Electrochemical Detection in HPLC
Analysis of Drugs and Poisons

RSC Chromatography Monographs

Series Editor: R.M. Smith, *Loughborough University of Technology, UK*

Advisory Panel: J.C. Berridge, *Sandwich, UK*; G.B. Cox, *Indianapolis, USA*, I.S. Lurie, *Virginia, USA*, P.J. Schoenmaker, *Eindhoven, The Netherlands*, C.F. Simpson, *Brighton, UK*; G.G. Wallace, *Wollongong, Australia*

Other titles in this series:

Applications of Solid Phase Microextraction
Edited by J. Pawliszyn, *University of Waterloo, Waterloo, Ontario, Canada*

Capillary Electrochromatography
Edited by K.D. Bartle and P. Myers, *University of Leeds, UK*

Chromatographic Integration Methods, Second Edition
N. Dyson, *Dyson Instruments, UK*

Cyclodextrins in Chromatography
By T. Cserháti and E. Forgács, *Hungarian Academy of Sciences, Budapest, Hungary*

HPLC: A Practical Guide
T. Hanai, *Health Research Foundation, Kyoto, Japan*

Hyphenated Techniques in Speciation Analysis
Edited by J. Scpunar and R. Lobinski, *CNRS, Pau, France*

Packed Column SFC
T.A. Berger, *Hewlett Packard, Wilmington, Delaware, USA*

Separation of Fullerenes by Liquid Chromatography
Edited by Kiyokatsu Jinno, *Toyohashi University of Technology, Japan*

How to obtain future titles on publication

A standing order plan in available for this series. A standing order will bring delivery of each new volume upon publication. For further information please contact:

Sales and Customer Care
Royal Society of Chemistry
Thomas Graham House
Science Park
Milton Road
Cambridge
CB4 0WF
Telephone: +44(0) 1223 420066, Fax: +44(0) 1223426017, Email: sales@rsc.org

RSC CHROMATOGRAPHY MONOGRAPHS

Electrochemical Detection in HPLC

Analysis of Drugs and Poisons

Robert J. Flanagan
Medical Toxicology Unit, Guy's and St Thomas' NHS Foundation Trust, London, UK

David Perrett
William Harvey Research Institute, Barts & The London, Queen Mary's School of Medicine and Dentistry, London, UK

Robin Whelpton
Department of Chemistry, Queen Mary, University of London, London, UK

advancing the chemical sciences

ISBN 0-85404-532-5

A catalogue record for this book is available from the British Library

Published by The Royal Society of Chemistry,
Thomas Graham House, Science Park, Milton Road,
Cambridge CB4 0WF, UK

Registered Charity Number 207890

For further information see our web site at www.rsc.org

Typeset by Alden Bookset, Northampton, UK
Printed by Athenaeum Press Ltd, Gateshead, Tyne & Wear, UK

Preface

The aim of this book is to give a balanced account of the application of HPLC with electrochemical detection (ED) in analytical toxicology and associated areas, including therapeutic drug monitoring, drug metabolism and pharmacokinetic studies. Many other reviews, book chapters and other sources of information on ED in HPLC tend to give an instrument manufacturer's view, not the view of the analyst, or are devoted to the analysis of easily oxidised species, such as catecholamines and related compounds. The emphasis in this monograph is on exogenous compounds, although catecholamines and other endogenous species are sometimes discussed to exemplify particular approaches or when such compounds have been used as drugs.

Although ED can give high sensitivity and selectivity and can be relatively inexpensive in operation, it is not an easy technique to use, the ever-present concerns being electrode deactivation or other more subtle factors that may act to influence response. The authors all have experience of using HPLC with both amperometric and coulometric detectors. This volume gives practical and useful information on the applications and limitations of the technique in the analysis of drugs and poisons in biological and related specimens, which tends to pose different problems to those encountered in the analysis of catecholamines and of pharmaceutical preparations.

Introductory chapters give information on basic electrochemistry and HPLC-ED, and on the specialised area of HPLC-ED of thiols. The major portion of the book is devoted to summary details of over 400 published HPLC-ED methods that are discussed in a standard format (column, eluent, internal standard, ED conditions, extraction procedure, limit of quantitation, *etc.*). These data are not always available *via* published abstracts and, wherever possible, sufficient information is given for the reader to decide whether a particular approach is worth pursuing. Chemical structures are given for most analytes and internal standards to ensure unambiguous identification and to illustrate possible electroactive moieties. Problems and pitfalls, and alternative techniques when appropriate, are emphasised throughout. Literature coverage is comprehensive up to the end of 2003.

Acknowledgements

We thank Steve Binder (Bio-Rad, Hercules, CA) and C.F.M. van Valkenberg (Antec, Leiden) for helpful comments on the manuscript, and the Series Editor, Prof Roger M. Smith, and the Royal Society of Chemistry for their help and encouragement.

Contents

Chapter 7   Applications of HPLC-ED in Toxicology and Related Areas   **211**

Appendix   Non-Aqueous Ionic Eluent Systems for Basic Drugs   **219**

Subject Index   **225**

List of Abbreviations

AASP	advanced automated sample processor
aq	aqueous
BAS	Bioanalytical Systems
BAL	British Anti-Lewisite (dimercaprol)
BALF	bronchoalveolar lavage fluid
CBA	carboxylic acid
CE	capillary electrophoresis
CoPC	cobalt phthalocyanine
CPE	carbon paste electrodes
CSF	cerebrospinal fluid
CV	cyclic voltammetry
CZE	capillary zone electrophotesis
DC	direct current
DEA	diethylamine
DHA	dihydroartemisinin
DHBA	dihydroxybenzylamine
DMAD	N,N-dimethylaminododecane
DOPA	3-(3,4-dihydroxyphenyl)alanine
DOPAC	3,4-dihydroxyphenylacetic acid
DTT	dithiothreitol
EC	electrochemical
ECD	electron capture detection
ED	electrochemical detection
EDTA	ethylenediaminetetra-acetic acid (or sodium salt)
ESA	Environmental Science Associates
FIA	flow-injection analysis
f.s.d.	full-scale deflection
GC	gas chromatography
GCE	glassy carbon (working) electrode

| GSH | reduced glutathione |
| GSSG | oxidised glutathione |

HFBA	heptafluorobutanoic acid
5-HIAA	5-hydroxyindoleactic acid
HIV	human immunodeficiency virus
HMDE	hanging mercury drop electrode
HPLC	high-performance liquid chromatography
HPLC-ED	HPLC with electrochemical detection

| i.d. | internal diameter |
| i.v. | intravenous |

LC	liquid chromatography
LoD	limit of detection
LoQ	limit of quantitation
LLE	liquid-liquid extraction
LLoQ	low limit of quantitation

MDA	methylenedioxyamphetamine (3,4-methylenedioxyamphetamine)
MDEA	methylenedioxyethylamphetamine
MDMA	methylenedioxymethylamphetamine
MS	mass spectrometry/spectrometric
MTBE	methyl *t*-butyl ether

NAC	*N*-acetyl-L-cysteine
NEM	*N*-ethylmaleimide
NAPM	*N*-(4-anilinophenyl)maleimide

ODS	octadecylsilyl
OPA	*o*-phthaldialdehyde
OSA	octanesulphonic acid

PAD	pulsed amperometric detection
PC	personal computer
PCA	perchloric acid
PFPA	pentafluoropropionic acid
PGE	porous graphite (working) electrode
PITC	phenylisothiocyanate

| RSD | relative standard deviation |

| SAM | self-assembled monolayers |
| SCE | standard calomel electrode |

SCX	strong cation-exchange
SDS	sodium dodecylsulphate
S/N	signal-to-noise
SPE	solid phase extraction
SPME	solid phase microextraction
TBA	tetrabutylammonium ion
TEA	triethylamine
THF	tetrahydrofuran
Tris	tris(hydroxymethyl)aminomethane
UV	ultraviolet

CHAPTER 1

General Introduction

1 Introduction

Electrochemical detection (ED) is used for the sensitive detection and measurement of electro-active analytes in many areas of analytical chemistry and biochemistry. These applications range from electrode sensor devices *via* flow injection analyses (FIA) to direct measurements of neurochemicals in brain tissue using *in vivo* cyclic voltammetry. In separation science, ED is used to detect and measure responsive analytes in flowing streams following analysis by high-performance liquid chromatography (HPLC) or capillary electrophoresis (CE). The use of ED with HPLC is by far and away the most important application of ED in flowing systems (Tables 1.1 and 1.2). The use of HPLC-ED grew by 500% between the 1980s and the 1990s. However, its popularity should be compared to some 20,000 papers that employed HPLC in combination with fluorescence detection and some 10,000 with MS detection (the most rapidly increasing combination at present). Most published HPLC methods use UV/visible detection, but the numbers are more difficult to quantify as this is not always made explicit in titles or abstracts.

Unlike UV or fluorescence detectors, ED does not exploit a physical property of an analyte, but an induced chemical change that results from an electrochemical reaction. ED must, therefore, be considered to be a type of post-column chemical reaction detector. ED differs, however, from other post-column reactors used in HPLC in that no reagents (other than electrons) or reaction devices are normally required to effect the chemical change in the analyte. In addition, the reaction kinetics are usually fast leading to minimal extra-column effects.

General aspects of electrochemistry have been covered in a number of books. In addition, the principles of ED when specifically applied to HPLC and/or CE have been reviewed.[1–7] A brief overview of this area is given in Chapter 3.

With UV detectors, selectivity is adjusted by varying the detection wavelength, lower wavelengths often giving enhanced sensitivity and a response from a wider range of analytes. A modest degree of selectivity is achieved by using UV detection in the aromatic region (240–270 nm) and traditionally 254 nm has proved popular. However, at lower wavelengths (200–210 nm), the absorption of the eluent, of other eluent constituents or of oxygen become limiting. Relatively few compounds show useful absorption at wavelengths higher than 340 nm (the lower limit of the

Table 1.1 *Publications on electrochemical detection used with different analytical modes*

Method	No of publications
ED + HPLC	9189
ED + CE	180
ED + FIA	111
ED + sensors	171
ED + nanotechnology	100

(Data based on a search of SCIRUS Database, December 2003 and a personal CE database (D. Perrett))

deuterium lamp emission). Generally, responses are usually independent of eluent conditions. In EC detection both sensitivity and selectivity are adjusted by varying the potential maintained between the working and reference electrodes, higher potentials, up to a local maximum, giving increased sensitivity. However, higher potentials usually induce a response from more compounds and therefore compromise selectivity. In oxidative mode, oxidation of eluent constituents becomes limiting at higher applied potentials, whilst in reductive mode, interference from dissolved oxygen can prove difficult to exclude. The response at the electrode is also very dependent on the eluent composition, especially its pH. Thus, as in all analytical methods, it is the signal-to-noise (S/N) ratio that is important and the detection conditions eventually adopted for a separation are a compromise between the electrochemical response of the analyte, the optimum eluent for both detection and elution, and interference from the sample matrix or from noise or drift from electronic or other sources.

Table 1.2 *Literature publications on applications of electrochemical detection, 1980–2003*

Applications	No of publications
Pharmaceuticals	3222
Clinical chemistry	1890
Neuroscience	1478
Biochemistry	1405
Chemical	2881
Environmental	1136
Industrial	829
Forensic	201

(Data based on a search of SCIRUS, December 2003)

In HPLC-ED the column eluate flows over the surface of an 'inert' electrode maintained at an appropriate positive or negative potential relative to a reference electrode. At the electrode surface analytes possessing electroactive functional groups undergo oxidation or reduction (oxidation being loss of electrons and *vice versa*). The electrons released (or donated) travel *via* the electrode and the change in current can be measured and related to the concentration of the analyte. Modern electronics allow the applied (working) potential to be held within very tight limits while at the same time measuring and amplifying the very small currents created. Hence these detectors can be very sensitive. A crude comparison of the sensitivity and applicability of the most common HPLC detectors towards favoured analytes under similar analytical conditions is given in Table 1.3. Both EC and fluorescence detectors can be at least 100 times more sensitive towards responsive compounds than a standard UV detector and are much more selective. Unfortunately, with time EC reaction products tend to accumulate at the electrode surface leading to loss of activity and hence loss of detector response – this is the major reason EC detection remains a relatively specialised field.

2 HPLC-ED in Analytical Toxicology

HPLC is widely used in analytical toxicology. UV/visible absorption (including diode-array and scanning instruments) and fluorescence detection remain of paramount importance, with pre- or post-column derivatisation sometimes being used to enhance sensitivity and/or selectivity. Modern UV detectors are in the main a considerable improvement on their predecessors. HPLC-MS and HPLC-MS-MS are being used increasingly in quantitative work, although the capital costs involved remain relatively high. Nevertheless, ED still finds a role in certain applications and there is a considerable body of literature associated with this topic. ED requires more care and thought in routine use than spectrophotometric detectors, principally because of the problems of electrode deactivation. On the other hand, running costs can be minimal and good sensitivity/selectivity can be attained with a number of analytes.

The aim of this volume is to give information to aid the use of HPLC-ED in the analysis of drugs and poisons in biological and related specimens. The available information (column, eluent, detection potential, extraction procedure, internal standard, sensitivity, *etc.*) is presented in a standard format in Chapters 6 and 7. These data are not always given in published abstracts and, wherever possible, sufficient information is given for the reader to decide whether a particular approach is worth pursuing. Chemical names or structural formulae are given to aid identification of electroactive moieties. The use of alternative techniques, including CE-ED, is emphasised as appropriate. Additional topics, such as analyte stability, are also discussed where relevant. Note that unambiguous details of the working and reference electrode combinations used in a particular application are not always given in published work – in such cases an informed guess as to the ED conditions actually used has had to be made.

Table 1.3 *Comparative performance of common HPLC detectors to appropriate analytes*

Detector	Type	Minimum sensitivity (nmol*)	Selectivity	Applicability	Ease of use/ interpretation	Typical price in UK**
UV/Visible absorption	Variable (single) λ	500	low	high	easy	<£4,000
	Diode array	700	low	high	moderate	<£8,000
Refractive index		10000	low	universal	easy	<£5,000
Mass detector (light scattering)		10000	low	universal	easy	<£6,000
Mass spectral	MS	10	moderate	universal	moderate	*ca.* £70,000
	MS-MS	10	very high	universal	skilled	*ca.* £130,000
	Ion-trap	10	very high	universal	skilled	*ca.* £130,000
Fluorescence		1	very high	limited	easy	<£7,000
Electrochemical	Single electrode	1	high	moderate	moderate	<£5,000
	Multi-electrode	1	very high	moderate	skilled	<£7,000

The details apply only to the detector
* Assuming standard column configuration and injection volume for a favoured analyte
** 2004 (US prices are usually significantly lower)

The drugs and other compounds reviewed are largely those given International Nonproprietary Names. Information on the dose, pharmacokinetics, metabolism and pK_a values for some of these compounds can be found in standard works such as those of Moffat *et al.*,[8] Baselt[9] and O'Neil *et al.*[10]

References

1 J.N. Barisci and G.G. Wallace, Electrochemical detectors for liquid chromatography, *Chim. Oggi.*, 1990, 9–13.
2 M.F. Cardosi, Electrochemical detection in HPLC: An overview, *Chromatogr. Anal.*, 1992, **21**, 9–11.
3 G.M. Greenway, Electrochemical detection for HPLC, *Chromatogr. Anal.*, 1989, 9–11.
4 WTh. Kok, Electrochemical techniques for detection in HPLC in *Selective Sample Handling and Detection in High-Performance Liquid Chromatography* (J. Chromatogr. Library 39A), R.W. Frei and K. Zech (eds), Chapter 6, Elsevier, Amsterdam, 1988, 309–363.
5 D.M. Radzik and S.M. Lunte, Application of liquid-chromatography electrochemistry in pharmaceutical and biochemical-analysis – a critical review, *CRC Crit. Rev. Anal. Chem.*, 1989, **20**, 317–358.
6 C.N. Svendsen, Multi-electrode array detectors in high-performance liquid chromatography: A new dimension in electrochemical analysis, *Analyst*, 1993, **118**, 123–129.
7 H. Parvez, M. Bastart-Malsot, S. Parvez, T. Nagatsu and G. Carpentier (eds), *Electrochemical Detection in Medicine and Chemistry*, Progress in HPLC-HPCE, Volume 2, VSP International Science Publishers, Zeist, 1987.
8 A.C. Moffat, M.D. Osselton and B. Widdop (eds), *Clarke's Analysis of Drugs and Poisons*, 3rd edn, Volumes 1 and 2, Pharmaceutical Press, London, 2003.
9 R.C. Baselt, *Disposition of Toxic Drugs and Chemicals in Man*, 7th edn, Foster City, Chemical Toxicology Institute, California, 2004.
10 M.J. O'Neil, S. Smith and P.E. Heckelman (eds), *The Merck Index*, 13th edn, Merck and Co, Whitehouse Station, New Jersey, 2001.

CHAPTER 2

Basic Electrochemistry for the Separation Scientist

Electrochemistry refers to the chemical changes that occur in reactions in which electrons are transferred from one species to another. In addition, it represents the interchange of chemical and electrical energies. However, the electrochemical (EC) process of most interest to the analyst is the transfer of electrons between ions in solution and an adjacent electrode surface.

In contrast to detection systems that are based on physicochemical properties (*e.g.* UV-visible absorption, fluorescence) and typically involve homogeneous solutions, it is clear that ED must be a heterogeneous process since interfaces are involved. Furthermore, electrochemical detectors (EDs) are reaction detectors and therefore the responses may be influenced not only by the amount(s) of electro-active analyte(s) present, but also by factors such as temperature and residence time.

By definition an electrical current is the movement of electrons and therefore electrochemistry can be seen to include two different types of processes:

1. The *production* of an electric current from an electron transfer (chemical) reaction
2. The *use* of an electric current to produce a chemical change

It is the first of these processes that is employed in ED. The second is found and used in electrolytic reactions.

1 Reversible Electrode Potentials

In the solid state the atoms in a metal are closely packed and the well-defined electron energy levels that are found in single atoms are not present. There is a continuum of levels and the available electrons fill the states from the bottom upwards, the highest level being known as the Fermi level. Hence, the electrons in metals are relatively mobile and metals are good conductors of electricity. When a metal electrode is dipped into a solution of the corresponding ions it will equilibrate:

$$M^{n+} + ne^- \rightleftharpoons M$$

and the potential on the electrode will be a function of the equilibrium position for the reaction. Electrodes of this kind are also referred to as 'electrodes of the first type'.

1.1 Redox Electrodes

In this electrode system, oxidised and reduced forms co-exist in solution, the electrons being donated or accepted by an inert electrode such as platinum. A representation of such a simple system is shown in Figure 2.1. Oxidation occurs when a species loses electrons, whilst when a species is reduced it accepts electrons. Such a pair of linked reversible reactions is called a *redox couple* and is expressed more generally as:

$$X_{ox} + ne^- \rightleftharpoons X_{red}$$

where X_{ox} represents the oxidised species
   X_{red} is the reduced species
   n is the number of electrons involved in the process.

Such a redox couple is exemplified by an aqueous solution of Fe^{2+} and Fe^{3+} salts:

$$Fe^{2+}(aq) \rightarrow Fe^{3+}(aq) + e^-(metal) \qquad \textit{oxidation}$$

$$Fe^{3+}(aq) + e^-(metal) \rightarrow Fe^{2+}(aq) \qquad \textit{reduction}$$

Each of the above combinations is referred to as a half reaction.

In the simple system the electrons are passing *via* the metal electrode and the electrode could be carrying an excess or deficit of electrons. This is called

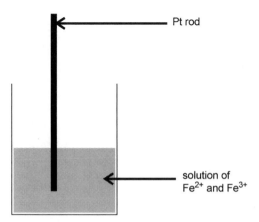

Pt rod

solution of
Fe^{2+} and Fe^{3+}

Figure 2.1   *A very simple electrochemical system.*

heterogeneous electron transfer. When the concentrations (activities) of Fe^{2+} and Fe^{3+} are unity and the heterogeneous electron transfer is fast, this is called the *standard electrode potential*, E^0, and is a characteristic of a particular redox couple (Table 2.1).

The equilibrium potential, E_{eq} is given by the Nernst equation:

$$E_{eq} = E^0 + \frac{RT}{nF} \ln \frac{[Ox]}{[Red]} \qquad (2.1)$$

where R is the gas constant ($8.315 \text{ J K}^{-1} \text{ mol}^{-1}$)

   T is absolute temperature (K)

   F is the Faraday constant ($96,485.3 \text{ C mol}^{-1}$)

(N.B. A number of slight variations on the value of the Faraday constant appear in the literature – the value given is that cited in the IUPAC Green Book[1])

At 25 °C (298 K) this becomes:

$$E_{eq} = E^0 + \frac{0.02567}{n} \ln \frac{[Ox]}{[Red]} \qquad (2.2)$$

or using logarithms to the base 10:

$$E_{eq} = E^0 + \frac{0.05913}{n} \log \frac{[Ox]}{[Red]} = E^0 + \frac{0.05913}{n} \log Q \qquad (2.3)$$

The reaction quotient (Q) is the mathematical product of the concentrations of the products of the reaction divided by the mathematical product of the concentrations of the reactants.

Table 2.1 *Some standard electrode potentials (25 °C)*

Half Reaction	*Potential* (V) *vs* SHE
$Pt^+ + e^- \rightarrow Pt$	+1.19
$Au^+ + e^- \rightarrow Au$	+1.68
$Ag^+ + e^- \rightarrow Ag$	+0.7996
$1/2Cu^{2+} + e^- \rightarrow 1/2Cu$	+0.3402
$1/2Hg^{2+} + e^- \rightarrow 1/2Hg$	+0.851
$Fe^{3+} + e^- \rightarrow Fe^{2+}$	+0.77
$H^+ + e^- \rightarrow H$	0.00
$H_2O + e^- \rightarrow H_2 + OH^-$	−0.83
$1/2Fe^{2+} + e^- \rightarrow 1/2Fe$	−0.44
$O_2 + e^- \rightarrow O_2^-$	−0.33
$Zn^{2+} + 2e^- \rightarrow Zn$	−0.7628

By connecting two half cells and measuring the potential difference between them it should be possible to measure an unknown electrode potential, providing the electrode potential of the other (reference) electrode is known. This is the dilemma – how does one ascertain the electrode potential of the very first reference electrode? As this is clearly impossible, the standard hydrogen electrode (Section 3) was arbitrarily chosen as the reference against which all other electrode potentials would be scaled.

From the Nernst equation, it is clear that if a potential different from E_{eq} is applied to the electrode, the position of the equilibrium in the EC reaction must change. If $E_{app} > E_{eq}$ the reduced species will be oxidised using electrons donated by the electrode and if $E_{app} < E_{eq}$ the electrode must donate electrons so the oxidised species will be reduced. The introduction of a second electrode allows the current to be measured. This is the basis of ED.

2 Theoretical Background Linking Electrochemistry to Electrochemical Detection

Faraday's Law states that 'the amount of substance consumed or produced at one of the electrodes in an electrolytic cell is directly proportional to the amount of electricity that passes through the cell'.

Since the current is formed from an electroactive chemical dissolved in the solution interacting with the adjacent electrode surface, it is not surprising that the probability of a molecule or ion reacting at the electrode surface is directly related to its concentration. Therefore, current density (*i.e.* current/electrode area) is normally directly proportional to analyte concentration. When all the analyte in a solution is electrolysed at an electrode then according to Faraday's Law, the measurable response is directly related to the molar amount of analyte:

$$Q = nFN \tag{2.4}$$

where Q is the total charge transferred (coulombs)
   n is the number of electrons transferred or equivalents per mole
   F is the Faraday constant
   N is the number of chemical equivalents of analyte

Thus the Faraday constant is the charge on the electron multiplied by Avogadro's number. By definition, an electrical current (i), measured in amperes (A), is determined as coulombs per second:

$$i = \frac{dQ}{dt} - nF\frac{dN}{dt} \tag{2.5}$$

Equation 2.5 shows that the rate at which electrons are moved across the electrode-solution interface (*i.e.* the current) is directly related to the rate of the reaction occurring at the surface interface. Therefore, when we measure the current, we are also measuring the rate of a chemical reaction.

Under conditions of complete (*i.e.* 100%) EC conversion of the analyte, the electrode is said to be operating in the *coulometric* mode and the total charge which passes depends on the total amount of analyte. Partial electrolysis is found in *amperometric* systems when the % conversion typically ranges between <1 and 5%. Since in ED systems the analyte is moving past the electrode the reaction is often incomplete and amperometry is more readily achieved than a coulometric response. In such cases the current is proportional to the analyte concentration.

Note that many steps are involved in an EC reaction, such as the electron transfer reaction, transport of molecules from the bulk solution to the electrode surface and chemical reactions coupled to the electron transfer reaction. As with any multi step reaction, the rate of the overall reaction is generally determined by the rate of the slowest step (the rate-limiting step), and it is important to identify this step. In the analytical electrochemistry of dissolved species, the limiting step is typically the transport of molecules to the electrode surface through the solution. However, there are many instances where this is not the case and where the rate of the heterogeneous electron transfer reaction is important, for example in corrosion electrochemistry.

The electrode surface area and the reactant concentration both affect the current. Other factors that also affect the current include the temperature and the viscosity of the solution and it is important to ensure that these parameters remain constant so the measured current is proportional to the reactant concentration. This is particularly important for quantitative analysis.

There are some other sources of current in addition to the Faradaic current due to the reaction of the analyte; these are collectively referred to as the *background* current. They include the electrolysis of impurities, the electrolysis of the electrolyte, the electrolysis of the electrode material and capacitive (or charging) current. The first three are Faradaic currents of the system, whereas the last one is a property of the interface between the electrode and the solution. This interface behaves like an electrical capacitor in that it can store charge and (to a first approximation) it obeys the equation:

$$Q = CV \qquad\qquad (2.6)$$

where Q is the charge stored
$\quad\quad$ C is the capacitance (in farads)
$\quad\quad$ V is the potential difference across the interface

An equation for the capacitive current (i_c) can be obtained by differentiating Equation 2.6:

$$i_c = \mathrm{d}Q/\mathrm{d}t = C(\mathrm{d}V/\mathrm{d}t) \qquad\qquad (2.7)$$

The total current is the summation of all currents, including the Faradaic currents for the analyte, the electrolyte and the electrode material, as well as the capacitive current. Currents other than that derived from the analyte are generally undesirable, and their magnitude should be minimised. In some cases, the background current can be subtracted from the total current to leave the signal of interest. However, in other cases, eliminating the background current is not so easy. In rare cases, the background and signal currents combine in a non-linear manner making it impossible to estimate the background current in the absence of the analyte. The capacitive current can be eliminated by operating the electrode at a fixed potential (*i.e.* $dV/dt = 0$). In HPLC-ED the background current can be essentially eliminated since: 1. the fixed potential eliminates the capacitive current as it is only present when the potential across the cell is changing and 2. the background Faradaic currents can be monitored between the peaks to form a chromatographic baseline. However, the capacitive current can change (even at a fixed potential) if the ionic strength or the solvent composition of the eluent changes – such changes can be seen, for example, at the 'solvent front'.

3 Reference Electrodes

As it is not possible to determine the 'absolute' potential of an electrode, the electrode potential must always be referred to an arbitrary zero point, defined by the potential of a chosen reference electrode. Thus, it is very important *always* to quote the type of reference electrode used in electrochemistry. Differences in operating potentials reported in the literature are often attributable to the use of different reference electrodes.

3.1 Standard Hydrogen Electrode

The standard hydrogen electrode (SHE) is the universal reference for reporting relative half-cell potentials, the equilibrium potential being arbitrarily defined as zero. A hydrogen electrode consists of platinised platinum foil (platinum black) immersed in a solution of hydrogen ions saturated with hydrogen gas. In the case of the standard hydrogen electrode the hydrogen ion concentration is unity (1 mol L^{-1}) and the gas is at a pressure of 1 atmosphere. Clearly this electrode would be very inconvenient to use and so other reference electrodes (*e.g.* calomel or silver/silver chloride electrodes) may be used instead, the measured electrode potentials being standardised by converting them to the 'hydrogen scale'.

3.2 Calomel Electrode

The calomel electrode is very similar in both construction and theory of operation to the silver/silver chloride electrode described below. The metal is mercury, the electrical connection being made by an inert metal wire and the salt is mercurous chloride. The equilibrium electrode potential is a function of the chloride concentration of the electrolyte. When the electrolyte is saturated potassium chloride, it is known as a saturated calomel electrode (SCE) producing an electrode potential of $+0.224$ V *vs* SHE. Potassium chloride is used because the ionic mobility ($m^2\ s^{-1}\ V^{-1}$) of K^+ (7.6×10^{-8}) is similar to that of Cl^- (7.9×10^{-8}).

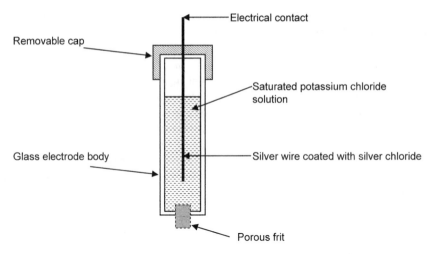

Figure 2.2 *A silver/silver chloride reference electrode.*

3.3 Silver/Silver Chloride Electrode

This is probably the most widely used reference electrode especially in HPLC-ED. The electrode assembly consists of a silver metal electrode in contact with solid silver chloride (usually coated on the silver wire) immersed in potassium chloride solution contained in glass tubing (Figure 2.2). The internal electrolyte may contain saturated silver chloride, which reduces the tendency of the silver chloride coating to be lost from the wire. The internal electrolyte and the external electrolyte, into which the whole assembly is immersed, are in ionic contact *via* a small porous ceramic plug ('frit' in Figure 2.2) sealed into the end of the glass tubing. Its potential is defined as +0.7996 V relative to a standard hydrogen electrode (0.0 V). As with the calomel electrode, and predicted by the Nernst equation, the equilibrium electrode potential is a function of the chloride concentration of the internal electrolyte (Table 2.2).

Table 2.2 *Effect of electrolyte concentration on the potential of reference electrodes*

Electrode type	Potassium chloride concentration (mol L^{-1})	Potential (V *vs* SHE)
Calomel	0.1	+0.334
	1.0	+0.280
	Saturated	+0.241
Silver/silver chloride	0.1	+0.288
	1.0	+0.235
	4.0	+0.222
	Saturated	+0.199

This is another source of variability in the operating potentials reported in the literature and it is not always possible to ascertain the precise nature of the silver/silver chloride reference electrode being employed. Depletion of potassium chloride will lead to an increase in the equilibrium electrode potential.

4 Voltammetry

The technique for studying the relationship between potential and current is known as voltammetry (*i.e.* the simultaneous recording of volts and amperes). Voltammetry can be performed in a variety of ways including:

Cyclic voltammetry
Square-wave voltammetry
Staircase voltammetry
Linear-sweep voltammetry
Fast cyclic voltammetry
Rotating disc voltammetry
Stripping voltammetry
Hydrodynamic voltammetry
Direct current (d.c.) polarography
Alternating current (a.c.) polarography
Pulse polarography

The majority of the procedures listed have been applied to HPLC-ED in research publications, although they are rarely employed in routine analysis. Therefore only those of relevance to HPLC-ED are discussed here.

4.1 Cyclic Voltammetry

Probably the most useful for initial investigations prior to ED is cyclic voltammetry (CV). CV is usually performed on dedicated equipment with the current being plotted against the potential of the working electrode rather than time as is normal on a chromatogram. In CV, a linearly increasing d.c. potential waveform of usually 1 V or less is applied to the working electrode over a few seconds and then the potential is exactly reversed (Figure 2.3).

A typical shape for the current response of a CV of an electrochemically reversible redox couple is shown in Figure 2.4. This is plotted according to the IUPAC convention with positive (oxidising) potentials increasing from left to right.

The anodic currents (oxidation) are plotted in the up direction and cathodic (reduction) in the downward direction. This is different from the frequently used polarographic convention (increasingly negative potentials from left to right and cathode currents in the up direction) but as in the majority of HPLC-ED applications the analyte is oxidised it makes the figure easier to understand. At point A the bulk solution contains only the reduced form (R) of the redox couple and there is no net conversion of R into the oxidised form (O). On increasing

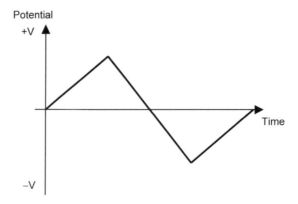

Figure 2.3 *Waveform applied in a cyclic voltammetry experiment.*

the potential across the cell, conversion of R to O starts to occur. This conversion results in a diffusion layer with concentration gradients set up for both R and O and, as the experiment is performed without stirring, the only mode of mass transport is by diffusion. The anodic peak (point B) occurs because the diffusion layer above the electrode has grown sufficiently that the flux of R to the electrode is no longer fast enough to satisfy the Nernst equation and the current, which is limited by the rate of mass transfer to the electrode surface, falls (point C). Upon reversal of the scan, the current flow is in the opposite direction (*i.e.* from electrode to solution) as the sweep converts O to R, and the cathodic current falls to a minimum at D.

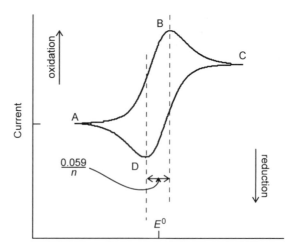

Figure 2.4 *Current-voltage curve for a reversible redox couple.*
                (plotted according to the IUPAC convention)

If a redox system remains in equilibrium throughout the potential scan, the EC reaction is said to be *reversible*. In other words, equilibrium requires that the surface concentrations of O and R are maintained at the values required by the Nernst equation. Under these conditions, the following parameters characterise the cyclic voltammogram of the redox process:

1. The positions of peak voltage do not alter as a function of voltage scan rate
2. The voltage separation between the current peaks is:

$$\Delta E = \frac{0.059}{n}$$

3. The peak currents are proportional to the square root of the scan rate

These relationships do not hold true for irreversible reactions, *i.e.* those for which heterogeneous electron transfer is slow relative to the voltage scan rate. The peak separation is no longer fixed, but varies as a function of the scan rate. Similarly the peak current no longer varies as a function of the square root of the scan rate. Generally, the greater the separation between the peaks the more irreversible is the electrode process.

Cyclic voltammetry is useful not only for ascertaining whether a compound is electroactive, but also for elucidating the species involved in the process, particularly intermediate species.

The most common format for presenting EC data is a plot of current *vs.* potential as shown in Figure 2.4. However, other representations appear in the literature. It is also necessary to remember that all potential values are quoted relative *to* a reference potential so there is no significance in the absolute values. Also note that there is no correlation between positive and negative potentials and oxidations and reductions; that is, reductions can occur at positive potentials and oxidations can occur at negative potentials.

4.2 Hydrodynamic Voltammograms

Cyclic voltammetry is not available to everyone and most analysts rely on some form of hydrodynamic voltammetry. As the analysis is to be performed in a flowing stream then this would appear to be most appropriate. Some commercially available instruments will allow the applied potential to be scanned and it is possible to construct current-voltage curves (*e.g.* Figure 2.5). However, optimum working potentials can also be ascertained by repeat injections of a solution of the analyte at differing detector voltages. Although this can be time-consuming, it does allow the analyst to determine electroactivity in actual chromatographic conditions.

In contrast to cyclic voltammetry, this technique is a steady state method in which the rate of mass transport is relatively high and occurs by both convection and diffusion.

Electrochemically reversible reactions produce steep curves whereas for irreversible reactions the slope may be much shallower, and it may be necessary to go to high potentials to reach the limiting current.

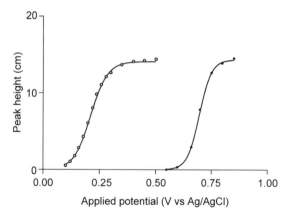

Figure 2.5 *Hydrodynamic voltammograms generated by repeat injections of physostigmine*
(•) and its hydrolysis product, the phenol eseroline (○). Chromatographic
conditions: Column: 250 × 4.6 mm (i.d.) Spherisorb S5W; Eluent: methanol-aq.
ammonium nitrate (1.0 mol L⁻¹, pH 8.5) (90 + 10); Flow rate: 0.75 mL min⁻¹;
Detection: glassy carbon electrode (GCE). The half-wave potentials (from the
fitted curves) were +0.69 V and +0.21 V vs Ag/AgCl.

5 Electroactivity of Functional Groups and Compounds

At high enough potentials most functional groups will exhibit some degree of
electroactivity, but for EC methods to be usable for analytical purposes the activity
must occur at a modest potential when the background currents should be low.
Figures 2.6 and 2.7 list some of the groups most commonly associated with either

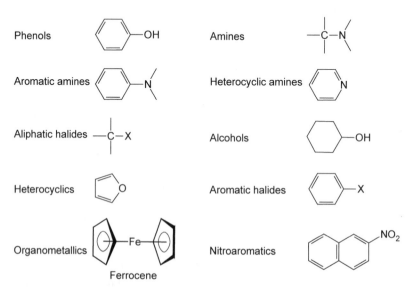

Figure 2.6 *Some electro-oxidisable functional groups.*

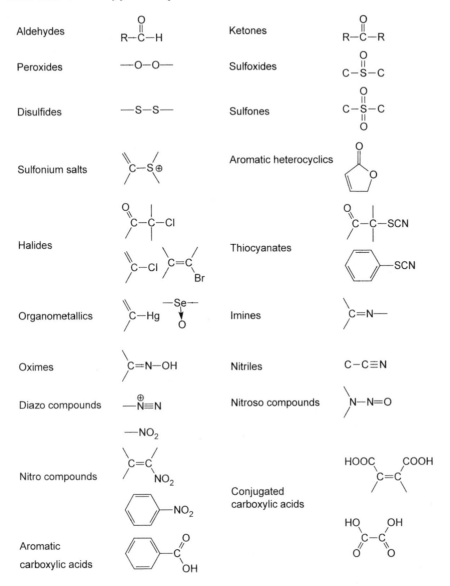

Figure 2.7 *Some electro-reducible organic functional groups.*

oxidation or reduction in electrochemistry. HPLC-ED methods for many molecules containing one or more of these groups in their structures have been developed. Later chapters in this book will illustrate how these groups are involved in the ED of drugs, *etc.* The figures are purely a guide to the type of groups to identify in a molecule when assessing if ED could be feasible as a detection method for HPLC or CE.

Electrochemically active compounds are typically aromatic and contain hydroxyl (phenols), methoxyl or amine moieties. Some aliphatic compounds, notably thiols and amines, are also electrochemically active. Ideally, for quantitative applications one should know the number of electrons transferred for a particular conversion. For instance, reduction of aldehyde to alcohol is a two-electron process, but reduction of nitro to amino requires six electrons. This difference would be reflected in the magnitude of the current.

5.1 Electrochemical Behavior of Some Specific Molecular Types

5.1.1 Catechol Containing Structures

The catechol (1,2-dihydroxybenzene) group is found in many neurochemicals, especially the catecholamines, and it has been one of the most studied electro-chemically active structures. Catechol-containing compounds, such as dopamine, adrenaline and noradrenaline, are readily oxidised at modest voltages (+0.7 V vs Ag/AgCl) at neutral pH. The reaction involves the two-electron oxidation of the hydroxyl groups to form an orthoquinone, as shown for dopamine (Figure 2.8).

5.1.2 Amino Acids and Peptides

Most of the common amino acids and their peptides are not oxidisable at commonly used electrodes. The exceptions are the amino acids tryptophan, tyrosine and cysteine. The oxidation of tyrosine is analogous to that of the catechols shown above. In the case of tryptophan and its metabolites serotonin and 5-hydroxyindoleacetic acid, the electrochemistry involves the oxidation of the carbon at position 2 on the indole ring with the formation of oxyindolalanine (Figure 2.9).

Thiol-containing amino acids, such as cysteine and penicillamine, are readily oxidised at mercury electrodes. The use of this reaction to measure them is discus-sed in detail in Chapter 5.

An unusual EC reaction applicable to the detection of most amino acids involves complexation with copper ions in solution. It was later shown that this also occurs at copper electrodes. Although sensitivity limits of 0.5 to 18 ng injected depending on the amino acid have been reported for this system, it is not widely used, probably because the difficulties of resolving 20+amino acids on reversed-phase columns. Most workers therefore use derivatisation of the amino acids prior to chromatography for their analysis. Even though many such derivatives, *e.g.* phenylisothiocyanate (PITC), are electroactive, most analysts favour fluorescence detection since it is more compatible with the necessary solvent gradients as well as being more selective.

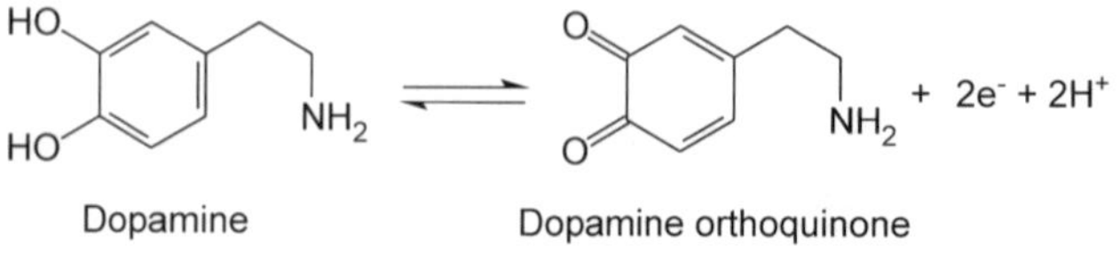

Figure 2.8   *The two-electron oxidation of a typical catechol.*

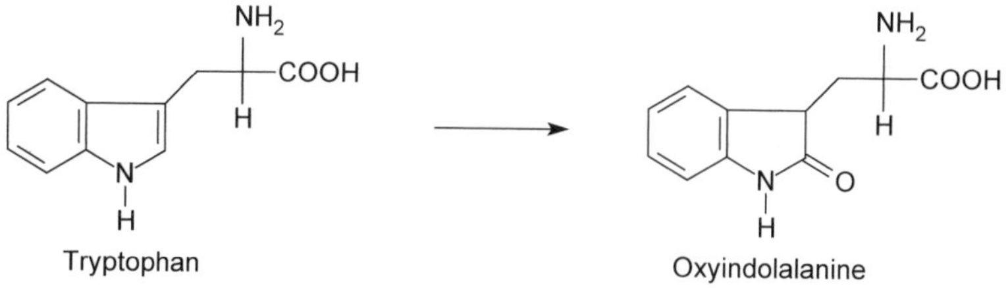

Figure 2.9 *Oxidation of tryptophan at a carbon electrode.*

5.1.3 Purine Rings

This heterocyclic ring structure is found not only in the purine bases of DNA and in its many metabolic end products, such as uric acid, but also in alkaloids such as caffeine. The ease of oxidation depends on the number of –OH groups attached to the main ring, with urate containing three –OH groups being the easiest to oxidise. However, unlike the oxidation of catechols, it is not the –OH groups themselves that are oxidised, but the C_4–C_5 that links the two rings of the molecule. In this case electrochemistry follows the biochemical catabolism of these molecules.

5.1.4 Sugars

Sugars are difficult to assay with any significant sensitivity by most physico-chemical methods, such as UV-visible detection, yet they are readily measured electrochemically. However, the optimum conditions for their reproducible

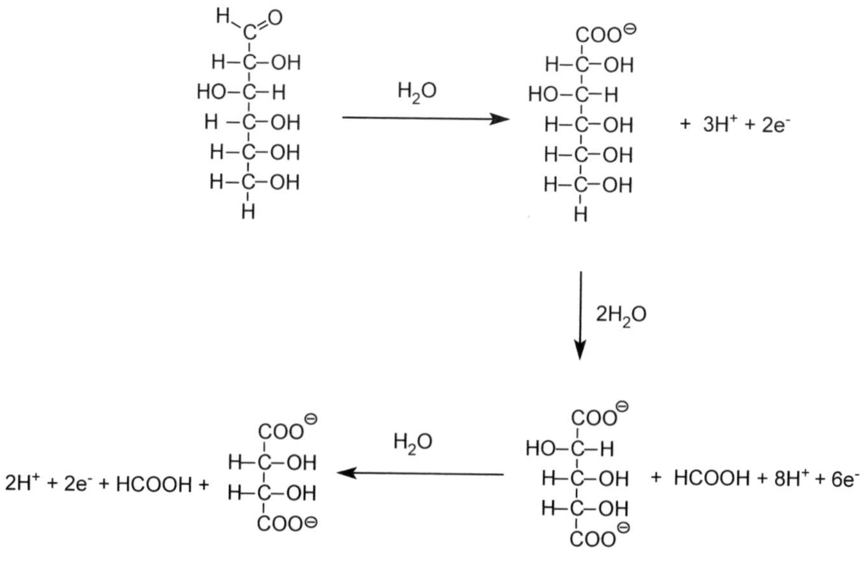

Figure 2.10 *Oxidation of glucose at gold electrodes under alkaline conditions, as proposed by Johnson and LaCourse.[2]*

electro-oxidation are somewhat unusual and require a strongly alkaline medium
and an inert electrode material, such as gold. The reaction pathway for a sugar may
involve a series of two-electron oxidations that sequentially cleave formic acid
from the terminal aldehyde of the sugar. This is shown for glucose in Figure 2.10.

References

1 I. Mills, T. Cvitas, K. Homann, N. Kallay and K. Kuchitsu, *Quantities, Units and Symbols in Physical Chemistry (The Green Book)*, Blackwell, London, 1993.
2 D.C. Johnson and W.R. LaCourse, Liquid-chromatography with pulsed electrochemical detection at gold and platinum-electrodes, *Anal. Chem.* 1990, **62**, A589–597.

Further Reading

A.J. Bard and L.R. Faulkner, *Electrochemical Methods: fundamentals and applications,* 2nd Edn., Wiley, New York, 2000.
D.R. Crow, *Principals and Application of Electrochemistry,* Blackie Academic and Professional, London, 1994.
D.B. Hibbert, *Introduction to Electrochemistry,* Macmillan, London, 1993.
H. Lund and O. Hammerich, *Organic Electrochemistry,* Marcel Dekker, New York, 2001.
P.T. Kissinger and A.W. Bott, Electrochemistry for the non-electrochemist, *Current Separations* 2002, **20**, 51–53. (Available *via* http://www.current separations.com/issues/20-2/20-2d.pdf).

CHAPTER 3

The Electrochemical Detector

1 Introduction

As noted in the previous chapters, when an EC detector is used in a flowing system, such as HPLC, it is simply one type of post-separation reaction detector. As with all post-column reaction detectors some knowledge of the chemistry involved in the detection process is essential in order to be able to use such detectors successfully. For maximum sensitivity with an EC detector, it is necessary to optimise conditions, such as the pH and composition of the reaction medium, the energy input, the time allowed for the reaction and the nature of the catalytic surface. However, when ED is used in conjunction with chromatography, the EC reaction conditions usually have to be a compromise with the chromatographic conditions necessary to achieve optimum resolution of the analytes, in particular the pH and the composition of the chromatographic eluent, as well as the need to maintain column stability.

2 Theoretical Background to ED in Flowing Liquid Streams

This section deals with the particular aspects of electrochemistry relevant to flowing systems. An introduction to the background of the theory was given in the previous chapter, and more detailed descriptions of the theoretical aspects can be found elsewhere.[1,2]

In outline, an EC reaction, at a solid surface electrode over which a liquid stream is flowing, is characterised by the following three separate and distinct stages:

1. Diffusion of the electroactive analyte to the electrode surface
2. Electron transfer as dictated by the electrochemical reaction
3. Diffusion of the reaction product(s) away from the electrode surface

During oxidation the EC reaction gives rise to the release of electrons into the electrode (stage 2). This flow of electrons (current) is subsequently converted to a signal, which can be monitored and recorded by suitable circuitry within the EC analyser or detector (Figure 3.1). The rate-limiting component in this scheme is the slowest of the necessary steps and in ED it is usually one of the mass transfer steps (stages 1 or 3). Due to flow-induced mixing, the analyte is present at a constant

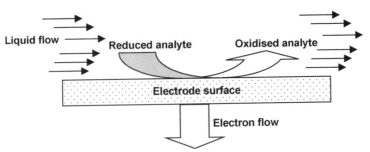

Figure 3.1 *Schematic diagram of the electrode processes occurring when, in this case, a reduced analyte flows over a static electrode held at an appropriate voltage.*

concentration throughout the bulk of the eluent except for a narrow layer (thickness δ) immediately above the electrode surface. Only analyte molecules in this relatively stationary layer can diffuse to the surface. Consumption of the analyte, either by oxidation or reduction, creates a concentration gradient (Figure 3.2). The thickness of the diffusion layer above the electrode surface is therefore critical to the electrochemistry.

The diffusion equation (Fick's second law) describes the rate of change of the analyte concentration above any planar surface. This analyte flux is proportional to the diffusion coefficient of the analyte (D). The flux therefore governs the limiting current (i_{lim}) and is directly proportional to the surface area of the electrode and analyte concentration (C) and inversely proportional to the thickness (δ) of the diffusion layer:

$$i_{lim} = \frac{nFADC}{\delta} \tag{3.1}$$

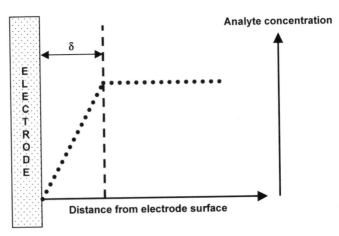

Figure 3.2 *Concentration profile (δ) for an analyte near an electrode surface with laminar hydrodynamic flow.*

Table 3.1 *Limiting current equations in a standard format for popular electrode designs*
(adapted from Hanekamp and van Nieuwkerk[3])

Type of cell	Limiting current
Thin-layer (rectangular channel)	$1.47\ nFC\ D^{2/3}U^{1/3}\ (A/d)^{2/3}$
Planar (parallel flow)	$0.68\ nFC\ D^{2/3}U^{1/2}\ (A/d)^{1/2}\ v^{-1/6}$
Planar (perpendicular flow)	$0.903\ nFC\ D^{2/3}U^{1/2}A^{3/4}v^{-1/6}$
Tubular	$1.61\ nFC\ D^{2/3}U^{1/3}(A/r)^{2/3}$
Wall-jet	$0.898\ nFC\ D^{2/3}U^{3/4}A^{3/8}v^{-5/12}a^{-1/2}$

Key to symbols: A = electrode area (cm^2), U = mean volume flow-rate (cm^3 s^{-1}), C = analyte concentration (mol L^{-1}), D = diffusion coefficient (cm^2 s^{-1}), F = Faraday constant, n = number of electrons, a = inlet diameter (cm), d = channel depth, *i.e.* gasket thickness (cm), r = radius of the channel (cm), v = kinematic viscosity

For most cell designs the limiting current has been calculated with respect to the inner geometry of the cell. Table 3.1 summarises the limiting current equations for popular electrode configurations in terms of their hydrodynamics. Diffusion coefficients (D) in aqueous solution for small molecules at room temperature vary from 10^{-5} to 10^{-6} cm^2 s^{-1} and are not usually under the control of the analyst. However, it is possible to increase D by raising the temperature, so decreasing the viscosity of the eluent (viscosity changes by 2–3% per °C). Thus, holding the cell at an elevated temperature might be useful in some cases. At an amperometric electrode the response (current) is proportional to molar concentration of analyte in the eluate.

3 The Construction of an Electrochemical Detector

3.1 Principles of Electrochemical Measurement

The simplest EC detector for an HPLC system would consist of a noble metal working electrode held at an appropriate potential using a battery. The eluate would flow past this electrode before reaching a metal return (counter or auxiliary) electrode. In practice such a simple circuit would be too unstable to be useful. A more stable circuit requires a second or reference electrode. In most EC detectors this electrode is a silver/silver chloride (Ag/AgCl) electrode of the type described in Chapter 2 with a potential defined as +0.7996 V relative to a standard hydrogen electrode (0.000 V). In the Environmental Science Associates (ESA) Coulochem detectors the reference electrode is palladium/hydrogen[4] (ESA Technical Note 70-4822P). These electrodes can be placed in close proximity to the working electrode, thus reducing non-compensatible potential drop (iR-drop) to a minimum (see below) and are operable to 6000 psi. Published operating potentials obtained with detectors using this type of reference electrode are some 0.2–0.3 V lower than most other EDs.

In practice even this two-electrode arrangement is not useful because the potential of the reference electrode depends on the current flowing through it. Polarisation of

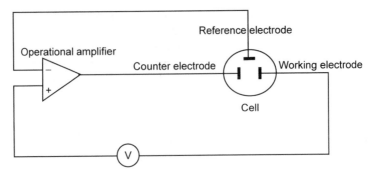

Figure 3.3 *Schematic diagram of a three-electrode electrochemical detector circuit.*

the mobile phase at the working electrode combined with the iR-drop due to conductivity of the eluent also causes the potential to differ from the applied voltage. So a two-electrode system would give a non-linear response and a small linear range. A third electrode, called the counter (or auxiliary) electrode, is therefore added specifically to collect the current flowing in the circuit so leaving the reference electrode at its reference potential. The counter electrode is often little more than the stainless steel outlet from the EC cell. Figure 3.3 shows a schematic diagram of this popular three-electrode configuration. The working electrode is usually grounded through the detector electronics and the voltage between the auxiliary and reference electrodes is adjusted to regulate the potential required for the EC reaction. Thus, for a positive (oxidative) potential difference to be applied to the working electrode it is necessary to apply a relative negative potential to the counter electrode.

3.2 Electrode Materials

For HPLC detection the working electrode material must:

1. Be electrically conducting
2. Give a low potential for the required EC measurement
3. Be robust enough to be mechanically formed into suitable shapes
4. Be able to be polished to a very smooth surface (thin-layer cells)
5. Withstand a constant flow or jet of eluent.

Additionally it must be inert towards the eluent, impurities in the eluent, other sample components, and any reactive species that might be formed by electrolysis of the analyte(s). In the event of electrode fouling, the working electrode assembly should be readily disassembled and the electrode easily cleaned or readily and cheaply replaced. Electrodes, such as the dropping mercury electrode, in which the electrode surface is continually renewed are not ideal for use in flowing systems since a more robust design is required to overcome the problems of being in a pumped flowing stream of eluent.

The most popular material for solid electrodes is carbon because it shows the least electrode deactivation in routine use. Some of the earliest EC detector cells used carbon or graphite pastes, both of which proved good materials for the working electrode, particularly for oxidative electrochemistry. Such electrodes in which solid particles are mixed with a binder are termed composite electrodes. Although inexpensive, they require considerable skill in preparation.[5,6] Chemically cleaned graphite particles (10–25 μm diameter) are combined with a binder, such as Nujol, in a ratio of approximately 1:25 by weight, packed firmly into the cavity of a plastic electrode block with a spatula and the surface smoothed to a glossy finish. Unfortunately carbon paste electrodes (CPE) are incompatible with many common HPLC solvents, such as acetonitrile, which dissolve the oil used to form the paste. However, under appropriate conditions and with careful use, carbon paste can give highly sensitive electrodes and many authors have claimed that such composite electrodes give the highest signal-to-noise (S/N) ratios.[7] Indeed, it has been claimed that using Kel-F wax as the binder gave more stable electrodes with higher sensitivity than either Nujol-graphite or glassy carbon electrodes.[8]

Solid electrodes, such as those formed from the noble metals, are much more suitable than pastes as working electrodes. Noble metal electrodes satisfy many of the requirements for good electrodes, but they often show rapid deterioration in response and are thus only employed in specific analyses especially where they exhibit electro-catalytic properties. Platinum and gold electrodes, for example, are used extensively in the pulsed amperometric detection (PAD) of thiols (Chapter 5) and carbohydrates. Copper electrodes have been used to measure amino acids specifically by exploiting the well-known complexation reaction between copper ions and amino acids.[9] Gold amalgam electrodes (*i.e.* gold electrodes with a coating of mercury) are used extensively for the measurement of thiols (Chapter 5).

In free-standing electrochemistry, liquid mercury has always been a common electrode material especially in the form of the dropping mercury electrode. Flow-through dropping-mercury detectors were first employed in the 1950s in a technique called chromato-polarography.[10] Rabenstein and Saetre[11] were the first to combine chromatography with a mercury-pool EC detector cell to monitor thiols eluting from a cation exchange column. They separated cysteine and other aminothiols and, using a working potential of +0.05 V *vs* SCE, could detect less than 1 nmol of most thiols tested. However, their detector was not commercially available and their system was not widely copied.

Others have attempted to improve selectivity by adsorbing compounds, such as metal ions, onto the electrode surface or by covering the surface with selective polymers, but such electrodes are not commercially available. Self-assembled monolayers (SAM) have become popular in many aspects of chromatography, including stationary phase modifications. 'SAM-modified' electrodes have also been reported. Usually gold electrodes are modified using a thiol, such as cysteine, which binds to the gold surface *via* a strong mercaptide bond, so functionalising the surface with an amino group.[12] A variety of analytes, such as hemoglobin, dopamine and most recently adrenaline,[13] have been analysed using cyclic voltammetry with surface-modified electrodes. However, they do not appear to have been used in flowing systems as yet.

Nowadays, the most commonly used electrode in HPLC-ED is the glassy carbon electrode (GCE). Glassy carbon was introduced as an electrode material in 1976.[14,15] Glassy carbon is formed by the controlled heating of pre-formed shapes of phenol-formaldehyde resin to *ca* 500 °C in an inert atmosphere. The nature of the final product is dependent on the nature of the starting material, the speed and duration of heating and the final temperature attained. The formation, chemical properties and uses of glassy carbon have been reviewed.[16] There are several manufacturers of glassy carbon and since the quality and chemical properties of their products differ, their EC behaviour also differs.

For amperometric detectors (Section 4), the shape of the glassy carbon pre-form is usually a rod of the required diameter, from which discs of suitable thickness are cut. The discs are mounted in an appropriate aperture in a plastic or glass mount, an electrical connection is made to the rear and the front surface is mechanically polished to a high sheen with a scratch-free finish using diamond and/or 0.1 μm alumina polishing paste on a flat, scratch-free, polishing plate. There is clearly scope for variation in this procedure, even in electrodes produced by a single manufacturer, and this may be an annoying source of variation even with a hitherto unused electrode. It has been suggested that the alumina normally used in the final polishing process may also influence the characteristics of the electrode, either by physical modification of the electrode surface[17] or by offering different polishing characteristics.[18] Since EC efficiency and noise depend on efficient diffusion of the analyte and the reaction products on to and off the electrode surface, the smoothness of the final surface is vital for high sensitivity and low noise. The physical quality of the junction between the electrode and the cell-block has also been claimed to affect the noise output of the detector.

The EC characteristics of a solid electrode can be chemically as well as physically modified so as to increase its stability, sensitivity, or most commonly, selectivity. An important example of selectivity modification is the formation of an amalgam on gold electrodes to specifically detect thiols (Chapter 5). Various authors have claimed increased sensitivity and/or increased selectivity by modifying carbon electrodes. Wang and Freiha[17] reported that the simple and routine act of polishing a GCE with 0.1 μm alumina paste changed the selectivity of the electrode towards ascorbate, adrenaline and oxalate by shifting the respective voltammograms to lower potentials. The effect was said to be due to electrocatalysis. However, such effects have, in general, not been confirmed; the initial observation may have been due to the type of glassy carbon employed or not allowing enough time for equilibration after polishing. Glassy carbon electrodes, however, have been known to change their optimum working potential markedly. In one case the same electrode used to detect catecholamines was initially operated at +1.0 V *vs* Ag/AgCl for maximum sensitivity, but after two years continuous use with frequent re-polishing was operable at +0.4 V *vs* Ag/AgCl with the same S/N ratio, even when using a new reference electrode. EC modification of a GCE by applying voltages from +5 to +10 V was reported to increase the sensitivity of the electrode towards timolol and oxprenolol by a factor of five.[19]

Glassy carbon electrodes have been modified with a hydrophilic, permeable film of horseradish peroxidase (HRP) covalently bound to a polyvinylpyridine polymer complexed with osmium to enhance the detection of hydrogen peroxide. Vreeke et al.[20] demonstrated that such a system could be used to quantify hydrogen peroxide (reduction at 0.0 V, *vs* SCE) produced from complex coupled reactions in their assay of NADH:

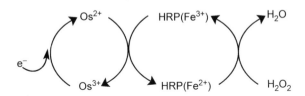

Huang et al.[21] found that a HRP-Os modified GCE (+0.1 V *vs* Ag/AgCl) was superior to Pt electrodes for the measurement of hydrogen peroxide generated from acetylcholine in post-column immobilised enzyme reactors (IMER, see Chapter 4, Section 2) and BAS have adopted this approach for their acetylcholine analysis kit. However, to obtain the sensitivity required to measure basal concentrations of acetylcholine in biological samples, it is necessary to use microbore columns and short elution times, under which conditions the acetylcholine peak is often masked by a very large choline peak. Immobilised enzyme reactors have been used to remove some of the interference from choline and other reactive species.[22] Osborne and Yamamoto[23] used disposable screen-printed film carbon electrodes that had been coated with an HRP-Os polymer and cross-linked glucose oxidase to measure glucose without the need for a separate post-column IMER. A disadvantage of HRP-Os GCEs is that the response to repeated injections of a standard solution decrease with time and the response to glucose showed a linear decrease of 37% after 40 injections.

All electrodes can be easily poisoned by exposure to complex matrices, such as urine, to certain classes of compounds, such as thiols and proteins, and by EC reaction products formed by operation at relatively high applied potentials. Attempts to limit protein adsorption have included coating the electrode surface with porous polymeric films. A film of cellulose acetate has been used with the aim of protecting platinum electrodes from deactivation by serum proteins in an assay for hydrogen peroxide.[24]

Other forms of solid carbon have been used in EDs although most have not found wide applicability in commercial systems. They include pyrolytic graphite,[25] low-temperature isotropic carbon,[26] reticulated vitreous carbon[27] and carbon cloth.[28] Flow-through electrodes can also be constructed from carbon. This is done either by compressing carbon granules to give a porous bed similar in form to a short HPLC column or by constructing a cell from pre-formed elements such as carbon gauze or hollow carbon rods. Although seemingly attractive to use in flowing systems, hollow tubular electrodes have proved unpopular in practice, although they were, in the past, employed in at least one commercial system.

3.3 Electrode Shielding and Grounding

Basic approaches to noise minimisation in ED are improved electrical shielding, efficient earthing, and thermostating of the detector, or at least of the cell. There has been much discussion about the need for a common earth and for Faraday shields for EC detectors but there appears to be little consensus. Common earthing (grounding) seems useful in some EC laboratories, particularly if the external electricity supply and the connection to the supply are less than optimal.[29] Common earthing is easily implemented and should include all parts of the system including stainless steel tubing carrying the column eluent. Even with the most sensitive assays we have found no benefit in constructing elaborate Faraday cages from aluminium foil or mesh with respect to noise or spike reduction, although other workers have claimed benefits. It is often difficult to decide if such cages are electrically beneficial or simply serve to reduce thermal variations (draughts).

3.4 Flow-Cell Designs

Of the various EC detector cell designs (Figure 3.4), the thin-layer and wall-jet configurations are commonly used in commercially available amperometric

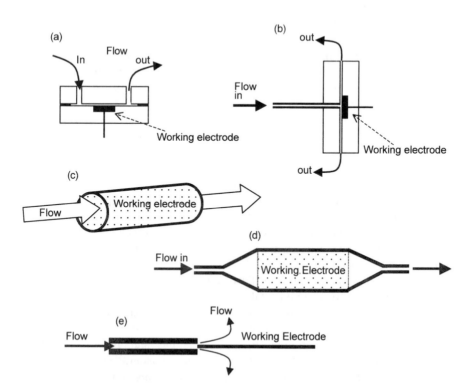

Figure 3.4 *Simplified designs of chromatographic ED cells: (a) Thin-layer cell; (b) Wall jet cell; (c) Tubular cell; (d) Porous flow-through cell; and (e) Single fibre electrode.*

detectors, although it is possible to work coulometrically using a porous cell provided a large enough surface area is used. However, the trend appears to be towards smaller surface areas to make the detectors compatible with microbore HPLC systems and to reduce noise, which is proportional to surface area. Of the porous flow-through designs the most frequently encountered are the porous graphite type used in several of the ESA Coulochem detector cells. Tubular electrodes have not been commercially exploited and the single fibre electrode is chiefly confined to use with miniaturised systems and CE.

3.5 Chromatographic Factors Affecting Flow-Cell Design

As well as the constraints imposed by the laws of electrochemistry, many chromatographic factors are also important in flow-cell design. Good mass transfer characteristics, electrical isolation and resistance to common HPLC solvents are important, but as with all HPLC systems extra-column band broadening must be minimal if full use is to be made of the column efficiency that can be attained. The total band broadening of an analyte peak is given by combining the individual variances (σ) caused by all the components in the flow stream, plus any other factors that cannot be defined (Equation 3.2).

$$\sigma^2_{total} = \sigma^2_{injector} + \sigma^2_{inlet\ tubing} + \sigma^2_{column} + \sigma^2_{detector\ path} + \cdots \qquad (3.2)$$

The maximum permissible cell volume (V_{opt}) in any type of HPLC detector is approximated by:

$$V_{opt} = \frac{qt_r u}{\sqrt{N}} \qquad (3.3)$$

where u is the flow velocity, and q is the maximum allowable relative degradation in efficiency (N) for the sharpest peak of interest eluting at time (t_r).

Typically q should not exceed 10% for the narrowest peak in the chromatogram. So for a 1 mm i.d. column packed with 3 μm particles and generating 25,000 theoretical plates, the maximum detector cell volume should be of the order of 1 μL.

Because of the low diffusion distances in ED, the swept detection volume of most types of amperometric flow-cell is very small. In fact, EC detector flow-cells have some of the lowest volumes of any type of HPLC flow-cell (normally <1 μL), and in this respect they easily comply with the criteria for minimising band broadening. Such characteristics make ED ideal, not just for standard HPLC detection, but also for micro-column LC (μLC), capillary LC and CE applications. Nevertheless, both manufacturers and users need to pay due attention to the overall design and assembly of their HPLC systems if they are to benefit fully from miniaturisation. The linking of the cell to the column should be *via* tubing of the minimum i.d. and shortest possible length. Ideally the column and cell should be mounted in the same thermostat-regulated housing and if possible directly coupled to each other. If dual detection is to be used then post-detection flow paths must also be very small. In some early cell designs, although the swept volume to the working electrode was less than 1 μL, the volume to the reference and counter electrodes could be up to 1 mL. Such systems would be totally unsuitable if the eluate was then to pass to another detector. Modern EC designs have very small swept volumes to the working electrode and more modest post-electrode volumes.

Two other design aspects need to be considered. The wetted parts of the EC detector must be constructed from materials inert to the solvents used in HPLC, such as methanol, acetonitrile and dilute aqueous buffers with pH values generally in the range 2–8. PTFE and stainless steel are the favoured materials of construction. Although most cells will not be subject to excessive pressures, they should be able to withstand the modest back pressure created by coupling a second detector in series or the use of a flow restrictor designed to prevent bubble formation.

3.6 Circuitry of an Electrochemical Detector

Although it is not necessary to understand electronics in order to use an EC detector, it is useful to have some idea of the circuitry involved, in the event that a detector is to be modified for a specific application. Simple circuit diagrams for potentiostats have been reported.[14,30–32] They have two elements, *viz.* potential control and current monitoring. A circuit for a self-metering detector built from available integrated circuits has also been described.[33] Circuits for dual and pulsed detectors will obviously be more complex. Since many modern EC analysers are computer controlled, it is no doubt possible to do the same for HPLC-ED detectors, but such detectors do not seem to be commercially available.

4 Amperometric Detection

4.1 Thin-Layer Cells

In the thin-layer cell, which was used in the first commercial HPLC-ED detector, the eluate enters the detector cell *via* a suitable micro-port and then flows through a narrow channel cut into a plastic gasket over a 'solid' working electrode. This is shown diagrammatically in Figure 3.4(a). The flow continues past the reference and counter electrodes. The body of the cell is formed from two solid blocks, one of which contains a force fitted disc or short rod of electrode material. The blocks are separated using a thin PTFE gasket as a spacer around the working electrode as well as the inlet and the outlet orifices of the cell. The block containing the working electrode is usually constructed from an inert material, such as PTFE or Kel-F. The second cell-block may be made of stainless steel and forms the counter electrode. Connection of these electrodes to the electronics of the system is usually *via* simple gold pins and suitable electrical connectors. This simple classic design was pioneered by Kissinger[34] and has formed the basis of the majority of cells in commercial instruments, including those supplied by BAS (Figure 3.5), Waters and Agilent.

The current produced in this type of electrode is described by the equation given in Table 3.1. From this, it is clear that for maximum current (sensitivity) thin-layer cells should be designed to:

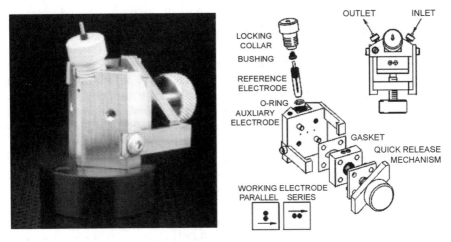

Figure 3.5 *Thin-layer cell design manufactured by BAS. The lower diagrams show how the two electrodes can be configured in the eluent stream.*

1. Maximise the electrode area
2. Use the thinnest possible gaskets
3. Operate at a high flow-rate

In practice, imperfections in the planarity of the electrode blocks limit the dimensions of the gaskets that can be employed. Increasing the surface area and the flow-rates will raise the background current, whilst in many cases flow-rates are limited by the chromatographic system. Typical electrode parameters are therefore an area of 15–20 mm^2 with gaskets of 20–100 µm giving a cell volume of less than 1 µL and capable of working with flow-rates of *ca* 1 mL min^{-1}.

Advantages of thin-layer cells are low cost of construction (only a means of clamping the two cell-blocks firmly together is required), ease of disassembly for maintenance (particularly polishing the working electrode), the variety of possible configurations (see below), the possible use of a variety of materials for the electrodes and cell body, rapid response times and easily adjustable (low) cell volumes giving minimal extra-column dispersion. Disadvantages include small electrode area, difficulty in achieving very small distances between the electrode blocks and the inability to withstand a back pressure when used in series with other detectors.

4.2 Wall-Jet Cells

In this type of cell the eluate is delivered, *via* a small orifice, into the cell and directly impinges as a jet at right-angles onto the face (wall) of the working electrode before flowing radially across the surface (Figure 3.4(b)). A wall-jet electrode was applied to flow injection analysis (FIA) in 1973 by Yamada and Matsuda[35] and very shortly afterwards applied to HPLC by Fleet and Little.[36] For many years, the HPLC cell designed by Fleet and Little was manufactured and sold by EDT in London. There are still several commercial instruments that use wall-jet electrodes (*e.g.* Antec, The Netherlands; BAS, USA; GBC Pty, Australia). The use of wall-jet detectors has been extensively reviewed.[37]

The signal is dependent on the size of the electrode and the diameter of the jet (Table 3.1). Since the flow of electrolysed eluate is away from the electrode, the design is intended to sweep the surface clean. High flow velocities are important. There are a number of other critical factors to consider. It must be assumed that the jet does not break up before it hits the electrode, an assumption that Fleet and Little[36] confirmed by showing that the jet remained intact for a distance of up to 10 mm. Secondly, it is important not to restrict the radial flow from the zone where the jet impinges on the electrode. Finally, the electrode diameter should be at least ten-fold greater than the jet diameter. In contrast to thin-layer designs, the jet need not be constrained, the cell can be unsealed and performance should not be degraded by trapped air-bubbles. However, it is unsuitable for coupling to a second detector and should therefore be last in line if serial detection is employed.

Other early designs used a ring counter (auxiliary) electrode, which surrounded the working electrode,[3,38] but these were not popular at the time. Mini- and micro-bore HPLC columns typically use eluent flow-rates of <100 µL min^{-1}. This means that the flow velocities in thin-layer cells would be reduced to the point where the electrochemistry became inefficient by not removing electrochemically generated products. A new design of wall-jet cell incorporates a flow-path in which the jet impinges onto a 3 mm electrode and travels across the surface to a radial outlet channel, which is connected to a solid state reference electrode composed of either Ag or Pd. This design, the BAS UniJet[39] (Figure 3.6), is claimed to offer a four-fold improvement in sensitivity over a standard thin-layer design when used with a 320 mm i.d. capillary column at an eluent flow of 100 µL min^{-1}.

4.3 Dropping Mercury Electrodes

Because of their fluidity, dropping mercury electrodes are not ideal in flowing systems because they change in shape, hence in electrical capacity, with variations

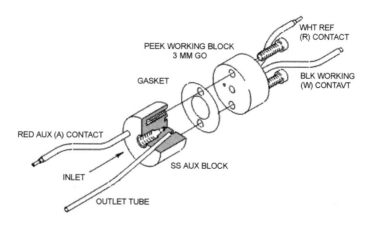

Figure 3.6 *BAS UniJet cell design.*

in eluent flow-rate, but at least one company (EG&G) manufactures such a detector particularly for reductive EC measurements. In this system a fine capillary is connected to a mercury reservoir *via* a precision micro-valve. A static mercury drop of fixed dimensions is formed either automatically or manually at the end of the capillary to create the electrode. The column eluate is directed at this drop. The drop is automatically re-generated every few seconds. The electrode can operate at +0.4 to −2.2 V depending on the eluent.

4.4 Operation of Amperometric Electrodes

Modern amperometric detectors possess a number of useful features. These include rapid response time, low cell volume, ease of access to the electrode surface for cleaning, ability to be used in series with other detectors, good sensitivity with suitable analytes, in-built facilities for scanning the detection potential and minimal running costs. Since amperometric electrodes are small it is possible to incorporate more than one into a thin-layer electrode block. Most commercial cells usually contain two electrodes with their necessary connections. At the simplest level this allows the rapid connection of the second electrode when the first becomes contaminated. The electrode connections are simply transferred to the other electrode pin without having to dismantle the cell. A discussion of the use of two or more amperometric electrodes for differential analysis is given below (Section 7).

As with all HPLC-ED, the principal disadvantage of amperometric detectors is that the working electrode can be easily deactivated as a result of accumulation of reaction products and possibly other contaminants on the electrode surface. The speed with which deactivation occurs varies depending on the nature of the electrode, the mode in which the electrode is used, the applied potential, the pH and composition of the eluent and the nature and amount of material injected. At higher applied potentials the risk of electrode deactivation is greater. This is because a higher proportion of the compounds reaching the electrode will undergo EC reactions at the electrode surface. Obviously at higher applied potentials the standing current will be greater and thus the risk of increased background noise/drift will be greater, hence the risk of interference will also be greater. Thus, in addition to enhanced selectivity, the decreased risk of electrode deactivation is a further factor favouring the use of as low an applied potential as possible. An alternative way of minimising electrode contamination is to arrange for the column eluate to by-pass the detector until the peaks of interest elute.[40] However, this is only possible if a stable baseline and reproducible response to the analyte(s) of interest can be obtained when flow to the detector is restored.

Because electrode deactivation is a constant worry when using ED, especially when working at higher applied potentials, frequent analysis of appropriate quality assurance specimens is necessary to ensure the reliability of quantitative data. Although the response of a deactivated electrode may sometimes be restored temporarily by using a slightly higher applied potential, the most reliable way of restoring the response is to disassemble the cell and to clean the surface of the electrode (Chapter 4, Section 1) unless the electrode can be cleaned electrochemically as in PAD.

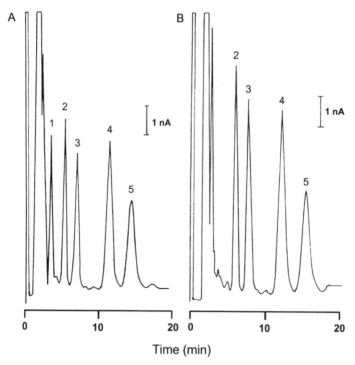

Figure 3.7 *Use of serial GCEs [electrode 1: (A) −0.8 V, (B) 0 V; electrode 2: (A) and (B) +0.85 V vs Ag/AgCl] to facilitate the ED of fluphenazine sulfoxide. Column: 300×3.9 (i.d.) mm μ Bondpak ODS-silica 10 μm. Eluent: methanol-formic acid (0.15 mol L^{-1}) containing dibutylamine (10 mmol L^{-1}), pH 3.2 (50 + 50); Flow-rate: 2.0 mL min^{-1}; Injection: lyophilised 3-methyl-1-butanol (isopentanol) extract of plasma containing added fluphenazine, fluphenazine sulfoxide and other fluphenazine metabolites (amounts and injection volume not given); Peaks: 1 = fluphenazine sulfoxide, 2 = 7-hydroxyfluphenazine, 3 = 8-hydroxyfluphenazine, 4 = fluphenazine, 5 = fluphenazine N-oxide.* With permission from Hoffman et al.[41]

In recent years, dual glassy carbon electrodes have been used either in parallel (to increase the surface area of the electrode and hence the response) or in series. In this instance several modes of operation become possible as discussed for coulometric operation below (Section 5). An example of the application of serial glassy carbon electrodes is given by the use of a negative potential (−0.8 V *vs* Ag/AgCl) at the first electrode to reduce fluphenazine sulfoxide to fluphenazine, which can then be detected at the second electrode (Figure 3.7) held at +0.8 V *vs* Ag/AgCl.

5 Coulometric Detection

In coulometric mode, in which EC conversion is considered to be complete, it can be shown that the steady state current (*i*) is dependent on the rate at which analyte is delivered to the electrode and is given by:

$$i = nFC\,u \tag{3.4}$$

where u is the volume flow-rate.

Thus increasing the flow-rate will increase the peak height but the peak area remains constant over a wide range of flow-rates. Coulometric detection is therefore an absolute method, such that peak area can provide a means of quantification by relating area directly to sample mass using Faraday's Law, provided the number of electrons being transferred is known.

When an electroactive analyte passes through a coulometric cell its concentration decreases with time according to first order kinetics:

$$\text{Rate of electrolysis} = \frac{dC_t}{dt} = -kC_t \tag{3.5}$$

So:

$$C_t = C_0 \cdot e^{-kt} = C_0/2^N \tag{3.6}$$

where C_t is the instantaneous concentration (at time t)
$\qquad$ C_0 is the initial concentration
$\qquad$ k is the reaction rate constant or cell constant
$\qquad$ N is the number of half times
$\qquad$ t is the time

To achieve coulometric operation it is necessary to have a very large rate constant. Since this reflects primarily mass transport in the cell it can be shown that:

$$k = \frac{AD}{Vd} \tag{3.7}$$

where A is the electrode surface area (cm^2)
$\qquad$ D is the diffusion coefficient ($cm^2\ s^{-1}$)
$\qquad$ d is the diffusion layer thickness (cm)
$\qquad$ V is the cell volume (cm^3)

Figure 3.8 shows the relationship between residence time and the percentage of the sample that has been electrolysed. Maximising k is achieved by optimising the area-to-volume ratio of a cell and minimising the diffusion layer. It has been estimated that a new ESA Model 5010 cell has a rate constant of 500 s^{-1} which, at a flow rate of 1 mL min^{-1} and a cell volume of 4.5 μL, equates to *ca* 200 half-times. A consequence of this is that up to 96% of the electrode surface can be fouled whilst only reducing the efficiency of electrolysis by 0.5%. Thus, electrode cleaning is needed less frequently.

One way of maximising the area-to-volume ratio is to create a porous working electrode with a very large active surface. This can be achieved in various ways, *e.g.* use of stacks of metal gauze, sintered metal tubes, reticulated vitreous carbon or packed beds along the lines of HPLC columns composed of conducting material, such as carbon particles or silver granules. One method of construction is to fuse

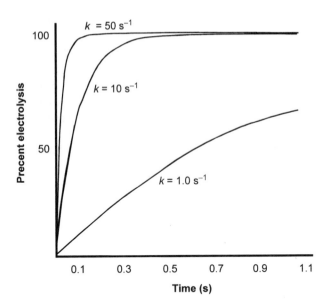

Figure 3.8 *Degree of electrolysis as a function of time for three values of cell constant. The large surface areas of PGEs produce* $k = 500\ s^{-1}$.
Redrawn from Coulochem II Operating Manual (ESA, Bedford MA, USA, 1992; Figure 5.5).

a mixture of carbon particles with a solid soluble salt, such as sodium chloride, at high temperature and then dissolve away the salt with water leaving a porous carbon form. For such a packed bed electrode the surface area can be derived in the same manner as for an HPLC column:

$$A = 6\frac{(1-e)}{d_{\mathrm{p}}} \qquad (3.8)$$

where d_{p} is the particle size and e is the bed porosity.

It can be calculated that for a cross-sectional area of 20 mm^2 and flow velocity of 8 mm s^{-1} then a bed length of *ca* 1 mm would give coulometric efficiency. Such a flow-through porous electrode design (porous graphite electrode, PGE) forms the basis of the cell in the ESA Coulochem detectors (Figure 3.9). In this latter cell there are two working electrodes, four counter electrodes and four reference electrodes. The electrodes are sealed within a flow-through housing. Full coulometric operation is available only in the standard cell that consists of two porous carbon electrodes each giving 100% EC conversion (see below for modes of operation). As the volumes of the flow through electrodes are less than 5 μL they should ensure minimal extra-column band broadening.

Variants of the standard cell include a version that offers one coulometric electrode with an amperometric analytical electrode with 5–10% EC conversion, and another with one coulometric cell and a removable wall-jet amperometric electrode. Since coulometric cells are flow-through they have some degree of resistance to flow and with use can develop a significant back pressure. To minimise

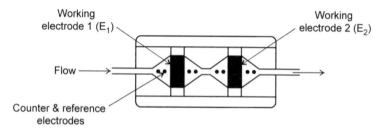

Figure 3.9 *Schematic diagram of the Coulochem Model 5010 Cell, showing the relative positions of the working electrodes (PGEs) and counter and reference (Pd) electrodes.*

such back-pressure changes, they need to be protected from particulate materials. Their intrinsic back pressure should also be borne in mind when connecting other types of HPLC detector cell in series.

5.1 Modes of Operation

At present the ESA Coulochem is the only commercially available 'coulometric' detector. Several different types of cell are now available, most of which contain two analytical electrodes in series. The standard cell (Model 5010) contains equal sized PGEs, but in the high sensitivity cell (Model 5011) the surface area of the second electrode is reduced with the aim of producing a better S/N ratio. Despite the argument that the increased surface area available in PGE systems only increases the noise in proportion to the increased signal,[42] such systems do appear to give an enhanced S/N ratio in certain applications, as compared to glassy carbon electrodes in thin-layer or wall-jet assemblies, and may offer more flexibility in routine use. Even better S/N ratios are claimed for the Model 5014 cell, which has been developed specifically for use with microdialysis samples.

Later ESA models, such as the Coulochem II and III, have the facility to operate in a pulsed mode (Section 6) and are available with a 'sugar' cell (Model 5040 with a gold working electrode) designed principally for the analysis of compounds containing aliphatic hydroxyl moieties. The Model 5040 cell can also be used with other working electrodes, including Pt, Ag and glassy carbon arranged in a thin-layer configuration. Additionally, there is a 'guard' cell (Model 5020) for pre-treating the eluent with the aim, for example, of minimising the background signal. The 'conditioning' cell (Model 5021) has been designed to be inserted been the column outlet and the detector, so that analytes can be oxidised or reduced prior to detection at an analytical cell[43] or interfering compounds can be oxidised at a lower potential before the analytical cell.

An operating potential can be selected by studying the response of a given quantity of analyte on the HPLC system at increasing potentials. Alternatively, the in-built potential scanning facility may be used. Because of coulometric conversion such a scan should be carried out with analyte flow through the cell. The easiest way to do this is to connect the detector directly to an injection valve fitted with a large (>2 mL) sample loop; 5 m of 0.05 inch (i.d.) tubing (*ca* 6 mL total

(a) Morphine   (b) Codeine

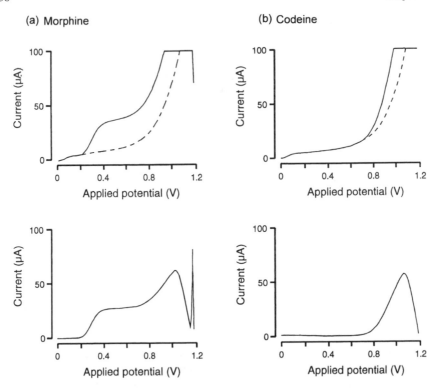

Figure 3.10 *Current-voltage plots (upper) for solvent (- - -) and drug + solvent and hydrodynamic voltammograms (bottom) [(signal given by drug + solvent) – (signal given by solvent)] for morphine and codeine obtained using the scanning facility on the ESA Coulochem (PGE). Both compounds (1 mg L^{-1}) were dissolved methanol-aq. ammonium nitrate buffer (0.1 mol L^{-1}, pH 9.5) (90 + 10).*

volume) is suitable. The eluent and eluent containing the analyte are then scanned (Figure 3.10). By recording the output signal digitally, the background due to the eluent can be subtracted to give a typical current-voltage curve. For morphine, the half-wave potential at approximately +0.3 V (*vs* Pd) is due to oxidation of the phenolic hydroxyl, whilst for both morphine and codeine the half-wave potential at approximately +0.9 V is due to oxidation of the tertiary amine moieties. Note that the optimum potential for oxidation of the tertiary amine moieties can only be measured after subtracting the signal due to the eluent. Scanning samples in this way is particularly useful for studying the effect of eluent composition (pH, ionic strength) on the response of a given analyte. To perform such a study by injecting samples onto an HPLC column would require days of work. A further use of scanning is to investigate candidate HPLC eluents to ascertain the useable range of potentials.

Coulochem cells with dual PGEs can be operated in three modes, often referred to as 'screen', 'difference' and 'redox'.[44] In *screen mode* the potential of the first electrode (E_1) in the analytical cell is set, typically, 0.2–0.3 V below that of the second electrode (E_2). (*N.B.* E_1 and E_2 will be used throughout to describe the first and second electrodes in Coulochem analytical cells. However, different nomenclature is sometimes used, including 'Cells 1 and 2' and 'T1 and T2'.) The aim is to reduce the concentration of any unwanted readily oxidisable components reaching the second electrode. This in turn should reduce the standing current and thus improve the S/N ratio obtained from the second electrode. Additional screening can be obtained by inserting a conditioning cell in line between the injection valve and the analytical column. The signals from the first and second electrodes used in 'screen' mode for the analysis of a mixture of prilocaine and a metabolite, 2-methylaniline, are shown in Figure 3.11 (a). The metabolite is readily oxidised at the first electrode and only a small proportion survives to reach the second. With serial dual electrode amperometric detection only a small proportion of the metabolite would have been oxidised at the first electrode. The higher applied potential of the second electrode is required to oxidise the aliphatic secondary amine moiety of prilocaine itself. The responses obtained from lignocaine (lidocaine) and a metabolite, 3-hydroxylignocaine, using the same electrode combination are shown in Figure 3.11 (b). The phenol moiety in the metabolite is oxidised at the first electrode, but both compounds respond at the second electrode as they contain the same aliphatic amine moiety.

A further way of reducing interference is to use *difference mode*. The Coulochem has three outputs: those for the two electrodes, E_1 and E_2, and the difference (E_1 minus E_2). If the electrodes are set to similar potentials, such that the analyte of interest is oxidised at the first electrode with little or no signal at the second, then the difference will contain a signal due to the analyte. If impurities produce similar

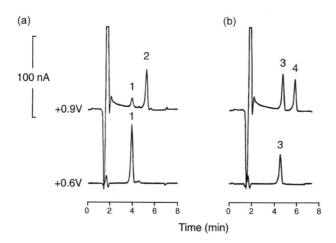

Figure 3.11 *Chromatograms obtained using the ESA Coulochem in 'screen' mode (PGEs, E_1: +0.6 V, E_2: +0.9 V vs Pd). Column: 150×4.6 (i.d.) mm Spherisorb S5CN cyanopropyl-modified silica; Eluent: acetonitrile-aq. phosphoric acid (14 mmol L^{-1}) (30+70). Flow-rate: 0.8 mL min^{-1}; Injections: (a) 50 μL aqueous 2-methylaniline (1) (0.01 mg L^{-1}) and prilocaine (2) (1 mg L^{-1}); (b) 50 μL aqueous 3-hydroxylignocaine (3) (0.1 mg L^{-1}) and lignocaine (4) (1 mg L^{-1}).*

responses at each electrode then their contribution to the signal in the difference channel will be small. The difference mode is especially useful for quantifying peaks on the solvent front or on a sloping base line.

The third mode of operation, *redox mode*, can be used if a product of oxidation or reduction is stable and it is possible to reduce or oxidise the reaction product at the second electrode. The oxidation and reduction signals produced at E_1 and E_2, respectively, for a plasma sample containing paracetamol (acetaminophen) are shown in Figure 3.12. Depending how other species respond at E_1 and E_2 either output may be used for quantification. As the reduction signal is negative with respect to the oxidation signal, the difference, E_1-E_2, in effect sums the two to produce a larger response. It is usually possible to use one or all three channels for quantification. In general, experience suggests that the RSDs for replicate analyses are lowest when the oxidation signal is used. However, the additional information provided by the reduction channel gives some indication of peak purity. Occasionally the oxidation signal may be lost beneath an impurity, but the reduction channel is unaffected and can thus be used in the analysis. This mode is particularly useful for catecholamine analysis; ascorbic acid added as a stabiliser does not undergo redox and so is not detected in the reduction channel.

A specialised form of redox mode is when the conditioning electrode is used to produce species that are then detected at the analytical electrode(s). The conditioning cell was developed primarily for catecholamine analysis and, like the analytical cells, is designed to minimise peak broadening. The guard cell, on the other hand, is

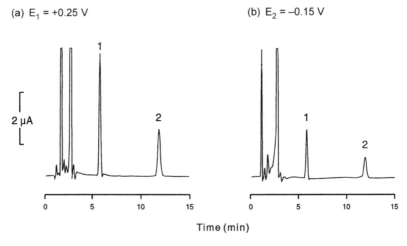

Figure 3.12 *Analysis of paracetamol using the ESA Coulochem in 'redox' mode (PGEs, E_1: +0.25 V, E_2: −0.15 V vs Pd). Column: 150×4.6 (i.d.) mm Ultrasphere ODS; Eluent: acetonitrile-aq. phosphoric acid (10 mmol L^{-1}) (7+93); Flow-rate: 1 mL min^{-1}; Injection: 50 μL supernatant obtained by mixing standard solution of paracetamol (5 mg L^{-1}) in neonatal calf plasma (100 μL) with aqueous trichloroacetic acid (5% v/v) containing N-propionyl-4-aminophenol (internal standard, 0.5 mg L^{-1}) (1 mL); Peaks: 1 = paracetamol, 2 = N-propionyl-4-aminophenol.*
After Whelpton et al.[45]

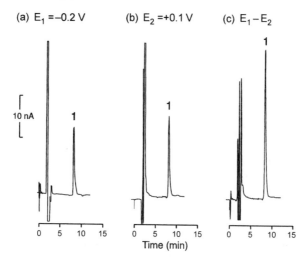

Figure 3.13 *Analysis of physostigmine using the ESA Coulochem [PGEs, guard cell (positioned between column outlet and analytical cell) +0.8 V, E_1: −0.2 V, E_2 + 0.1 V vs Pd]. Column: 150×4.6 (i.d.) mm Spherisorb S3W silica; Eluent: acetonitrile-methanol-aq. ammonium nitrate buffer (0.1 mol L^{-1}, pH 9.5) (45 + 45 + 10); Flow-rate: 0.75 mL min^{-1}; Injection: 50 µL methanol containing 1 ng physostigmine (1).*

designed for insertion between the pump and the injection valve in order to oxidise impurities in the eluent. Despite this, guard cells can often be used as conditioning cells with little noticeable effect on the efficiency of the system. The utility of this approach can be seen from Figure 3.13. Physostigmine was oxidised at a guard cell (+0.8 V) positioned between the outlet from the column and the analytical cell. The species produced was reduced (−0.2 V) and re-oxidised (+0.1V).[46] This gave a marked improvement in S/N ratio when compared to the use of screen mode described previously (graphite electrodes, E_1: +0.45 V, E_2: +0.65 V vs Pd).[47]

A further advantage of PGEs is that stoichiometric conversion can be performed under appropriate conditions. Sometimes this can be exploited to improve non-ED detection. For example, amoxicillin can be electrochemically oxidised post-column and then detected by fluorimetry.[48] This approach may be especially useful when the standing current/background noise is too high to permit direct ED. A further development of the Coulochem was the use of serial PGEs each operating at slightly increasing applied potentials – a multi-electrode array (Section 7).[49,50]

6 Pulsed Detection

Pulsing the applied potential at the working electrode using simple waveforms has long been employed in EC analysis and was also proposed in some of the earliest patents for EC detectors for liquid chromatography. In the early 1980s, Johnson and co-workers at Iowa State University began an extensive series of studies on pulsed amperometric detection at platinum electrodes in a flow injection system of simple

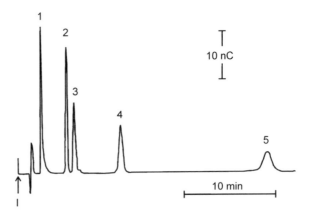

Figure 3.14 *Detection of carbohydrates using PAD following isocratic separation by liquid chromatography. Peaks: (1) glucitol, (2) glucose, (3) fructose (4) sucrose and (5) maltose.*
(Redrawn from LaCourse and Johnson)[53]

alcohols[51] and of carbohydrates.[52] However, ED for HPLC with PAD capabilities only became commercially available in the mid-1980s with the introduction of electronically controlled pulsed detection by Dionex. Initially this was developed for the measurement of carbohydrates at gold electrodes in very alkaline eluents (Figure 3.14). Modern electronics are capable of reproducibly generating a variety of waveforms. A typical pulse sequence is shown in Figure 3.15. The initial

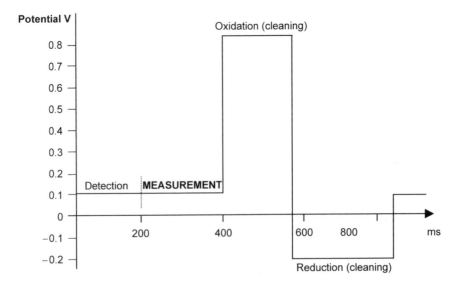

Figure 3.15 *A pulse sequence of the form typical for the measurement of aldehydes and alcohols.*

detection potential is held for a few hundred milliseconds, this is followed by a short period in which the response is sampled to give the measured output. Although the electrode itself may not be able to respond as rapidly, its response will be reproducible so allowing quantitative performance. During this detection stage the noble metal electrode is fouled when being used to detect aliphatic compounds, such as sugars. The next step of the waveform represents anodic (oxidative) desorption generating a oxide layer on the electrode plus complete oxidation of the analyte residues to water and CO_2. The next step reduces the electrode surface back to the noble metal. Finally the detection potential is re-established.

Subsequently pulsing profiles have been employed to maintain electrode performance for some other difficult analytes, such as thiols (Chapter 5), and inorganic ions, such as sulfite in foods. A detailed description of the use and applications of PAD has been provided.[54]

7 Multiple Detection Systems

Unlike Coulochem detectors (Section 5) in which the electrodes are arranged in series, amperometric detectors, containing two or more electrodes, can be used either in parallel or in series. In the first approach, the eluate simultaneously passes over two electrodes, which can be held at different working potentials. The benefits of this approach are that the ratio of currents at each electrode can help provide peak purity and identification information. Oxidation and reduction reactions can be carried out simultaneously. Signals from reactions requiring both low and high potentials can be monitored at different sensitivities. A difference signal can be used to remove a common interference.

In the dual series configuration the eluate first passes over one electrode held at one potential, and then over a second electrode possibly held at a different potential. In thin-layer cells the two electrodes are contained in the same cell-block. The benefits are that the ratio of currents at each electrode can again help provide peak purity and identification information. Selectivity can be enhanced by partial oxidation at the first electrode. Oxidation/reduction reactions can be carried out in series, as for disulfides, which can first be reduced followed by detection of the thiol at the second electrode (Chapter 5). Signals from low and high potential reactions can be monitored at different sensitivities. A difference signal can be plotted to remove a common interference.

7.1 Array Detectors

Dual electrodes in series are used relatively frequently. Matson *et al.*[55] extended this approach to an *n*-electrode, three-dimensional HPLC-ED system. Using coulometric flow-through electrodes, up to 15 such electrodes were incorporated into one housing. Each electrode was held at incremental working voltages from 0 to 650 mV and a series of chromatograms were obtained simultaneously. Each electrode was monitored every 50 ms and the output from each electrode was

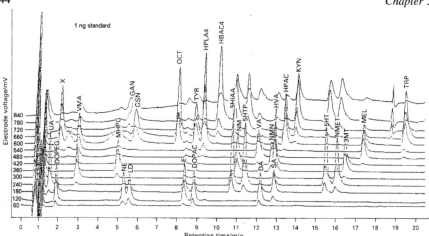

Figure 3.16 *HPLC and EC resolution of a wide range of neuroactive substances using a 16-electrode coulometric array detector.*[49] *Conditions: gradient elution 1% methanol to 40% in phosphate buffer (0.1 mol L⁻¹, pH 3.4); Injection: 30 component standard solution (1 ng).*
(Figure courtesy ESA)

Compound name	Compound abbreviation	Oxidation Pontential (mV)	Retention Time (min)
Dihydroxyphenylacetic acid	DOPAC	150	8.91
Dihydroxyphenylethyleneglycol	DOPEG	180	1.95
l-Dopa	LD	150	5.61
Dopamine	DA	150	12.13
Epinephrine	E	180	8.35
Guanine	GAN	700	5.65
Guanosine	GSN	840	5.89
Homovanillic acid	HVA	450	12.97
Hvdroxybenzoic acid	HBAC	700	9.95
Hydroxyindoleacetic acid	HIAA	180	10.75
Hydroxyphenylacetic acid	HPAC	650	13.32
Hydroxyphenyllactic acid	HPLA	650	9.32
Hydroxytryptophan	HTP	300	11.42
Kynurenine	KYN	800	13.82
Melatonin	MEL	600	17.36
Metenephrine	MN	480	10.98
Methoxyhydroxyphenyl glycol	MHPG	450	4.95
Methoxytyramine	MT	450	16.48
N-Methylserotonin	NMET	300	12,13
Norepinephrine	NE	180	5.16
Normetanephrine	NMN	480	12.87
Salsolinol	SA	180	12.98
Octopamine	OCT	620	7.83
Serotonin	HT	180	15.46
Tryptophan	TRP	600	19.11
Tyrosine	TYR	650	8.94
Uric acid	UA	300	1.62
Vanillic acid	VA	480	12.03
Vanillylmandelic acid	VMA	300	2.97
Xanthine	X	70	2.02

captured and displayed by a computer in a 3-D format together with time and working potential (Figure 3.16). Since the electrodes were in series and were coulometric, some analytes were totally removed at appropriate electrodes so leading to an apparent improvement in the chromatograms obtained. The electrode array design is commercially available from ESA Coulochem with 8–16 electrodes in series, each independently under full computer control. The overall concept of the electrode array system is similar to that of the diode array detector. However, in a diode array detector a single beam of light passes through a *single* sample cell and is split into appropriate wavelengths prior to impinging on the diode array, rather than there being a series of individual electrodes.

Array systems have been applied to the analysis of neurotransmitters in brain and other biological samples[49,55] and of neurotransmitters, amino acids, drugs and drug metabolites (apomorphine, hydralazine, isoprenaline (isoproterenol), methoxamine, morphine and morphine-3-glucuronide and phenylephrine) in microdialysis fluid obtained *in vivo*.[56] In addition, a gradient elution method for the analysis of 27 aromatic amines, phenols and phenylurea herbicides, including 2,4,5-trichlorophenol, 2,4-dinitrophenol, 1,2-phenylenediamine, diuron and linuron, in drinking water has been described.[57] An ODS-modified silica analytical column was used with an electrode array (four cell modules each containing four PGEs arranged in series, E_1: 0 V, E_2: +0.08 V, then in increments of 0.08 V to E_{16}: +1.20 V *vs* Pd). The analysis time was 63 min. Sample preparation was by SPE (ODS-modified silica) and LoDs were 0.5 ng L^{-1} or better (1 L sample).

7.2 Rapid Scanning Detectors

Rapid scanning voltammetry forms the basis of a type of EC detector in which the working potential at a single amperometric electrode is varied with time and the resulting current is measured as a function of analysis time and potential.[58] Effectively it is a system for continuously obtaining cyclic voltammograms. Such systems were originally developed for dropping mercury electrodes in static polarography systems. To gain useful information from such a system, when it is being used to monitor a chromatographic eluate requires that the potential range to be scanned is approximately 1 V with a scan rate of *ca* 1 V s^{-1}. A number of waveforms can be used, *i.e.* square-wave, normal pulse and staircase. The staircase form is considered the most suitable for use with solid electrodes because the equal voltage steps lead to more consistent background currents, which can then be subtracted from the signal. The resulting chromato-voltammogram is the ED equivalent of the rapid scanning optical detector. The most obvious advantage of rapid scanning detection is that a voltammogram for each chromatographic peak is obtained. Although not such a true physicochemical measurement as a spectrum, the result is still information rich especially with respect to confirming peak purity and helping with identification.

7.3 Use in Series With Other Detectors

The most common dual arrangement with ED is in series with a UV detector. The order in which the detectors are placed will be dictated by the requirements of the assay and by the design of the detectors. Theoretically, the detector with the smallest cell volume, *i.e.* lowest dispersion, should be placed first. The working volume of EC detectors is usually very small, but the dead volume of many EC flow-cells after the working electrode is often high because of the presence of large, poorly swept volumes around the reference electrode. Secondly, unlike many EC cells, most UV flow-cells can withstand modest backpressures, *e.g.* up to 400–500 psi, and similar back pressures are often inadvertently created when coupling flow-cells. Even so, as in all coupled systems care must be taken to keep pressures low to avoid leaks or cracked flow-cell windows. This is even more important if a fluorescence detector is employed since their flow-cells can often only withstand a few bar before leaking.

Special care should be taken when coupling detectors with PGEs. The Coulchem guard cell is designed to withstand back pressures of 6000 psi, but contamination of analytical PGEs can lead to rapid increases in resistance and back pressure. Consequently, it is recommended that PGE cells are first in any series of cells. If the ED is operated in the coulometric mode, the production of new chemical species may have a marked effect on the spectral characteristics of the parent compound(s) and thus of the response. This may, as in the case of amoxicillin (Section 5.1), be exploited to produce a species with improved spectral characteristics, or it may lead to a loss of signal at subsequent detectors.

8 Micro-Detectors

The combination of low diffusion distances and small cell volumes make ED ideal for use with small bore (1 or 2 mm i.d.) and capillary HPLC columns. The only change usually required may be to reduce the length and bore of the column-to-detector connections. Changing from 'standard' 4.6 mm i.d. columns to 2.1 mm or 1 mm columns will theoretically increase detector sensitivity by 4.8- and 21-fold, respectively. In practice the measured improvements are somewhat lower (Figure 3.17), but even then are still appreciable.

Miniaturised EC detectors have been developed based on the designs already described. Baur and Wightman[59] built a carbon fibre micro-cylinder electrode with a volume of only a few pL. Decreasing the surface area of the fibre reduces the background current hence the detector is less susceptible to flow-rate fluctuations. Using fine wire or carbon fibre electrodes can result in further miniaturisation. Neurochemists led the way in the application of carbon fibre electrodes for *in vivo* analysis of neurotransmitters.[60,61] The first application of similar devices in the separation sciences came much later, notably when White *et al.*[62] used a carbon fibre electrode and a step waveform to detect 0.1 μmol L^{-1} of hydroquinone. An additional advantage of such carbon fibre designs are their low cost and therefore ready interchangeability. Carbon fibres are manufactured in bulk for a variety of industrial applications and are readily obtained from a variety of manufacturers. At present the major drawback with such microelectrodes is fabrication, which for most workers requires sealing the fibres in glass capillaries by either drawing out the capillaries so trapping the fibre or by the use of epoxy resin to hold the fibre.

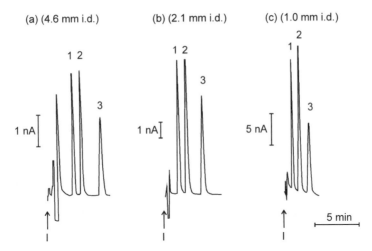

Figure 3.17 *Effect of column diameter on assay sensitivity of noradrenaline (1) and adrenaline (3): (a) 100 mm×4.6 mm (i.d.); (b) 100 mm×2.1 mm (i.d.); (c) 100 mm×1 mm (i.d.). Column: 5 μm ODS Hypersil; Eluent: potassium dihydrogen orthophosphate (40 mol L⁻¹) + citric acid (40 mol L⁻¹) + OSA (0.5 mol L⁻¹) + Na₂EDTA (1 mol L⁻¹) in methanol-water (1 + 9); Flow-rates: (a) 1 mL min⁻¹; (b) 0.5 mL min⁻¹; (c) 0.2 mL min⁻¹; Detector: BAS GCE +0.7 V vs Ag/AgCl; Peak 3: dihydroxylbenzylamine (internal standard).*

Although such electrodes have continued to be used as implantable micro-electrodes in neurochemical applications and are commercially available,[63] they are not specifically manufactured for use with HPLC or CE.

8.1 Detectors for Capillary Systems

8.1.1 Carbon Fibre Electrodes

Wallingford and Ewing[64] were the first to show that ED was feasible with capillary electrophoresis (CE) when they demonstrated that it was possible to measure the concentrations of neurochemicals in a single neuron. To achieve this it was necessary to overcome the intrinsic problem that CE works with very high voltages whilst ED works with very low voltages and that CE capillaries are only 25–75 μm i.d. To achieve their exceptional result, they employed a carbon fibre micro-electrode and decoupled the high voltage of the CE system from the detector by using a porous conductive break in the capillary immediately prior to the detector.

ED is finding increasing use in CE. The usual approach is to earth (ground) the very high running voltages used in CE prior to the EC detector.[65] Many other research groups have developed ED systems for CE, but so far none have been commercialised. Also of relevance is that although the electrodes are very small, the devices required for their operation, such as xyz manipulators and microscopes, are bulky.

8.1.2 Electrochemical Detectors 'on a Chip'

Chip-based and other nanotechnology applications continue to excite much interest in analytical chemistry. Miniaturised devices offer the potential to shrink 'normal' analytical systems with advantage of speed, initial cost, portability, sample size, lower consumables usage and disposability. ED would seem to be a natural choice for such systems since EC effects work fundamentally at a nanoscale.

Possibly the first EC detector on a chip was that designed by Slater and Watt.[66] ED on a chip becomes especially relevant when an electrophoretic separation is also being developed on the device. Complete self-contained micro-analytical systems have been realised that have one-chip potential control, as described by Woolley et al.[67] from Mathies' group (University of California, Berkeley). Using photolithography they placed a detecting electrode just apart from a capillary electrophoretic separation channel (Figure 3.18) in order to decouple the high voltage separating potential from the low voltage detector. The electrode material was platinum and the separation/detection of both catecholamines and DNA restriction digest fragments (nucleic acids) was demonstrated.

It was soon shown that more direct on-chip coupling of separation and ED was possible. Wang et al.[68] described a micromachined CE chip in which the liquid flow was jetted onto a screen-printed thick-film working electrode mounted at right angles to the CE chip. The separation gap was 50 μm. Hilmi and Luong[69] used an 'electroless' (sputtered) deposited gold film on the capillary channel's outlet from the chip. Further integration of the chip and detector was described by Martin et al.[70]: the carbon paste detector was placed within a channel of a poly(dimethylsiloxane) (PDMS) layer outside the exit from the separation channels. End-channel amperometric detection with a single electrode was used to detect amino acids derivatised with naphthalene dicarboxaldehyde. Two electrodes were placed in series and used to detect copper (II) peptide complexes. An advantage of carbon paste was said to be that catalysts could be easily incorporated into the electrode. Carbon paste chemically modified with cobalt phthalocyanine has been used to detect thiols following CE (Chapter 5).

Clearly the integration of ED onto chips is an area of rapid development that will doubtless yield many new devices in the near future (Table 3.2). Rossier et al.[71]

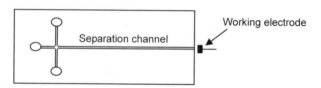

Figure 3.18 *Schematic diagram of a chip-based CE system with an external electro-chemical detector.*

Table 3.2 *Selected applications of on-chip nanoseparations combined with ED*

Analyte(s)	Working electrode	Electrochemical mode	Applied potential	Reference
Amino acids	Gold film	Amperometric	+0.8 V	73
Neurochemicals	Platinum	Amperometric	0.8 V	67
	Carbon	Amperometric	+0.7 V	68
	Carbon	Amperometric	+0.78 V	74
Chlorophenols	Carbon	Amperometric	+1 V	75
Hydrazines	Platinum film	Amperometric	+0.5 V	76
Organophosphate pesticides	Carbon	Amperometric	−0.5 V	77
Nitroaromatic explosives	Carbon	Amperometric	−0.7 V	68
	Gold	Amperometric	−0.8 V	78
DNA	Platinum	Amperometric	+0.8V	67

and Wang[72] have comprehensively reviewed the, so far, relatively small literature on ED for microscale analytical systems.

9 Comparative Performance of EC Detectors

There has been much discussion as to which mode of ED, *i.e.* amperometric or coulometric, is 'best', and secondly which manufacturer's instrument, in either configuration, is 'best'. In nearly all cases 'best' was judged by sensitivity towards a selected analyte or group of analytes, usually catecholamines. Such comparisons can sometimes be made, for example, with UV detectors under standard test conditions. However, for EC detectors such comparisons are rather simplistic due to the chemical nature of the ED process.

Some of the comparisons are described below, but the results must be interpreted with due caution particularly in view of continuing improvements in detector design. One report compared two coulometric detectors for catecholamine analysis and concluded that they were equivalent to current amperometric detectors.[79] Another study[80] compared wall-jet and thin-layer cell configurations. Other more comprehensive studies have been controversial. Forzy *et al.*[81] compared 11 detectors for the analysis of 5-hydroxyindoleactic acid (5-HIAA). They used the same HPLC system with each detector to determine linearity, repeatability, absolute sensitivity, limit of detection and stabilisation time. Driebergen and Benders[82] evaluated 10 detectors with respect to their suitability for routine use in a pharmaceutical company using tetramethylbenzidine as the test compound. Both of these reports found similar relative results with respect to sensitivity and ease of use but stressed the importance of matching instrument to application.

References

1 S.G. Weber, Detection based on electrical and electrochemical measurements in *Detectors for Liquid Chromatography*, E.S. Yeung (ed.), Wiley, New York, Chapter 7, 1986.

2 W.Th. Kok, Electrochemical techniques for detection in HPLC in *Selective Sample Handling and Detection in High Performance Liquid Chromatography* (J. Chromatogr. Library 39A), R.W. Frei and K. Zech (eds), Chapter 6, Amsterdam, Elsevier, 1988, 309–363.

3 H.B. Hanekamp and H.J. van Nieuwkerk, Theoretical considerations on the performance of electrochemical flow-through detectors, *Anal. Chim. Acta*, 1980, **121**, 13–22.

4 W.R. Matson, Electrochemical detection system and method of analysis, *US patent 4,404,065*, 1983.

5 P.T. Kissinger, C. Refshauge, R. Dreiling and R.N. Adams, An electrochemical detector for liquid chromatography with picogram sensitivity, *Anal. Lett.*, 1973, 6, 465–477.

6 P.T. Kissinger, L.J. Felice, R.M. Riggin, L.A. Pachla and D.C. Wenke, Electrochemical detection of selected organic components in the eluate from high-performance liquid-chromatography, *Clin. Chem.*, 1974, **20**, 992–997.

7 D.E. Tallman and D.E. Weisshaar, Carbon composite electrodes for liquid-chromatography electrochemistry – Optimizing detector performance by tailoring the electrode composition, *J. Liquid Chromatogr.*, 1983, **6**, 2157–2172.

8 J. Wangsa and N.D. Danielson, Electrochemical detection for high-performance liquid chromatography using a Kel-F wax-graphite electrode, *J. Chromatogr. A*, 1990, **514**, 171–178.

9 W.Th., Kok, H.B. Hanekamp, P. Bos and R.W. Frei, Amperometric detection of amino acids with a passivated copper electrode, *Anal. Chim. Acta*, 1983, 142, 31–45.

10 W. Kemula, Chromato-polargraphic studies, I, General considerations and description of the set of apparatus, *Rocz. Chem.*, 1952, **26**, 281–287.

11 D.L. Rabenstein and R. Saetre, Mercury-based electrochemical detector for liquid chromatography for the detection of glutathione and other sulfur-containing compounds. *Anal. Chem.*, 1977, **49**, 1036–1039.

12 D.D. Schlereth, E. Katz and H.L. Schmidt, Surface-modified gold-electrodes for electrocatalytic oxidation of NADH based on the immobilization of pheno-xazine and phenothiazine-derivatives on self-assenbled monolayers, *Electro-analysis*, 1995, **7**, 46–54.

13 S.-F. Wang, D. Du and Q.-C. Zou, Electrochemical behavior of epinephrine at L-cysteine self-assembled monolayers modified gold electrode, *Talanta.*, 2002, **56**, 687–692.

14 J. Lankelma and H. Poppe, Design and characterization of a coulometric detector with a glassy carbon electrode for high-performance liquid chroma-tography, *J. Chromatogr. A*, 1976, **125**, 375–388.

15 U.R. Tjaden, J. Lankelma and H. Poppe, Anodic coulometric detection with a glassy carbon electrode for reversed phase high-performance liquid chromatography. Determination of blood levels of perphenazine and fluphenazine, *J. Chromatogr. A*, 1976, **125**, 275–286.

16 W.E. van der Linden and J.W. Dieker, Glassy carbon as electrode material in electro-analytical chemistry, *Anal. Chim. Acta*, 1980, **119**, 1–24.

17 J. Wang and B. Freiha, Liquid chromatography with detection by α-alumina modified glassy carbon electrodes, *Anal. Chem.*, 1984, **56**, 2266–2269.

18 D.E. Weisshaar and T. Kuwana, Considerations for polishing glassy-carbon to a scratch-free finish, *Anal. Chem.*, 1985, **57**, 378–379.

19 M.R. Gregg, Increased electrochemical detector sensitivity by electrode surface modification., *Chromatographia*, 1985, **20**, 129–133.

20 M. Vreeke, R. Maidan and A. Heller, Hydrogen peroxide and β-nicotinamide adenine dinucleotide sensing amperometric electrodes based on electrical connection of horseradish peroxidase redox centers to electrodes through a three-dimensional electron relaying polymer network, *Anal. Chem.*, 1992, **64**, 3084–3090.

21 T. Huang, L. Yang, J. Gitzen, P.T. Kissinger, M. Vreeke and A. Heller, Detection of basal acetylcholine in rat brain microdialysate. *J. Chromatogr. B*, 1995, **670**, 323–327.

22 T. Kato, J.K. Liu, K. Yamamoto, P.G. Osborne and O. Niwa, Detection of basal acetylcholine release in the microdialysis of rat frontal cortex by high-performance liquid chromatography using a horseradish peroxidase-osmium redox polymer electrode with pre-enzyme reactor, *J. Chromatogr. B*, 1996, **682**, 162–166.

23 P.G. Osborne and K. Yamamoto, Disposable, enzymatically modified printed film carbon electrodes for use in the high-performance liquid chromatographic-electrochemical detection of glucose or hydrogen peroxide from immobilized enzyme reactors, *J. Chromatogr. B*, 1998, **707**, 3–8.

24 G. Sittampalam and G.S. Wilson, Surface-modified electrochemical detector for liquid chromatography, *Anal. Chem.*, 1983, **55**, 1608–1610.

25 R.M. Wightman, E.C. Paik, S. Borman and M.A. Dayton, Evaluation of the basal plane of pyrolytic graphite as an electrochemical detector for liquid chromatography, *Anal. Chem.*, 1978, **50**, 1410–1414.

26 B.R. Hepler, S.G. Weber and W.C. Purdy, The behaviour of an electrochemical detector used in liquid chromatography and continuous flow voltammetry, Part 2, Evaluation of low-temperature isotropic carbon for use as an electrode material, *Anal. Chim. Acta*, 1978, **102**, 41–59.

27 A.N. Strohl and D.J. Curran, Controlled potential coulometry with the flow-through reticulated vitreous carbon electrode., *Anal. Chem.*, 1979, **51**, 1050–1053.

28 J.E. Girard, Ion chromatography with coulometric detection for the determination of inorganic ions, *Anal. Chem.*, 1979, **51**, 836–839.

29 J.A. Helpern, J.R. Ewing and K.M.A. Welch, Increased signal-to-noise ratio through grounding and the proper selection of a pump for high-performance

liquid chromatography with electrochemical detection, *J. Chromatogr.*, 1982, **240**, 491–492.

30 M.W. White, Determination of morphine and its major metabolite, morphine-3-glucuronide, in blood by high-performance liquid chromatography with electrochemical detection, *J. Chromatogr.*, 1979, **178**, 229–240.

31 L.C. Blank, Dual electrochemical detector for liquid chromatography, *J. Chromatogr.*, 1976, **117**, 35–46.

32 H.W. van Rooijen and H. Poppe, Noise and drift phenomena in amperometric and coulometric detectors for HPLC and FIA, *J. Liquid. Chromatogr.*, 1983, **6**, 2231–2254.

33 J.R. Russell, A self-metering circuit for a high performance liquid chromatography detector, *Lab. Pract.*, 1986, **January**, 99–101.

34 P.T. Kissinger, Amperometric and coulometric detectors for high-performance liquid chromatography, *Anal. Chem.*, 1977, **49**, 447A–456A.

35 J. Yamada and H. Matsuda, Limiting diffusion currents in hydrodynamic voltammetry, *J. Electroanal. Chem.*, 1973, **44**, 189–198.

36 B. Fleet and C.J. Little, Design and evaluation of electrochemical detectors for HPLC, *J. Chromatogr. Sci.*, 1974, **12**, 747–752.

37 H. Gunasingham, Analytical applications of the wall-jet detector, *Trends Anal. Chem.*, 1988, **7**, 217–221.

38 J.C. Hoogvliet, F. Elferink, C.J. van der Poel and W.P. van Bennekom, Design and characterization of an electrochemical ring-disk flow-through detector for liquid chromotography, *Anal. Chim. Acta*, 1983, **153**, 149–159.

39 C.E. Bohs, M.C. Linhares and P.T. Kissinger, The UniJet: A new electrochemical detector for microbore chromatography, *Current Sep.*, 1994, **12**, 181–186.

40 A. Greischel, K. Beschke, H. Rapp and W. Roth, Quantitation of the new hypoglycaemic agent AG-EE 388 ZW in human plasma by automated high-performance liquid chromatography with electrochemical detection, *J. Chromatogr.* 1991, **568**, 246–252.

41 D.W. Hoffman, R.D. Edkins and S.D. Shillcutt, Human metabolism of phenothiazines to sulfoxides determined by a new high performance liquid chromatography-electrochemical detection method, *Biochem. Pharmacol.* 1988, **37**, 1773–1777.

42 D.M. Radzik and S.M. Lunte, Application of liquid chromatography/ electrochemistry in pharmaceutical and biochemical analysis: A critical review, *CRC Crit. Rev. Anal. Chem.* 1989, **20**, 317–358.

43 G.V. Melzi d'Eril, G. Achilli and G.P. Cellerino, A new microprocessor controlled multi-mode multi-electrode electrochemical detector for HPLC: Sensitivity and selectivity enhancements in neurochemical measurements, *Intl. LabMate*, 1992 (February), 17–18.

44 R. Whelpton, An evaluation of a two-electrode coulometric detector in *Drug Determination in Therapeutic and Forensic Contexts*, E. Reid and I.D. Wilson (eds), Plenum, London 1984, 189–190.

45 R. Whelpton, K. Fernandes, K.A. Wilkinson and D.R. Goldhill, Determination of paracetamol (acetaminophen) in blood and plasma using high performance

liquid chromatography with dual electrode coulometric quantification in the redox mode, *Biomed. Chromatogr.*, 1993, **7**, 90–93.

46 P.R. Hurst and R. Whelpton, Solid phase extraction for an improved assay of physostigmine in biological fluids, *Biomed. Chromatogr.*, 1989, **3**, 226–232.

47 R. Whelpton and T. Moore, Sensitive liquid chromatographic method for physostigmine in biological fluids using dual-electrode electrochemical detection, *J. Chromatogr.*, 1985, **341**, 361–371.

48 H. Mascher and C. Kikuta, Determination of amoxicillin in plasma by high-performance liquid chromatography with fluorescence detection after on-line oxidation, *J. Chromatogr.*, 1990, **506**, 417–421.

49 C.N. Svendsen, Multi-electrode array detectors in high-performance liquid chromatography: A new dimension in electrochemical analysis, *Analyst*, 1993, **118**, 123–129.

50 I.N. Acworth, M. Naoi, H. Parvez and S. Parvez (eds), Coulometric electrode array detectors for HPLC in *Progress in HPLC-HPCE*, Vol. 6., Zeist, VSP, 1997.

51 S. Hughes, P.L. Meschi and D.C. Johnson, Amperometric detection of simple alcohols in aqueous solutions by application of a triple-pulse potential waveform at platinum electrodes, *Anal. Chim. Acta*, 1981, **132**, 1–10.

52 S. Hughes and D.C. Johnson, Amperometric detection of simple carbohydrates at platinum electrodes in alkaline solutions by application of a triple-pulse potential waveform, *Anal. Chim. Acta*, 1981, **132**, 11–22

53 W.R. LaCourse and D.C. Johnson, Optimization of waveforms for pulse amperometric detection (p.a.d.) of carbohydrates following separation by liquid chromatography, *Carbohydr. Res.*, 1991, **215**, 159–178.

54 W.R. LaCourse, '*Pulsed Electrochemical Detection in High Performance Liquid Chromatography*', New York, Wiley, 1997.

55 W.R. Matson, P. Langlais, L. Volicer, P.H. Gamache, E. Bird and K.A. Mark, n-Electrode three-dimensional liquid chromatography with electrochemical detection for determination of neurotransmitters, *Clin. Chem.* 1984, **30**, 1477–1488.

56 I.N. Acworth, J. Yu, E. Ryan, K.C. Gariepy, P. Gamache, K. Hull and T. Maher, Simultaneous measurement of monoamine, amino acid and drug levels using high performance liquid chromatography and coulometric array detection, *J. Liquid Chromatogr.* 1994, **17**, 685–705.

57 G. Achilli, G.P. Cellerino, G.V. Melzi d'Eril and F. Tagliaro, Determination of illicit drugs and related substances by high-performance liquid chromatography with an electrochemical coulometric-array detector, *J. Chromatogr. A*, 1996, **729**, 273–277.

58 C.E. Lunte, Voltammetric detection for liquid-chromatography, *LC-GC*, 1989, **7**, 492–498.

59 J.E. Baur and R.M. Wightman, Microcylinder electrodes as sensitive detectors for high-efficiency, high-speed liquid chromatography, *J. Chromatogr. A*, 1989, **482**, 65–73.

60 F. Gonon, R. Cespuglio, J.-L. Ponchon, M. Buda, M. Jouvet, R.N. Adams and J.-F. Pujol, in vivo continuous electrochemical determination of dopamine

release in rat neostriatum, C.R. Acad. Sci. Hebd. Seances Acad. Sci. D, 1978, **286**, 1203–1206.

61 M. Armstrong-James and J. Millar, Carbon fibre microelectrodes, *J. Neurosci. Meth.*, 1979, **1**, 279–287.

62 J.G. White, R.L. St-Claire and J.W. Jorgenson, Scanning on-column voltammetric detector for open-tubular liquid-chromatography, *Anal. Chem.*, 1986, **58**, 293–298.

63 J. Millar, Extracellular single and multiple unit recording with microelectrodes In *Monitoring Neuronal Activity: A Practical Approach*, J.A. Stamford (ed), Chapter 1, IRL Press, London, 1991, 1–27.

64 R.A. Wallingford and A.G. Ewing, Capillary zone electrophoresis with electrochemical detection, *Anal. Chem.*, 1987, **59**, 1762–1766.

65 J.M. Mesaros, A.G. Ewing and P.F. Gavin, Electrochemical detection in microcolumn separations. *Anal. Chem.*, 1994, **66**, 527A-527A.

66 J.M. Slater and E.J. Watt, On-chip microband array electrochemical detector for use in capillary electrophoresis, *Analyst*, 1994, **119**, 2303–2307.

67 A.T. Woolley, K.Q. Lao, A.N. Glazer and R.A. Mathies, Capillary electrophoresis chips with integrated electrochemical detection, *Anal. Chem.*, 1998, **70**, 684–688.

68 J. Wang, B.M. Tian and E. Sahlin, Micromachined electrophoresis chips with thick film electrochemical detectors, *Anal. Chem.*, 1999, **71**, 5436–5440.

69 A. Hilmi and J.H.T. Luong, Electrochemical detectors prepared by electroless deposition for microfabricated electrophoresis chips, *Anal. Chem.*, 2000, **72**, 4677–4682.

70 R.S. Martin, A.J. Gawron, B.A. Fogarty, F.B. Regan, E. Dempsey and S.M. Lunte, Carbon paste-based electrochemical detectors for microchip capillary electrophoresis/electrochemistry, *Analyst*, 2001, **126**, 277–280.

71. J. Rossier, F. Reymond and P.E. Michel, Polymer microfluidic chips for electrochemical and biochemical analyses, *Electrophoresis*, 2002, **23**, 858–867.

72 J. Wang, Electrochemical detection for microscale analytical systems, A review, *Talanta*. 2002, **56**, 223–231.

73 J. Wang, M.P. Chatrathi and B. Tian, Micromachined separation chips with a precolumn reactor and end-column electrochemical detector, *Anal. Chem.*, 2000, **72**, 5774–5778.

74 R.S. Martin, A.J. Gawron and S.M. Lunte, Dual-electrode electrochemical detection for poly(dimethylsiloxane)-fabricated capillary electrophoresis microchips, *Anal. Chem.*, 2000, **72**, 3196–3202.

75 J. Wang, M.P. Chatrathi and B. Tian, Capillary electrophoresis microchips with thick-film amperometric detectors: separation and detection of phenolic compounds, *Anal. Chim. Acta*, 2000, **416**, 9–14.

76 J. Wang, M.P. Chatrathi, B. Tian and R. Polsky, Capillary electrophoresis chips with thick-film amperometric detectors: Separation and detection of hydrazine compounds, *Electroanalysis.*, 2000, **12**, 691–694.

77 J. Wang, M.P. Chatrathi, A. Mulchandani and W. Chen, Capillary electrophoresis microchips for separation and detection of organophosphate nerve agents, *Anal. Chem.*, 2001, **73**, 1804–1808.

78 A. Hilmi and J.H.T. Luong, Micromachined electrophoresis chips with electro-chemical detectors for analyis of explosive compounds in soil and ground water, *Environ. Sci. Technol.*, 2000, **34**, 3046–3050.

79 C. Bunyagidj and J.E. Girard, A comparison of coulometric detectors for catecholamine analysis by LC-EC, *Life Sci.*, 1982, **31**, 2627–2634.

80 M. Patthy, R. Gyenge and J. Salat, Comparison of the design and performance-characteristics of the wall-jet type and thin-layer type electrochemical detectors – Separation of catecholamines and phenothiazines, *J. Chromatogr.*, 1982, **241**, 131–139.

81 G. Forzy, J.L. Dhondt and J.M. Hayte, Comparison of 11 electrochemical detectors used in high performance liquid chromatography, *Ann. Biol. Clin. (Paris)*, 1988, **46**, 793–793.

82 R.J. Driebergen, and A.J.T.C. Benders, Comparative performance of electro-chemical detectors for FIA and HPLC. *Chromatogr Anal* 1990 (February): 13-4. Response: Andrews RW. *Chromatogr Anal* 1990 (June): 17.

CHAPTER 4

Development and Troubleshooting of HPLC-ED Methods

Optimisation of an HPLC system for a specific analysis can be a complex task and many additional factors must be considered when using EC detection. These can be divided into system dependent variables and analyte dependent variables. System dependent variables include the nature of the working, reference and auxiliary electrodes and associated electronics, and the applied potential as discussed in Chapters 2 and 3. In addition, there may be influences from the eluent (solvent composition, pH, ionic strength), temperature and flow-rate. Analyte-dependent variables are specifically the presence of one or more electroactive moieties or functional groups on the compound(s) of interest. With compounds which are not inherently electroactive it may be possible to derivatise them, or to chemically modify them in some other way, to impart electrochemical properties.

1 System Optimisation

As with all HPLC method development, it is important to start work with a new column and fresh eluent at an optimum flow rate. Frequently, the HPLC separation is initially optimised using a UV or other detector that relies on a physical property of the analyte and is therefore independent of the performance of the ED. Transferring the analysis to an electrochemical detector and then to real samples often brings problems. Firstly, the operational sensitivity may be completely different and it may be necessary to compare separations of standards performed at maximum sensitivity using a UV detector with those done with an ED at almost minimal sensitivity. Although the selectivity of the ED using real sample extracts will probably be better than with a UV detector, there is likely to a substantial frontal peak caused by electrode disturbance and/or the presence of other electroactive compounds in the extract, such that the HPLC conditions have to be adjusted.

1.1 Hydrodynamic Voltammograms

One of the first tasks is to ascertain whether the compound(s) of interest are electroactive and to define the optimum working potentials. Some guidance may be found in the literature, but when attempting to reproduce published HPLC-ED conditions it must be remembered that the reported applied potential may be apparent rather than real. This is because factors such as the composition, surface condition and cleanliness of the working electrode, and the nature and conductivity of the eluent, may act to influence the effective potential at the surface of the electrode. A further factor is that the applied potential is always measured relative to a reference electrode, the exact nature, physical dimensions and positioning of which differ between manufacturers, as discussed in Chapter 3. Therefore, to avoid misleading results, working with clean electrodes, new reference electrodes (where relevant) and fresh eluent is highly recommended.

When using EC detection, it is important to ascertain the optimum working potential for the analyte using the HPLC system chosen for the analysis. A plot of ED response versus applied potential in a flowing system is referred to as a hydrodynamic voltammogram or simply voltammogram (Figure 2.5). Such voltammograms are highly system dependent. Their shape and voltage relationships may vary with eluent pH and composition, the electrode material, electrode contamination, the nature and condition of the reference electrode, and the electrochemical process (Chapter 2, Section 4).

1.1.1 Using the HPLC-ED

A voltammogram is readily obtained by re-injecting the analyte at a suitable concentration at different working potentials. It is important that the HPLC-ED system is stable and reproducible before attempts are made to ascertain optimal working potentials. Since it takes longer to establish a stable electrode response at higher potentials, *i.e.* > 0.8 V, it is best to allow the electrode to stabilise at the highest voltage useable with the appropriate eluent flowing through the system. Depending on the detector, the sensitivity used and the nature of the eluent it may take some time to stabilise, possibly overnight. The amount of analyte being injected should be such that the detector sensitivity is set in mid-range. Once stabilised, the peak responses (height or area) for repeat injections of a standard analyte solution are measured. This is repeated at successive 0.1 V reductions in the applied potential. After each change it is necessary to allow time for the detector output to re-stabilise, but this stabilisation time will be shorter for reduced voltages than if the voltage was successively increased. For trace analysis, it is also important to measure the baseline noise when the detector is operating at high sensitivity. At each voltage, the peak-to-peak baseline noise should be measured with the detector at a suitably high sensitivity level. The output signals and the noise can then be plotted in a voltammogram and the S/N ratio calculated (Figure 4.1).

1.1.2 Using Flow Injection Analysis

When there is a limited number of analytes, and when the HPLC conditions have been pre-determined, the approach described above is convenient and has

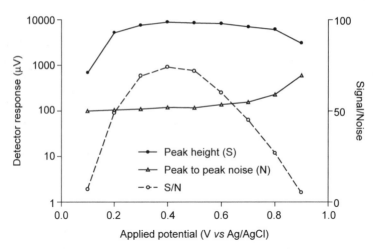

Figure 4.1 *Comparison of a typical voltammogram with changes in signal (•) background noise (△) and signal to noise ratio (○) for noradrenaline (0.5 pmol injected). Column: 5 μm ODS Hypersil (100×4.6 mm i.d.); Eluent: phosphate/citrate buffer (50 mmol L⁻¹, pH 3.3)–octanesulfonic acid (OSA) (3 mmol L⁻¹) in methanol-water (15 + 85); Flow rate: 1 mL min⁻¹; Detection: BAS GCE thin layer cell.*

the advantage that the optimum working potentials are ascertained under the HPLC conditions (eluent, flow-rate, *etc.*) to be used for the assay. However, for pure analytes, a voltammogram can be more speedily constructed using flow injection analysis (FIA). In order to use the FIA approach on an HPLC system, all that is usually necessary is to replace the analytical column with a length of narrow bore stainless steel tubing, *e.g. ca* 2 m of 0.1 mm i.d. tubing. This creates sufficient back pressure to allow most HPLC pumps to continue to pump reproducibly whilst also broadening the injection bolus so that a reasonable response is obtained. It will be necessary to use a relatively low eluent flow-rate (*ca* 0.3 mL min⁻¹) to minimise baseline noise. Injecting a sample *via* the HPLC loop will cause a peak to appear in *ca* 30 s. Since a column is not employed, not only is the method fast, but it also allows a wider range of potential eluent conditions (*e.g.* alkaline buffers) to be investigated. One problem with this method is that artifacts due to the solvent used and impurities in the analyte or solvent can give false data. So such procedures are best employed to rapidly explore conditions for use as a basis on which to build more detailed studies with the column in place as described above.

Some EC detectors include software to perform simple cyclic voltammetry on analytes held at the electrode. This can be performed by either FIA, as above, or by catching a peak from the column in the flow-cell by rapidly stopping the flow. This is best done by quickly disconnecting the column to cell tubing since the de-pressurisation of the system after switching off the pump usually sweeps the peak from the detector. From the cyclic voltammogram the optimum applied potential can be ascertained. However, most of the restraints mentioned for the FIA approach apply.

1.1.3 Operating Potentials

Using either of the above procedures the optimum working potential for maximum sensitivity can be ascertained from the voltammogram. It is usual to work on the first part of the plateau if maximum stability is required at moderate sensitivity with some selectivity. If maximum sensitivity is required the best working potential will be that giving the maximum S/N ratio – this is often about 0.1 V below the plateau. If complex extracts are to be injected it is possible that neither of these two voltages will be suitable since the number of electroactive species found in a sample usually increases with applied voltage. If chromatographic selectivity cannot be improved then for overall selectivity it is best to derive voltammograms for the pure standard analyte and also for peaks in the sample matrix that elute at or near the analyte. Since these will usually be unknown it is necessary to inject a typical sample at each working potential studied and observe the number and size of the peaks seen. Equally important is that higher working potentials can lead to the appearance of late eluting peaks from endogenous compounds or impurities under some conditions, which will increase analysis times and hence decrease throughput.

1.2 Eluent Composition

1.2.1 Purity

In EC detection the eluent must contain a supporting electrolyte. The degree of conductivity required is dependent on detector design, but provided the electrodes are in close proximity, weakly buffering the eluent will provide the necessary conductivity. It is important that neither the eluent solvent nor the buffer salts contribute unduly to the standing current or the baseline noise if high sensitivity is to be attained. Water should be distilled and/or de-ionised and of the highest possible quality. Organic modifiers, such as methanol, and other additives, such as ion-pairing agents and buffer salts, should also be of the highest quality available and should be tested to ensure that they do not add unduly to the background current.

HPLC solvents are normally purified for use with UV detectors and are not necessarily ideal for ED work. Antioxidants, for example, are readily oxidised and thus eluents and extraction solvents containing such compounds should be either avoided or purified before use. Ethers, such as diethyl ether, di-isopropyl ether and tetrahydrofuran are likely to contain up to 0.1% (w/v) pyrogallol or quinol (hydroquinone) as stabiliser. If the stabiliser is removed, peroxides will form and their concentration will increase with time unless the solvent is stored under nitrogen. Not only do peroxides present a hazard from explosion, but they may also oxidise susceptible analytes. Methyl t-butyl ether (MTBE), on the other hand, is stable to oxidation. To test whether a high background signal arises from the eluent, the standing current at the working potential should be monitored with the flow switched off. If the current drops sharply this indicates the presence of electroactive compounds. Periodic monitoring of the standing current provides a simple means of monitoring eluent contamination and/or electrode deactivation during normal use.

Sodium, potassium or ammonium orthophosphate, acetate, perchlorate or citrate salts are commonly used for buffering the eluent and should be of the highest possible grade (AR or Aristar) – experience suggests that HPLC grade reagents may have high UV purity, but not EC purity. It is important to remember that perchlorates are strong oxidising agents and appropriate caution must be exercised in their use. Ammonium salts provide buffering capacity at higher pH values, but oxidation of ammonia may become a limiting factor. Amine modifiers, such as dimethyloctylamine or triethylamine, added in relatively high concentration to improve the peak shapes of basic drugs on conventional reversed-phase HPLC columns, are relatively easily oxidised and their use may preclude the use of ED for many basic drugs.

Eluents containing salt concentrations of 10 mmol L^{-1} or so are generally suitable for use with ED. At higher ionic strengths an increase in the standing current may reduce the linear range of coulometric detectors. For some applications higher ionic strengths may be necessary with certain amperometric detectors to maintain stability and so the effect of ionic strength of the eluent should be considered when adapting an amperometric method for coulometric use.

ED is sometimes particularly susceptible to increases in background noise caused by metal ions present in the eluent, introduced from the samples, eluent or from components of the HPLC system. Some analysts add EDTA to their eluents, typically at concentrations of 30–150 mg L^{-1} (0.1–0.5 mmol L^{-1}), with the aim of chelating iron and possibly other metal ions. EDTA may also enhance the stability of analytes, such as catecholamines, during sample storage and analysis.

Oxygen is a potential source of interference, particularly in reductive electrochemistry. Depending on the conditions and in an acid environment, it is reduced to either hydrogen peroxide or water. For work at high negative potentials, eluents should be oxygen-free and this can be achieved by sparging with helium or oxygen-free nitrogen, although there is the risk that too much sparging will alter the eluent solvent composition. MacCrehan *et al.*[1] deoxygenated the eluent by using a platinum catalyst held in a cartridge and placed in the eluent prior to the injection valve. Regardless of the deoxygenating process, the eluent must then be pumped *via* stainless steel tubing to the column and thence to the detector. This is because Teflon tubing is porous to air and the eluent can be reoxygenated during pumping through the HPLC system. Samples may also need deoxygenation by sparging.

Air in the eluent can cause bubbles to form in the detector cell, but degassing prior to use will reduce this, as will recycling the eluent. Back pressurisation of ED cells to minimise the risk of bubble formation is usually not advisable since many of them readily leak.

Since EC detection is a destructive technique, recycling the eluent is often effective in removing reactive components. Once stable isocratic chromatography has been achieved, the eluent can be recycled from a vessel covered to minimise solvent evaporation and dust collection. It is usual to recycle the eluent for anything up to a few weeks and, provided that the solvent reservoir is sealed, the eluent composition only changes slowly with time. This enables the detector to remain switched on continuously so achieving maximal stability. The column performance will also be more stable. Recycling is best for eluents containing 10% or more organic modifier since this inhibits bacterial growth. The background current and column performance should be monitored to decide when fresh eluent should be prepared.

1.2.2 Compatibility with the HPLC System

One aspect of eluent compatibility with EC detection is that there should be no effect on the components of the detector. Detector cell bodies are now routinely constructed of PTFE, other fluoroplastics, glass or stainless steel, and seem stable to most eluents. Nevertheless electrodes are vulnerable to chemical attack. Problems with the longer term use of some eluents at potentials around +1 V *vs* Ag/AgCl have been experienced. For example, ammonium acetate buffers have caused flaking of the surface of glassy carbon electrodes held at as little as +0.1 V for one batch of electrodes. Noble metal electrodes are easily contaminated by a number of eluents unless the electrode is cleaned by pulsing the applied voltage as in carbohydrate analysis (Chapter 3, Section 6).

Ag/AgCl reference electrodes contain electrolytes that can either diffuse into, or be contaminated by ions diffusing from, the mobile phase. The greater the difference between the concentration of salts in the eluate and that in the reference cell, then the faster will be the contamination of the reference electrode. This leads to a gradual deterioration in the performance of the electrode. It is important that the cell design permits the reference electrolyte to be renewed regularly. If that is not possible, the glass parts of the electrode can be soaked overnight in a saturated potassium chloride solution. For longer term storage, reference electrodes should be enclosed in saturated potassium chloride. For some eluents, such as those containing lithium salts (as used in some vitamin measurements), it is necessary that the reference electrolyte be changed to the lithium form prior to use. The use of perchlorates can lead to deposition of potassium perchlorate on contact with potassium chloride, which is commonly used as the electrolyte in Ag/AgCl reference electrodes. The reference potential may be altered as a result.

1.2.3 pH

Since EC detection involves ionisation of the analyte, it is necessary to understand how eluent pH affects the response. In HPLC-ED eluent pH also often affects analyte retention. This is particularly true when reversed-phase ion-pair systems are used. It is sometimes useful to ascertain the effect of pH on EC response independently of the chromatographic system. Again the FIA approach to studying system variables in isolation can be usefully employed. It is preferable to use buffer salts at or near their pK_a for maximum system stability. The pH of the eluent should be measured and defined (in the method) for the aqueous component in the absence of the organic modifier, preferably without even ion-pairing agents, because the presence of organic solvents will change not only the measured pH, but also the apparent pH of the reference electrode in the pH meter. When, as is often the case, a 'final eluent' pH is quoted, it should be remembered that it is not only an 'apparent pH', but also that it is very pH meter dependent.

Aliphatic amines cannot be detected under strongly acidic conditions since only the fully-protonated (non-oxidisable) species is present (Figure 4.2). In contrast, phenols can be oxidised at most pH values. Usually, acidic conditions are preferred because this reduces the oxidation of other compounds that may be present.

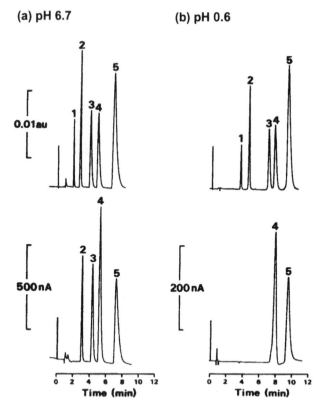

Figure 4.2 *Influence of eluent pH on the EC detection of secondary and tertiary aliphatic amines. Column: 125×5 (i.d.) mm Spherisorb S5W silica; Eluent: (a) Methanolic ammonium perchlorate (10 mmol L^{-1}, apparent pH 6.7). (b) Methanolic perchloric acid (approximately 1 mmol L^{-1}, apparent pH 0.6); Flow-rate: 2.0 mL min^{-1}; Detection: serial UV (254 nm) and EC (V25 grade GCE, +1.2 V vs Ag/AgCl); Injection: 20 μL methanolic solution containing amphetamine (1) (100 mg L^{-1}), nortriptyline (2), amitriptyline (3), imipramine (4) and methdilazine (5) (all 10 mg L^{-1}). (With permission from Flanagan and Jane[2])*

Aliphatic alcohols, on the other hand, require alkaline pH values to oxidise them – this is, in part, the basis of the pulsed amperometric detection of carbohydrates. Because most silica-based columns are incompatible with aqueous buffers exceeding *ca* pH 7.5, it may be necessary to use a high pH stability column or to employ post-column pH modification.

1.3 Temperature

Ideally all HPLC-ED components, *i.e.* injector, eluent, column and detector cell should be held at a constant temperature. In many systems the column, injector and cell are held in a temperature control unit. At a very minimum the column and cell should be shielded from draughts and placed in the same compartment. There is no optimum temperature for EC detection and the best working temperature should be found by plotting changes in S/N for the analyte(s) of interest against varying temperatures.

To obtain coulometric conversion not only must the appropriate operating potential and other conditions be used, but also the analyte must be present in the cell for enough time for the reaction to proceed to completion. Higher temperatures increase the reaction and diffusion rates[3] and the original Coulochem had built-in heaters to maintain a temperature of 40 °C. However, such heaters generally produced an unacceptable increase in baseline noise. A better way to work at elevated temperatures is to place the column and EC cell in a suitable oven.

1.4 Flow Rate

The EC detector, as a reaction detector, is very sensitive to pressure/flow fluctuations and the flow rate stability of the HPLC pump is very important particularly in high sensitivity work. Noise depends on both the magnitude of the flow rate fluctuation and the standing current. The flow rate fluctuation varies with the design of the HPLC pump,[4] but in all systems pulse damping can be introduced independently. Most pumps with low speed pistons, whether single or dual headed will require additional pulse damping when used with ED at anything other than the lowest sensitivity settings. For moderate sensitivity, flow fluctuations of dual headed pumps can often be reduced by introducing long coils of narrow bore stainless steel tubing between the pump and injector. More efficient pulse damping is achieved with mechanical pulse dampeners, such as the Shodex DP-1 (with an iron core) or the liquid membrane dampeners incorporated into some HPLC pumps and in the Antec ED. For some very high sensitivity applications it may be necessary to use more than one pulse dampener.

Sometimes the effect of pump noise can be minimised by operating the detector amplifier at a slower sampling rate (time constant), for example 2 or even 10 s, if the chromatographic conditions allow. The flow from most HPLC pumps is controlled by changing the rate of pumping and this can place a lower limit on flow rate as the frequency becomes too low for the resultant noise to be filtered out electronically. Even a single piston pump, with a Shodex pulse dampener, is suitable for flow rates down to 0.8 mL min^{-1}. If pump noise is thought to be giving rise to unacceptable baseline noise, turning the time constant down (*e.g.* to 0.1 s) will confirm this. In extreme cases pulse-free piston pumps, or high frequency pumps, have been used to reduce background pulsation noise.

1.5 Electrode Surface

It has already been stressed that the electrode surface should be clean and very smooth for efficient EC detection to occur. Even brand new electrodes can sometimes function less than optimally due to the conditions of storage since manufacture. Manufacturers quote various 'optimal' methods for cleaning electrodes.

1.5.1 Cleaning Thin-Layer Electrodes

In the case of solid electrodes that can be dismantled, such as the BAS cells, then the electrodes can be cleaned by polishing using jewellery grade alumina dust and suitable gem polishing pads. A small amount of alumina is placed on the pad and then wetted with distilled or deionised water and the electrode cleaned manually using a rotary motion for a few minutes. The electrode should be briefly washed with pure water and inspected under a low powered magnifying glass until it looks smooth and clean. It can then be briefly sonicated in deionised or distilled water to remove the final traces of alumina. The whole cell can then be re-assembled having checked for any damage to the spacer. Damaged spacers should be replaced. Solid metal electrodes can, in addition to the polishing process, also be cleaned by a brief exposure to a strong acid such as nitric acid.

1.5.2 Maintaining and Cleaning Porous Graphite Electrodes

Successful use of PGEs is dependent on preventing blockage and keeping the electrode clean. Before storage, cells should be washed with water and then methanol to prevent bacterial growth. Eluents should be filtered through a 0.22 µm filter before use. Appropriate in-line carbon filters are supplied for use with the Coulochem instrument and any increase in back pressure can usually be traced to blockage of a filter. Should the working electrode(s) become contaminated, as indicated by a reduced response, then applying a potential of +1 V (*vs* Pd) for 30 min (or −0.4 V for 10 min) may restore the response. Eluent should not be re-circulated during such treatment.

Washing with a series of solvents of decreasing polarity (*e.g.* water, methanol, acetonitrile and tetrahydrofuran) and then reversing the procedure may serve to remove lipophilic contaminants. If contamination by silica is suspected then the cell should be washed with aqueous sodium hydroxide (2 mol L^{-1}) or aqueous ammonium hydroxide (1 mol L^{-1}). If the cell appears to be partially blocked then flushing with phosphoric acid (60% w/v) is recommended. Flushing with nitric acid (6 mol L^{-1}), which was once recommended by the manufacturers, should only be used as a last resort. Of course, these procedures may not work and use of appropriate sample purification procedures, high quality solvents, appropriate filters, low electrode potentials, *etc.*, should be beneficial in the long term. If cell replacement does become necessary then a new hydrodynamic voltammogram should be constructed since the new cell may not be exactly comparable to that used previously.

1.6 Gradient Elution

The instability of EC detectors towards changes in eluent ionic strength implies that isocratic elution is to be preferred, especially for applications requiring high sensitivity. Gradient elution can, however, be used in conjunction with HPLC-ED if the stability of the baseline and of the response is acceptable, as exemplified by some of the reports discussed in Chapter 6.[5-10] Either step or continuous gradients can be used at low to moderate sensitivities, and *ortho*-phthaldialdehyde (OPA) derivatives of amino acids have been analysed in this way. For such work, Joseph and Davies[11] stressed the need to reduce metal ion contamination and recommended the replacement of the metal sintered frits on the pump eluent lines with sintered glass. With such changes they were able to operate at 50 nA full-scale deflection (f.s.d.). Under these conditions the baseline change during the gradient was limited to *ca* 10% f.s.d. With dual electrode coulometric systems it may be possible to oxidise the analyte at E_1 and to subtract the baseline signal at E_2 to reduce the slope of the baseline in the resulting chromatogram.

2 Analyte-Related Variables

EC detection relies on the presence and accessibility of specific electroactive moieties or functional groups (Figures 2.6 and 2.7).[12] However, if the structure of the molecule facilitates delocalisation of the added charge then a response will probably be obtained at a lower applied potential than if charge delocalisation cannot occur. An example here is tranylcypromine [(±)-*trans*-2-phenylcyclopropylamine] where the juxtaposition of the primary aliphatic amine and cyclopropyl moieties means that a good response can be obtained at +0.8 V *vs* Ag/AgCl[13] – normally at least +1.6 V (*vs* Ag/AgCl) is needed to oxidise primary aliphatic amines (Table 4.1). On the other hand, the presence of other substituents may detract from electroactivity. The accessibility of the electroactive group(s) and of course the effect of eluent pH on protonation/ionisation of the group(s) are further factors which may influence EC response.

Table 4.1 *Optimal electrochemical detection conditions for certain functional groups*

Functional group	*Optimum oxidation voltage (V)*	*Approximate detection limit (ng)*	*Typical standing current (nA)*
Phenol, aromatic amine	0.7	0.1	2–5
Phenothiazine sulfur	0.8	0.1	5–10
Imidazoyl nitrogen, indole	0.9	0.2	10–20
Tertiary aliphatic amine	1.0	0.5	50–100
Secondary aliphatic amine	1.2	2	200–500
Primary aliphatic amine	1.6	20	>2000
Pyridyl nitrogen, quaternary ammonium compound	>1.6	–	–

Column: 125×5 (i.d.) mm Spherisorb S5W silica; Eluent: Methanolic ammonium perchlorate (10 mmol L^{-1}, apparent pH 6.7 – see Chapter 4, Section 1); Flowrate: 2.0 mL min^{-1}; Working electrode: V25 Grade glassy carbon; Reference electrode: Ag/AgCl; Auxiliary electrode: stainless steel;[13] Maximum useable applied potential for biological extracts: +1.2 V

Published values for half-wave potentials,[12,14,15] are useful for assessing whether specific compounds or functional groups are amenable to ED, but normally a higher applied potential (0.1–0.2 V) than in a static system is needed to maximise the EC response in HPLC. Cyclic voltammetry performed under conditions as similar as possible (electrode composition, solvent composition, pH) to those likely to be encountered in the course of the HPLC analysis provides a simple means of suggesting an appropriate potential for detection.

2.1 Electrochemical Oxidation

Phenols, particularly catechols and hydroquinones,[16] and aromatic amines are easily oxidised as evidenced by the colours sometimes produced in solutions of these compounds on storage. In general, compounds containing such moieties are oxidised at carbon electrodes at potentials of +0.6 V *vs* Ag/AgCl. In contrast, secondary and tertiary aliphatic amines usually require oxidation potentials in the range +1.0 to +1.2 V *vs* Ag/AgCl.[13,17,18] The higher potentials required to oxidise the corresponding primary amines, amides and quaternary ammonium compounds usually renders ED unsuitable (Table 4.1).

EC oxidation is commonly employed in the analysis of some basic drugs, especially morphine and related opioids (Chapter 6, Section 1). Even if a compound is not amenable to EC oxidation, its metabolites may be. Phase I metabolism of aromatic xenobiotics often proceeds with aromatic hydroxylation. Hydrolysis of phenolic esters, reduction of diazo double bonds to primary amines and other reactions also occur.[19] EC detection is not widely used to measure acidic or neutral compounds, such as salicylate or paracetamol after overdosage, since these compounds, although easily oxidised, are normally present at relatively high concentrations and UV detection is adequate. However, EC methods may be useful in measuring the plasma concentrations of these compounds attained after a single oral dose.[20]

Thiols are easily oxidised using a mercury electrode, but the analysis of thiol drugs poses special problems and is discussed in Chapter 5. Solutions of pheno-thiazines, particularly those without an electron withdrawing substituent at the 2-position, become coloured on storage due to oxidation. EC oxidation of these compounds at carbon electrodes is possible at +0.8 V *vs* Ag/AgCl. Dihydro-pyridines and other reduced aromatic compounds are also often oxidizable at moderate potentials and HPLC-ED methods for nifedipine,[21] other 1,4-dihydro-pyridines,[22,23] and pirprofen,[24] a pyrroline derivative which can be oxidised to a pyrrole at +0.8 V *vs* Ag/AgCl, have been described.

2.2 Electrochemical Reduction

Compared to oxidation, EC reduction has not been not widely used with HPLC. Reducible groups include hydroxylamines, nitrosamines, *N*-oxides, peroxides, quinones, aromatic nitro compounds and disulfides. The antibiotic chloramphenicol can be assayed in blood by EC reduction using a mercury film electrode.[25]

A pendent mercury drop electrode has been used to reduce the 4,5-azomethine group in 1,4-benzodiazepines.[26–28] There were additional contributions from partial reduction of nitro moieties, as in nitrazepam, and *N*-oxides, as in chlordiazepoxide. The 5-lipoxygenase inhibitor methyl 2-[(3,4-dihydro-3,4-dioxo-1-naphthalenyl)amino]benzoate (CGS 8515) has been measured in ethyl acetate extracts of plasma from animals treated with the drug using an ODS-modified silica column with methanol-aq. acetate buffer (0.15 mol L,$^{-1}$ pH 4.5) (60 + 40) as eluent and EC detection (GCE, −0.15 V *vs* Ag/AgCl).[29] No internal standard was used. The LoD was 5 mg L^{-1} (0.4 mL sample).

The endoperoxides arteether, artemether and their *O*-dealkylated metabolite dihydroartemisinin (DHA) compounds, which are thermally labile and possess no UV chromophores or oxidisable groups, have been measured in plasma using HPLC with reductive ED (GCEs, 1.0 V *vs* Ag/AgCl).[30,31] Eluents were deoxygenated by bubbling with argon (2 h, 35 °C) and other precautions to exclude oxygen were necessary to maintain good sensitivity. Karbwang *et al.*[32] used similar conditions to measure artemether and DHA with pure helium (99.99%, 1 mL min^{-1}, 2 h, 35 °C) being used to deoxygenate the eluent. Stainless steel tubing was used throughout the system and sample extracts were also deoxygenated before injection.

Jastrebova *et al.*[33] described the use of an on-line deoxygenator (NovaTech, Westtown, PA) placed between the analytical column and the detector in the analysis of DHA and related compounds (Au electrode, −0.6 V *vs* Ag/AgCl). Knitted membrane tubing (300 cm × 4.8 mm i.d.) was employed in the deoxygenator, which was purged with nitrogen (inlet pressure 0.2 bar). Both lower eluent flow and lower acetonitrile content were associated with lower backgound currrents. Prior deoxygenation of either the sample or the eluent was unnecessary. Band broadening due to the deoxygenator was not a problem, but sample analyses were not reported.

EC reduction has been used extensively in the detection of residues of nitro-aromatic, nitramine, and nitrate explosives at pendent Hg, Au amalgam or glassy carbon electrodes – the forensic evaluation of this subject has been reviewed.[34] The methodology described may be applicable to drugs and other compounds containing these functional groups.

2.3 Common Interferences in Biofluid Analyses

Commonly used drugs, dietary constituents and food additives may be electroactive and consequently are potential sources of interference when biological samples are analysed. Quinine may originate from tonic water, for example, caffeine from coffee and caffeinated soft drinks and some proprietary stimulants, nicotine and cotinine from tobacco smoke, chloroquine and related compounds from malaria prophylaxis, and pholcodine and other opiate analogues from cold cures. Many such compounds and their metabolites will show an EC response at carbon electrodes under appropriate conditions and can be sources of confusion if unrecognised.

Paracetamol, for example, is a very widely used drug and is easily oxidised. The detection of paracetamol after therapeutic administration at concentrations of 9 mg L^{-1} and above was an incidental finding from a study designed to measure monoamine metabolites, such as 5-HIAA in cerebrospinal fluid (CSF) in children.[35] Paracetamol was detected at a GCE (+0.75 V *vs* Ag/AgCl). Plasma concentrations of paracetamol will be at or above the concentrations reported in CSF. Lignocaine (lidocaine) used as a topical anaesthetic may be encountered in plasma at concentrations up to 2 mg L^{-1} or so and is easily extracted into organic solvents under basic conditions showing a good response at carbon electrodes at potentials of +1.0 V *vs* Ag/AgCl.[36]

2.4 Sample Preparation and Choice of Internal Standard

Aspects of sample preparation have been reviewed.[36,37] Phenols and aromatic amines, although amenable to EC oxidation detection, tend to be relatively water soluble and thus polar extraction solvents such as ethyl acetate may have to be used if liquid-liquid extraction (LLE) is to be employed. Solid-phase extraction (SPE), using disposable cartridges or plastic syringe barrels packed usually with chemically-modified silica materials (Table 4.2) analogous to those used in HPLC columns, but with greater porosity, provides an alternative and may be more effective in extracting relatively hydrophilic analytes. On-line sample preparation procedures such as prior to EC detection have also been described.

Internal standards are commonly used in chromatographic analyses and some aspects of the rationale for, and use of, internal standardisation have been

Table 4.2 *Some alkylsilyl-modified silica column packing materials for use in SPE and HPLC*

Bonded Phase	Name	Bonded Phase	Name
$-(CH_2)_{17}CH_3$	n-Octadecyl (ODS, C18)	$-(CH_2)_3NH_2$	Aminopropyl
$-(CH_2)_7CH_3$	n-Octyl (C8)	$-(CH_2)_3CN$	Cyanopropyl, nitrile (CN)
$-(CH_2)_5CH_3$	n-Hexyl (C6)	$-CH_2COOH$	Carboxymethyl (WCX)
$-CH_2CH_3$	Ethyl (C2)	$-(CH_2)_3 - \overset{\oplus}{N}(CH_3)_3$	Trimethylaminopropyl (SAX)
$-CH_3$	Methyl (C1)	$-(CH_2)_3-N(CH_2CH_3)_2$	Diethylaminopropyl (WAX)
	Phenyl		Cyclo-hexyl
$-(CH_2)_3OCH_2\underset{CH_2OH}{\overset{|}{CHOH}}$	Diol		4-Sulfophenylpropyl (SCX)

discussed.[38] When choosing internal standards for use with HPLC-ED, the EC properties of the molecule should be considered. Ideally, not only the extraction and chromatographic characteristics should be as similar as possible, but so should the EC properties. If derivatisation or post-column reaction is to be used then the internal standard should also undergo the same reaction as the analyte(s). The best internal standards are thus likely to be homologues or close structural analogues of the analyte, having almost identical hydrodynamic voltammograms. If suitable compounds are not commercially available, remember that homologues will often have been synthesised during drug development and may still be available. Failing that, it may be necessary to synthesise an internal standard. This need not be daunting as putative metabolites may prove useful starting materials and HPLC can be used to monitor the progress of the reaction. Because of the high sensitivity of EC detection only small amounts of internal standard are required and it may be possible to purify sufficient material by thin-layer chromatography.

2.5 Derivatisation

In the HPLC of drugs and poisons, derivatisation reactions, although sometimes used to stabilise an analyte, are seldom needed to achieve satisfactory chromatography except when performed to permit the separation of enantiomers; for example Srinivas and Igwemezie[39] list a number of derivatisation reagents that have been used to facilitate enantiomer separations by HPLC-ECD. However, derivative formation is chiefly used to enhance the selectivity and sensitivity of detection. Fluorescent derivatives are the most popular, but derivatisation has also been used to enhance UV or EC detection characteristics.[40,41]

Many reagents can be used in either pre- or post-column mode. Primary amines may be derivatised using OPA in the presence of a thiol (2-mercaptoethanol, *N*-acetylcysteine, 3-mercaptopropionate, *t*-butylthiol) or sodium sulfite and by other methods developed primarily for amino acid analysis. EC oxidation or, more usually, fluorescence detection following post-column reaction with OPA/2-mercaptoethanol may be used for primary aliphatic amines, although some compounds, such as tocainide, do not react under all conditions.[13] The OPA/2-mercaptoethanol derivative of histamine can be oxidised at a potential of +0.4 V *vs* Ag/AgCl, whereas histamine itself requires more than +1.2 V.[42,43] The derivative formed by reacting the γ-amino acid baclofen with OPA/*t*-butyl thiol (Figure 4.3) was oxidised at +0.6 V *vs* Ag/AgCl using a GCE.[44] Lovell and Corran[45] used *N*-acetylcysteine with OPA for their analysis of L-thyroxine. The derivative was not fluorescent, but could be detected at +0.73 V *vs* Ag/AgCl.

2,3-Naphthalenedialdehyde (NDA) in the presence of cyanide ion is an alternative to OPA.[46,47] The methodology was developed to produce fluorescent derivatives, but products derivatised at more than one site may not be fluorescent due to quenching. However, they are oxidisable at approximately +0.75 V *vs* Ag/AgCl. Sherwood *et al.*[10] derivatised over 40 α-amino acids using phenylisothiocyanate, a particularly useful reagent as imino acids, such as proline, also react. The derivatives were analysed on an ODS-modified silica column with

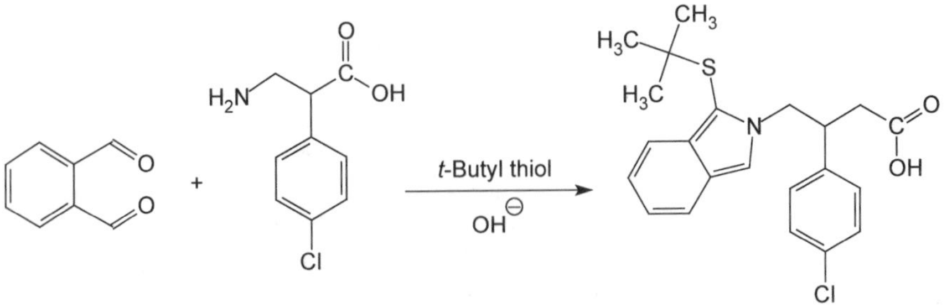

Figure 4.3 *Derivatisation of baclofen with ortho-phthaldialdehyde/t-butylthiol under alkaline conditions.*[44]

gradient elution and were detected at a GCE (+1.1 V *vs* Ag/AgCl). Shimada *et al.*[48] have developed several ferrocene derivatives for pre-column reaction with primary amines to give electroactive products. Of these, *N*-succinimidyl-3-ferrocenyl propionate was shown to be the most valuable when phenylethylamine was the test compound.

Substituted phenacyl bromides are derivatising agents for carboxylic acids. Munns *et al.*[49] synthesised 1-(4-hydroxyphenyl)-, 1-(2,4-dihydroxyphenyl)- and 1-(2,5-dihydroxyphenyl)-2-bromoethanone derivatives of quinoxaline-2-carboxylic, benzoic and salicylic acids, and various penicillins (Figure 4.4). LoDs as low as 1 pmol on column were claimed. Derivatisation of alcohols and ketones is exemplified in two methods employing aromatic nitro-derivatives. In the first example digoxin was assayed as the 3,5-dinitrobenzoyl ester.[50] The derivative was reduced at −0.8 V *vs* Pd using a glassy carbon wall-jet electrode, which gave better sensitivity (0.98 ng) than a gold electrode. Even more sensitivity (0.39 ng) was reported using a Coulochem model 5010 cell in the redox mode (first electrode potential −0.8 V, second electrode potential +0.8 V, both *vs* Pd). In the second report, 17-ketosteroid sulfates were reacted with 4-nitrophenylhydrazine and the hydrazone was oxidised at +0.8 V *vs* Ag/AgCl[51] – the electrode material used was not stated (Yanagimoto VMD 101 ED), but was probably glassy carbon.

The extent to which derivatisation will be used to enhance ED in future is uncertain. Several of the examples cited above used novel reagents. Unless these reagents are made easily available, it is difficult to believe that these techniques

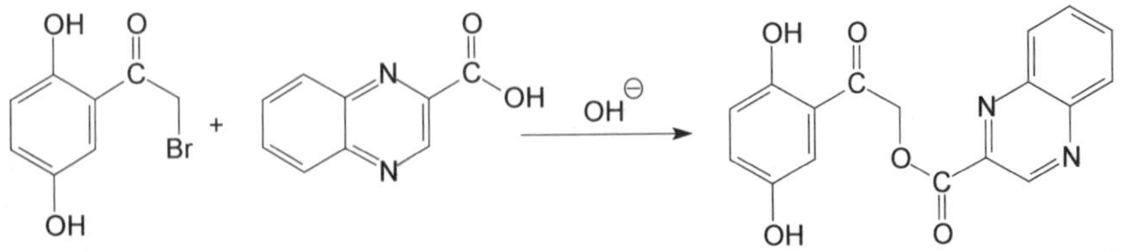

Figure 4.4 *Derivatisation of quinoxaline-2-carboxylic acid with 1-(2,5-dihydroxyphenyl)-2-bromoethanone under alkaline conditions.*[49]

will be adopted widely. Derivatisation with compounds containing chromophores composed in part of aromatic nitro moieties is well established for UV detection and might prove useful for ED of acids, alcohols and ketones, but there is the danger that large quantities of 'background' materials will react and that the selectivity of the method will be lost. However, reductive ED is generally more difficult because of the need to exclude oxygen (Section 2). EC detection is usually applied when analyte concentrations are very low and there may be problems with the reaction chemistry at these low concentrations. Furthermore, newer techniques, such as PAD coupled with polymer columns suitable for eluents of more extreme pH, may extend the range of groups that can be detected directly.

2.6 Hydrolysis/Photolysis/Enzymic Reactions

Post-column pH change can also be used to raise the eluent pH to facilitate PAD of carbohydrates. Phenolic esters may be amenable to post-column hydrolysis to reveal the phenolic hydroxyl. Krause,[8] for example, used a post-column reaction coil with aqueous sodium hydroxide (0.1 mol L^{-1}) at 100 °C to hydrolyse aryl *N*-methylcarbamate residues. The *N*-methylcarbamate physostigmine has been hydrolysed electrochemically and then quantified using the redox mode of the Coulochem.[52] Boppana *et al.*[53] developed a post-column enzyme reactor containing β-glucuronidase on glass beads to hydrolyse fenoldopam glucuronides to the parent compound, which was detected at a GCE (+0.6 V *vs* Ag/AgCl). Obviously esters may also be hydrolysed pre-column, although this may require two assays, particularly for glucuronides; in some instances the hydrolysis product may be unstable.

Post-column photolysis to give electroactive products has been used in the measurement of trichothecene mycotoxins, such as deoxynivalenol (vomitoxin) in grain extracts,[54] penicillins and cefoperazone[55] and spironolactone[56] in pharmaceutical preparations, and aspoxicillin (N^4-methyl-D-asparaginylamoxicillin) in BAL fluid.[57]

In the analysis of inorganic anions using acetonitrile-water (3+7) containing citric acid and hexadecyltrimethylammonium hydroxide as eluent, Wheals[58] reported that bromide, cyanide, dithionite, ferrocyanide, iodide, metabisulfite, nitrite, perborate, peroxide, sulfide, thiocyanate and thiosulfate respond at a GCE (+0.75 V *vs* Ag/AgCl). Dou and Krull[59] have further shown that post-column irradiation (low pressure Hg arc discharge lamp) can facilitate the EC detection of chromate, dichromate and perchlorate and give improved sensitivity for thiocyanate. Despite this work, however, the application of EC detection to the analysis of these ions in biological systems has not been reported as yet.

Tagliaro *et al.*[60,61] have described the HPLC of ethanol and methanol in plasma standards after 1+9 dilution in eluent (aq. potassium phosphate (10 mmol L^{-1}, pH 7.5)). The analytical column was a 2-hydroxyethyl methacrylate/ethylene dimethacrylate co-polymer. Detection was *via* post-column reaction with immobilised alcohol oxidase and monitoring of released hydrogen peroxide (Pt electrode, +0.50 V *vs* Ag/AgCl). No internal standard was used, methanol gave a nonlinear response above 0.5 mg injected and the analysis of authentic samples was not reported. LoDs were 0.1 mg injected for both compounds. A layer of Nafion 117, a perfluorinated cation-exchanger, was deposited on the working electrode prior to use to minimise the risk of deactivation. A similar approach, but using immobilised acetylcholine esterase and choline oxidase has been employed to detect acetylcholine and choline after HPLC.[62] The use of modified GCEs to detect hydrogen peroxide was discussed in Chapter 3 Section 2.

Eisenberg *et al.*[63] have described the analysis of allopurinol by EC detection *via* post-column reaction with immobilised xanthine oxidase to give oxypurinol. This facilitated EC detection at a GCE (+0.85 V *vs* Ag/AgCl). Oxypurinol (a metabolite of allopurinol) was resolved on the chromatographic system employed.

3 Maintenance of HPLC-ED Systems

There is no doubt that EC detectors are more prone to operational difficulties than most other HPLC detectors, and it is also more difficult to take corrective action. The close inter-relationship between the chromatograph and the detector makes fault diagnosis and correction even more difficult. A good understanding of the basic principles underlying EC detection is essential to a logical and hopefully successful approach to system maintenance. Nevertheless some problems associated with EC detection can be difficult to resolve even for the most experienced workers. In this short review it will be assumed that the chromatographic system is working satisfactorily and only problems with the EC detector will be discussed. Table 4.3 lists some of the possible causes of noise in HPLC-ED.

Table 4.3 *Sources of noise and their possible solution in EC detection*

Symptom	Possible cause	Action
A. Noise mainly electrical in origin		
Increased baseline noise	Detector sensitivity too high	Reduce if possible
	Working potential too high	Check voltammogram; reduce to lowest value consistent with required sensitivity; optimise S/N from voltammogram
	Poor electrode surface	Polish electrode or replace if badly damaged
	Electrode type inappropriate	Change to more appropriate electrode type, *e.g.* metal
	Electrostatic effects	Common ground the system; use Faraday cage
	Time constant too low	Select better setting but do not over-damp
	Dirty/poor electrode contacts	Clean or re-make connections
	Eluent pH incorrect	Check pH of eluent

(Continued)

Table 4.3 *(Continued)*

Symptom	Possible cause	Action
	Pump	Pump valves not functioning correctly, clean; flow rate too high, check flowrate; add more pulse dampening
Baseline drift	System unstabilised	Allow cell longer to stabilise before use; replace eluent
Sharp spikes on baseline	Regular spikes can come from other equipment in the lab	Use clean mains for detector – use computer isolation block Move equipment to different location – away from ovens, spectrometers, *etc.*
Loss of sensitivity	Change in voltammogram Electrode fouled Change in eluent Change in reference electrode Detector/integrator wrongly set Injector blocked Pump not pumping	Check optimum voltage Clean electrode Replace eluent; check buffer pH Check voltammogram; replace reference electrode Correct settings Unblock; install correct loop; correct injector setting Check

B. Noise mainly of eluent origin

Symptom	Possible cause	Action
Excessive baseline noise	Grade of reagents employed	Use HPLC grade reagents; use better source of water; re-cycle eluent; investigate use of other buffer salts; add 1–2 mmol L^{-1} EDTA to eluent
Baseline drift	Leaks in system Electroactive material leaching from system components Eluent grossly contaminated	Seal leaks Clean column and fittings such as frits; passivate stainless steel components Replace with fresh eluent
Spikes in baseline	Bubbles in system Closed electrical circuit in buffer Oxygen in sample	De-gas; use continuous degasser; recycle eluent Let solvent drip from end De-gas before injection

C. Miscellaneous sources of noise

Symptom	Possible cause	Action
Increased baseline noise	Flow noise	Select column dimensions and flow-rate giving minimum noise
	Pulsations from pump	Use a better pump, use more pulse dampeners
	Metal fittings in system	Remove or replace metal sinters with glass/plastic sinters; incorporate EDTA in buffers
Baseline drift	Electrode temperature changes	Thermostat cell along with column or shield cell from draughts

References

1 W.A. MacCrehan, S.D. Yang and B.A. Benner, Catalytic oxygen scrubber for liquid-chromatography, *Anal. Chem.,* 1988, **60,** 284–286.

2 R.J. Flanagan and I. Jane, High-performance liquid chromatographic analysis of basic drugs on silica columns using non-aqueous ionic eluents – I. Factors influencing retention, peak shape and detector response, *J. Chromatogr.,* 1985, **323,** 173–189.

3 R. Whelpton, An evaluation of a two-electrode coulometric detector in *Drug Determination in Therapeutic and Forensic Contexts,* E. Reid and I.D. Wilson, (eds), Plenum, London, 1984, 189–190.

4 D. Perrett, Comparative studies on different electrochemical detector systems in *Drug Determination in Therapeutic and Forensic Contexts, Methodological Surveys in Biochemistry,* E. Reid and I.D. Wilson, (eds), Plenum, New York, 1984, **14,** 111–121.

5 G. Achilli, G.P. Cellerino, G.V. Melzi d'Eril and F. Tagliaro, Determination of illicit drugs and related substances by high-performance liquid chromatography with an electrochemical coulometric-array detector, *J. Chromatogr. A,* 1996, **729,** 273–277.

6 V. Rizzo, G.M. d'Eril, G. Achilli and G.P. Cellerino, Determination of neuro-chemicals in biological fluids by using an automated high-performance liquid chromatographic system with a coulometric array detector, *J. Chromatogr.,* 1991, **536,** 229–236.

7 E. Kirchmann, R.L. Earley and L.E. Welch, The electrochemical detection of penicillins in milk, *J. Liq. Chromatogr.,* 1994, **17,** 1755–1772.

8 R.T. Krause, High-performance liquid chromatographic determination of aryl *N*-methylcarbamate residues using post-column hydrolysis electrochemical detection, *J. Chromatogr.,* 1988, **442,** 333–343.

9 D.L. Mount, L.C. Patchen, P. Nguyen-Dinh, A.M. Barber, I.K. Schwartz and F.C. Churchill, Sensitive analysis of blood for amodiaquine and three metabolites by high-performance liquid chromatography with electrochemical detection, *J. Chromatogr.,* 1986, **383,** 375–386.

10 R.A. Sherwood, A.C. Titheradge, D.A. Richards, Measurement of plasma and urine amino acids by high-performance liquid chromatography with electro-chemical detection using phenylisothiocyanate derivatization, *J. Chromatogr.,* 1990, **528,** 293–303.

11 M.H. Joseph and P. Davies, Electrochemical activity of *o*-phthalaldehyde-mercaptoethanol derivatives of amino acids. Application to high-performance liquid chromatographic determination of amino acids in plasma and other biological materials, *J. Chromatogr.,* 1983, **277,** 125–136.

12 W.F. Smyth, *Voltammetric Determination of Molecules of Biological Significance,* Wiley, Chichester, 1992.

13 I. Jane, A. McKinnon and R.J. Flanagan, High-performance liquid chromato-graphic analysis of basic drugs on silica columns using non-aqueous ionic eluents – II. Application of UV, fluorescence and electrochemical oxidation detection, *J. Chromatogr.,* 1985, **323,** 191–225.

14 N.L. Weinberg and H.P. Weinberg, Electrochemical oxidation of organic compounds, *Chem. Rev.,* 1968, **68,** 449–523.

15 V.D. Parker, Anodic oxidation of amines in *Organic Electrochemistry,* M.M. Baizer, (ed). Dekker, New York, 1973, 509–529.

16 E. Cadenas and L. Ernster, Quinoid compounds: high-performance liquid chromatography with electrochemical detection, *Methods Enzymol.,* 1990, **186,** 180–196.

17 G. Musch, M. de Smet and D.L. Massart, Expert system for pharmaceutical analysis. I. Selection of the detection system in high-performance liquid chromatographic analysis: UV versus amperometric detection, *J. Chromatogr.,* 1985, **348,** 97–110.

18 G. Musch and D.L. Massart, Expert system for pharmaceutical analysis. II. Relative contribution of and rule validation model for amperometric detection (oxidation mode), *J. Chromatogr.,* 1986, **370,** 1–19.

19 S.M. Lunte, D.M. Radzik and P.T. Kissinger, An introduction to the study of xenobiotic metabolism using electroanalytical techniques, *J. Pharm. Sci.,* 1990, **79,** 557–567.

20 R. Whelpton, K. Fernandes, K.A. Wilkinson and D.R. Goldhill, Determination of paracetamol (acetaminophen) in blood and plasma using high performance liquid chromatography with dual electrode coulometric quantification in the redox mode, *Biomed. Chromatogr.,* 1993, **7,** 90–93.

21 H. Suzuki, S. Fujiwara, S. Kondo and I. Sugimoto, Determination of nifedipine in human plasma by high-performance liquid chromatography with electro-chemical detection, *J. Chromatogr.,* 1985, **341, 341**–347.

22. N.D. Huebert, M. Spedding and K.D. Haegele, Quantitative analysis of the dihydropyridines, 3-(2-furoyl)-5-methoxycarbonyl-2,6-dimethyl-4-(2-nitro-phenyl)-1,4-dihydropyridine and nifedipine, by high-performance liquid chromatography with electrochemical detection, *J. Chromatogr.,* 1986, **353,** 175–180.

23 Y. Iida, Y. Kinouchi, Y. Takeichi, T. Imai and M. Otagiri, Simultaneous determination of a new dihydropyridine calcium antagonist (MPC-1304) and its metabolite in dog plasma by high-performance liquid chromatography with electrochemical detection, *J. Chromatogr.,* 1991, **571,** 277–282.

24 L. Zecca, P. Ferrario, R. Pirola, F. Zambotti, N. Zonta and F. Fraschini, Analysis of pirprofen in cerebrospinal fluid, plasma, and synovial fluid by high-performance liquid chromatography with electrochemical detection, *J. Pharm. Sci.,* 1989, **78,** 776–779.

25 D.M. Radzik and S.M. Lunte, Application of liquid chromatography/electro-chemistry in pharmaceutical and biochemical analysis: A critical review, *C.R.C. Crit. Rev. Anal. Chem.,* 1989, **20,** 317–358.

26 M.R. Hackman and M.A. Brooks, Differential pulse amperometric detection of drugs in plasma using a dropping mercury electrode as a high-performance liquid chromatographic detector, *J. Chromatogr.,* 1981, **222,** 179–190.

27 J.B.F. Lloyd and D.A. Parry, Detection and determination of common benzodiazepines and their metabolites in blood samples of forensic science interest. Microcolumn cleanup and high-performance liquid chromatography

with reductive electrochemical detection at a pendent mercury drop electrode, *J. Chromatogr.,* 1988, **449**, 281–297.

28 J.B.F. Lloyd and D.A. Parry, Forensic applications of the determination of benzodiazepines in blood samples by microcolumn cleanup and high-performance liquid chromatography with reductive mode electrochemical detection, *J. Anal. Toxicol.,* 1989, **13**, 163–168.

29 S.K. Kuwahara, W.F. Brubaker, E. Williams, S. Tripp, F.L. Douglas and A.N. Kotake, Determination of the 5-lipoxygenase inhibitor (CGS 8515) in plasma by high-performance liquid chromatography using reductive electrochemical detection, *J. Chromatogr.,* 1990, **534**, 260–266.

30 V. Melendez, J.O. Peggins, T.G. Brewer and A.D. Theoharides, Determination of the antimalarial arteether and its deethylated metabolite dihydroartemisinin in plasma by high-performance liquid chromatography with reductive electrochemical detection, *J. Pharm. Sci.,* 1991, **80**, 132–138.

31 N. Sandrenan, A. Sioufi, J. Godbillon, C. Netter, M. Donker and C. van Valkenberg, Determination of artemether and its metabolite, dihydroartemisinin, in plasma by high-performance liquid chromatography and electrochemical detection in the reductive mode, *J. Chromatogr.,* 1997, **691**, 145–153.

32 J. Karbwang, K. Na-Bangchang, P. Molunto, V. Banmairuroi and K. Congpuong, Determination of artemether and its major metabolite, dihydroartemisinin, in plasma using high-performance liquid chromatography with electrochemical detection, *J. Chromatogr.,* 1997, **690**, 259–265.

33 J. Jastrebova, L. Nyholm, K.E. Markides and Y. Bergqvist, On-line deoxygenation for reductive electrochemical detection of artemisinin and dihydroartemisinin in liquid chromatography, *Analyst,* 1998, **123**, 313–317.

34 J.B.F. Lloyd, HPLC of explosives materials in *Advances in Chromatography,* J.C. Giddings, (ed). Dekker, New York, 1992, **32**, 173–261.

35 F.X. Walsh, P.J. Langlais and E.D. Bird, Liquid-chromatographic identification of acetaminophen in cerebrospinal fluid with use of electrochemical detection, *Clin. Chem.,* 1982, **28**, 382–383.

36 R.J. Flanagan, HPLC of psychotropic and related drugs, in *High Performance Liquid Chromatography in Neuroscience Research,* R.B. Holman, A.J. Cross and M.H. Joseph (eds), Wiley, Chichester, 1993, 321–356.

37 R.M. Smith, Before the injection – modern methods of sample preparation for separation techniques, *J. Chromatogr. A,* 2003, **1000**, 3–27.

38 R. Whelpton, Standards (d) Internal Standards in *Encyclopedia of Analytical Science,* A. Townshend, P. Worsfold, R. Macrae, S. Haswell, I. Wilson and H. Werner, (eds), Academic Press, London, 1995, 4785–4792.

39 N.R. Srinivas and L.N. Igwemezie, Chiral separation by high performance liquid chromatography. I. Review on indirect separation of enantiomers as diastereomeric derivatives using ultraviolet, fluorescence and electrochemical detection, *Biomed. Chromatogr.,* 1992, **6**, 163–167.

40 K. Imai, Derivatization in liquid chromatography. in *Advances in Chromatography,* J.C. Giddings, E. Grushka and P.R. Brown, (eds), Dekker, New York, 1987, **27**, 215–245.

41 H. Lingeman and W.J.M. Underberg, (eds), *Detection Orientated Derivatization Techniques in Liquid Chromatography*, Dekker, New York, 1990.

42 L.G. Harsing, H. Nagashima, D. Duncalf, E.S. Vizi and P.L. Goldiner, Determination of histamine concentrations in plasma by liquid chromatography/ electrochemistry, *Clin. Chem.*, 1986, **32**, 1823–1827.

43 L.G. Harsing, H. Nagashima, E.S. Vizi and D. Duncalf, Electrochemical determination of histamine derivatized with *o*-phthalaldehyde and 2-mercapto-ethanol, *J. Chromatogr.*, 1986, **383**, 19–26.

44 G.M. Wall and J.K. Baker, Determination of baclofen and alpha-baclofen in rat liver homogenate and human urine using solid-phase extraction, *o*-phthalaldehyde-*tert*.-butyl thiol derivatization and high-performance liquid chromatography with amperometric detection, *J. Chromatogr.*, 1989, **491**, 151–162.

45 G. Lovell and P.H. Corran, Determination of L-thyroxine in reference serum preparations as the *o*-phthalaldehyde-*N*-acetylcysteine derivative by reversed-phase liquid chromatography with electrochemical detection, *J. Chromatogr.*, 1990, **525**, 287–296.

46 S.M. Lunte, T. Mohabbat, O.S. Wong and T. Kuwana, Determination of desmosine, isodesmosine, and other amino acids by liquid chromatography with electrochemical detection following precolumn derivatization with naphthale-nedialdehyde/cyanide, *Anal. Biochem.*, 1989, **178**, 202–207.

47 W.F. Kline and B.K. Matuszewski, Improved determination of the bisphosphonate alendronate in human plasma and urine by automated precolumn derivatization and high-performance liquid chromatography with fluorescence and electrochemical detection, *J. Chromatogr.*, 1992, **583**, 183–193.

48 K. Shimada, T. Oe, M. Tanaka and T. Nambara, Sensitive ferrocene reagents for derivatization of amines for high-performance liquid chromatography with electrochemical detection, *J. Chromatogr.*, 1989, **487**, 247–255.

49 R.K. Munns, J.E. Roybal, W. Shimoda and J.A. Hurlbut, 1-(4-Hydroxyphenyl)-, 1-(2,4-dihydroxyphenyl)- and 1-(2,5-dihydroxyphenyl)-2-bromoethanones: new labels for determination of carboxylic acids by high-performance liquid chromaography with electrochemical and ultraviolet detection, *J. Chromatogr.*, 1988, **442**, 209–218.

50 L. Embree and K.M. McErlane, Electrochemical detection of the 3,5-dinitro-benzoyl derivative of digoxin by high-performance liquid chromatography, *J. Chromatogr.*, 1990, **526**, 439–446.

51 K. Shimada, M. Tanaka and T. Nambara, Studies on steroids. CC. Determination of 17-ketosteroid sulphates in serum by high-performance liquid chromatography with electrochemical detection using pre-column derivatization, *J. Chromatogr.*, 1984, **307**, 23–28.

52 P.R. Hurst and R. Whelpton, Solid phase extraction for an improved assay of physostigmine in biological fluids, *Biomed. Chromatogr.*, 1989, **3**, 226–232.

53 V.K. Boppana, K.-L.L. Fong, J.A. Ziemniak and R.K. Lynn, Use of a post-column immobilized beta-glucuronidase enzyme reactor for the determination of diastereomeric glucuronides of fenoldopam in plasma and urine by

high-performance liquid chromatography with electrochemical detection, *J. Chromatogr.*, 1986, **353**, 231–247.

54 W.L. Childress, I.S. Krull and C.M. Selavka, Determination of deoxynivalenol (DON, vomitoxin) in wheat by high-performance liquid chromatography with photolysis and electrochemical detection (HPLC-hv-EC), *J. Chromatogr. Sci.*, 1990, **28**, 76–82.

55 C.M. Selavka, I.S. Krull and K. Bratin, Analysis for penicillins and cefoperazone by HPLC-photolysis-electrochemical detection (HPLC-hv-EC), *J. Pharm. Biomed. Anal.*, 1986, **4**, 83–93.

56 W.J. Bachman and J.T. Stewart, HPLC-photolysis-electrochemical detection in pharmaceutical analysis: application to the determination of spironolactone and hydrochlorothiazide in tablets, *J. Chromatogr. Sci.*, 1990, **28**, 123–128.

57 T. Yamazaki, T. Ishikawa, H. Nakai, M. Miyai, T. Tsubota and K. Asano, Determination of aspoxicillin in broncho-alveolar lavage fluid by high-performance liquid chromatography with photolysis and electrochemical detection, *J. Chromatogr.*, 1993, **615**, 180–185.

58 B.B. Wheals, Ion chromatography of inorganic anions on a dynamically modified polystyrene-divinylbenzene packing material and its application to anion screening by sequential ultraviolet absorbance and electrochemical detection, *J. Chromatogr.*, 1987, **402**, 115–126.

59 L. Dou and I.S. Krull, Determination of inorganic anions by flow injection analysis and high-performance liquid chromatography combined with photolytic-electrochemical detection, *J. Chromatogr.*, 1990, **499**, 685–697.

60 F. Tagliaro, D. De Leo, M. Marigo, R. Dorizzi and G. Schiavon, Use of enzymatic reactors in the high performance liquid chromatographic determination of ethanol and methanol with electrochemical detection, *Biomed. Chromatogr.*, 1990, **4**, 224–228.

61 F. Tagliaro, R. Dorizzi, S. Ghielmi and M. Marigo, Direct injection high-performance liquid chromatographic method with electrochemical detection for the determination of ethanol and methanol in plasma using an alcohol oxidase reactor, *J. Chromatogr.*, 1991, **566**, 333–339.

62 Y. Ikarashi, H. Iwatsuki, C.L. Blank and Y. Maruyama, Glassy carbon precolumn for direct determination of acetylcholine and choline in biological samples using liquid chromatography with electrochemical detection, *J. Chromatogr.*, 1992, **575**, 29–37.

63 E.J. Eisenberg, P. Conzentino, G.G. Liversidge and K.C. Cundy, Simultaneous determination of allopurinol and oxypurinol by liquid chromatography using immobilized xanthine oxidase with electrochemical detection, *J. Chromatogr.*, 1990, **530**, 65–73.

CHAPTER 5

Thiols, Disulfides and Related Compounds

1 Introduction

Thiols, also known as sulfydryl compounds, mercapto compounds and mercaptans, are a group of organic compounds resembling alcohols, but where the oxygen of the hydroxyl group is replaced by sulfur. They are found in many pharmaceuticals and endogenous compounds. Together with their disulfides (R-S-S-R') they have an extensive and interesting chemistry.[1,2] Reactions of thiols include oxidation to and reduction from disulfides, metal chelation and thiazolidine formation. The reactivity of thiols makes their analysis challenging. The most common approaches to the assay of thiols are summarised in Table 5.1 – it should be noted that EC techniques are particularly important.[3,4] In order to distinguish between naturally occurring thiols, many methods rely on HPLC, hence HPLC-ED is a particularly valuable technique for thiol/disulfide analysis.[5–7]

1.1 Biologically Important Thiols

Naturally occurring thiols have important roles in areas as diverse as cellular metabolism, maintaining protein structure and maintaining intracellular oxidation status.[8,9] Cysteine, homocysteine and reduced glutathione (L-glutamyl-L-cysteinyl-glycine, GSH) are the most important biological thiols. Since they possess three ionisable groups (namely –SH, –NH$_2$, and –COOH with pK_a values of ca 10, 8 and 2, respectively) at physiological pH they are present as zwitterions. L-Cysteine along with its disulfide, L-cystine, occurs free in body fluids and cells as well as being a constituent of most proteins. GSH and its disulfide (GSSG) are found in relatively high concentrations in the cells of most species where they fulfil many roles.[10] Depletion of GSH and variations in the oxidative status marker (the GSH/GSSG ratio) are major topics in toxicology. Homocysteine is found at low concentrations in humans as the free thiol, as its disulfide, homocystine, and as mixed disulfides and has received much recent attention as a marker of cardiovascular disease.[11]

Table 5.1 *Approaches to the assay of thiols, disulfides and related compounds*

Principle	Details/Examples
Spectrophotometric	Complex formation with metals (*e.g.* ferric ions)
	Colorimetric reaction (*e.g.* nitroprusside reaction)
Electrochemical	Differential pulse polarography
	Cathodic stripping voltammetry
Chromatographic	Cation-exchange with post-column detection (*e.g.* amino acid analysis)
Immunoassay	Homocysteine analysis
Gas Chromatography	Derivatisation (*e.g.* with trimethylsilyl)
HPLC	Post-column derivatisation (*e.g.* Ellman's reagent)
	Non-specific pre-column derivatisation (*e.g.* of amino acids to form chromophores or fluorophores)
	Pre-column derivatisation to form fluorophores (*e.g.* with bimanes, maleimides)
	Oxidative and reductive electrochemical detection
Capillary Electrophoresis	Direction detection at low wavelengths
	Derivatisation followed by appropriate detection
Mass Spectrometry	LC-MS/MS
Nuclear Magnetic Resonance Spectrometry	Intracellular GSH and GSSG measurement

1.2 Thiol-Containing Drugs

The oldest thiol-containing drug is the anti-arsenical British Anti-Lewisite (BAL, dimercaprol, 2,3-dimercaptopropan-1-ol). D-Penicillamine, a degradation product of penicillin, is used as a chelating agent in treating Wilson's disease and sometimes in treating heavy metal poisoning, as a thiol-disulfide exchanger in the treatment of the in-born error of metabolism cystinuria and as a second line agent in the treatment of rheumatoid arthritis. A number of compounds structurally related to cysteine are used as mucolytics. Captopril (D-3-mercapto-2-methyl-propanoyl-L-proline) was the first marketed angiotensin converting enzyme (ACE) inhibitor and is an antihypertensive agent. Modified nucleoside bases, such as 6-mercaptopurine and thioguanine, have been used in chemotherapy since the 1960s either as the free thiol or in pro-drug form, as in azathioprine. *N*-acetyl-L-cysteine is widely used as a mucolytic and as an antidote in paracetamol poisoning. Thiols have also been studied extensively as radioprotective agents. Some thiol-containing drugs and naturally occurring thiols are listed in Table 5.2.

Table 5.2 *Naturally occurring thiols and some thiol containing drugs*

Naturally occurring thiols

Thiol	Disulfide
Hydrogen sulfide	
Cysteine	Cystine
Homocysteine	Homocystine
Reduced glutathione (GSH)	Oxidised glutathione (GSSG)
Co-enzyme A	Lipoic acid
Propanethiol	
Ethanethiol	
Ergothionine	
Mercaptoethanol	
Cysteamine	

Some thiol-containing drugs

Clinical use	Chemical name	Proprietary or trivial name(s)
Metal chelation	Dimercaprol	BAL
	D-Penicillamine	Distamine, Pendramine
	Dimercaptopropansulfonate	DMPS
	Dimercaptosuccinic acid	DMSA
Mucolytics	N-Acetyl-L-cysteine	Airbron, Fabrol
	Carbocysteine	Mucodyne, Mucolex
	Methylcysteine	Visclair
Antithyroid	Carbimazole	Neomercazole
	Propylthiouracil	
Antihypertensive	Captopril	Capoten
Anthelmintic	Levamisole	
	Thiobendazole	Mintezol
Immunosuppressant	Azathioprine	Imuran
	6-Mercaptopurine	Puri-Nethol
	Thioguanine	Lanvis
Antirheumatoid	D-Penicillamine	Distamine
	5-Thiopyridoxine	
	Pyrithioxine	
	Thiopronine	

(Continued)

Table 5.2 *(Continued)*

Some thiol-containing drugs

Clinical use	Chemical name	Proprietary or trivial name(s)
	Mercaptopropylglycine	Thiola
	Sodium aurothiomalate	Myocrisin
	Aurothioglucose	Solganol
	Aurothiopropanol sulfonate	Allochrysine
Radioprotective	Cysteamine	
	WR2721	Amifostine
Chemoprotective	*N*-Acetyl-L-cysteine	NAC, Parvolex

1.3 Biochemical Background to Thiol Analysis

The metabolism of sulfur-containing amino acids is well documented,[8,9] but their relative concentrations in tissues and physiological fluids remain uncertain. Because of the reactivity of the –SH group in the presence of oxygen and/or trace quantities of heavy metals, thiols readily oxidise to disulfides. Disulfides with homocysteine and thiolactate are also possible. With an excess of the free thiol, the homodisulfide also forms in quantity. The ability of most thiols to form disulfides or undergo thiol-disulfide exchange is dependent on their relative concentrations, equilibrium constants and redox potentials. Free thiols as well as disulfides can be detected in plasma, but most thiol-containing drugs are present largely bound to proteins as disulfides. Measured values are often dependent on biochemical oxidation status and the sample preparation procedure used. Different analytical approaches are therefore needed to assay free thiols, low formula weight disulfides, protein-bound thiols, disulfides contained within the structures of proteins and possibly thiol-metal chelates or nitroso-thiols, either individually or in groups.

2 Sample Preparation

Most thiols and thiol-containing drugs oxidise in solutions, such as physiological fluids, at neutral pH. Careful sample preparation is thus essential to ensure that the measurement accurately reflects true thiol concentrations.

2.1 Standard Solutions

The rate of oxidation of thiols to homodisulfides in deionised water at room temperature is dependent on the structure of the analyte. For example, cysteine can be completely oxidised in a few minutes, especially when traces of heavy metal ions are present, whereas steric hindrance means that the mercaptopurines and D-penicillamine oxidise only slowly. Thiols can be stabilised in solution by lowering the pH to <2 and/or by adding disodium EDTA to chelate trace metals. It is best to prepare working calibration standards daily by dilution in an acidic solution.

2.2 Blood and Tissues

Endogenous thiols, such as cysteine and thiol-containing drugs, are rapidly lost from plasma/serum *in vitro* through protein binding (for example to the –SH residues at the surface of proteins, such as albumin) and disulfide formation with endogenous thiols. This loss may be minimised by the collection of whole blood into disodium EDTA-containing tubes, followed by rapid centrifugation at 4 °C using pre-cooled centrifuge buckets. Plasma proteins should then be precipitated immediately. Trichloroacetic, perchloric, metaphosphoric and sulfosalicylic acids have all been used because, after mixing and centrifugation, the resulting supernatant solution can be analysed directly by HPLC. Provided the above steps are completed within 3–5 min recoveries should be high. Recoveries from plasma of 104% for D-penicillamine[12] and 95% for thiomalate[13] have been reported. GSH can be measured in tissues by homogenisation in a suitable acidic protein precipitant. Mills *et al.*,[14] however, suggested that ultrafiltration is the optimum method for the accurate measurement of GSH in whole blood. Definitive measurement of GSH and GSSG in intact cells may only be possible using NMR spectroscopy.[15] Despite flash-freezing, acidification and storage at −20 °C, further loss of thiols to disulfides usually occurs. It is clear therefore that free thiol measurements should be performed as soon as possible after sample collection.

An amino acid analyser with post-column ninhydrin detection can give specific and accurate measurement of sulfur-containing amino acids, such as cystine in plasma and urine. However, analysis times are long (hours) and thiol-containing amino acids, such as cysteine, cannot be measured since the necessary elevated column temperatures lead to their rapid oxidation. In an attempt to circumvent this problem, Brigham *et al.*[16] derivatised cysteine with iodoacetate to give the stable analyte *S*-carboxymethylcysteine. Subsequently, *N*-ethylmaleimide and iodoacetamide have been used to derivatise amino-thiols prior to amino acid analysis.

2.3 Urine

Thiols disappear from urine in contact with air, forming both homo- and mixed-disulfides. Rapid losses have been reported for cysteine,[17] D-penicillamine,[18] and thiomalate.[13] It is therefore essential to acidify urine immediately after it has been voided. Collection of urine into 10 mL of 6 mol L^{-1} hydrochloric acid per litre of urine inhibits thiol oxidation for at least 12 h at room temperature. Although addition of EDTA and storing at −20 °C inhibits thiol degradation, further loss is still possible. Dithiothreitol (DTT) can be used to maintain thiols in the reduced form prior to analysis, but this procedure can give erroneous results since DTT might release thiols from mixed disulfides. Furthermore, DTT often interferes with HPLC-ED methods.

Disulfides can be quantified with adequate precision after acidification alone, provided suitable precautions are taken to stop disulfide formation from thiols.

If oxidation of thiols is a problem they can be protected by reaction with thiol-derivatising reagents, such as NEM.[19] When complex sample preparation procedures are needed, an internal standard should be used. For thiols being measured by HPLC-ED, the only suitable compounds are other thiols. The internal standard is thus likely to be unstable, with an *in vitro* half-life not necessarily the same as that of the analyte and mixed disulfides may form. Nevertheless, Drummer *et al.*[20] used *N*-acetylcysteine as the internal standard in an assay for D-penicillamine.

3 Measurement of Free Thiols using HPLC-ED

As well as the problems associated with sample collection and analyte stability, thiol assay presents other difficulties. UV detection has low sensitivity since simple thiols do not possess strong chromophores or fluorophores. Traditionally, thiol-containing amino acids were analysed using cation-exchange resins. In HPLC, these compounds can be analysed using ODS-modified silica packings, for example, using either low pH eluents to suppress ionisation of the acidic moieties, or by adding ion-pairing agents. Low pH eluents have the advantage of reducing the rate of disulfide formation, which may be the cause of the multiple peaks often observed with some thiols at neutral pH, most commonly with ODS-modified columns.[21–24] The ion-pairing agents commonly employed for the resolution of amino-thiols on ODS columns are heptane- and octane-sulfonic acids. However, cysteine and cystine are very poorly retained.[21,23,25,26] The same ion-pairing systems, however, give good resolution of D-penicillamine and its disulfide. Detailed studies of the chromatography of thiols were reported by Allison *et al.*,[25] who considered that the ion-pair systems gave the best selectivity.

Peak efficiencies can also be improved by adding of 1–2 mmol L^{-1} EDTA to the eluent.[27] EDTA prevents thiol complexation with metal ions either in the eluent or released by the components of the HPLC system including the column packing. Perrett and Rudge[27] observed an unusual carry-over phenomenon when analysing thiols on ODS-modified silica packings. Apparently thiols were adsorbed onto the packing and then displaced by other thiols, even though injections of eluent alone did not give such ghost peaks. Such situations can arise when thiol-containing drugs are being assayed in physiological fluids containing relatively large amounts of endogenous thiols, such as cysteine. Injecting a strongly displacing thiol, such as D-penicillamine, and regular washing with methanol, can clean columns. Jonkers *et al.*[28] reported that dithioerythritol improved peak shape and recovery in the RPLC of thiopurines. Presumably other reducing agents, such as DTT and cyanide ions, could have the same effect when included in the eluent, though care would be needed to ensure the additives did not produce cleavage of disulfides or interfere with detection.

4 Electrochemistry of Thiols

The EC activity of thiols was discovered by Dixon and Quastel,[29] who reported the anodic oxidation of cysteine. Of more analytical relevance was the characterisation of the electrochemistry of cysteine at the dropping mercury electrode.[30] It was suggested that the wave observed corresponded to a specific one-electron reaction (Reaction 5.1). With respect to thiols the reaction is indirect since the current measured is derived from the oxidation of mercury and this can also occur in the presence of other species, such as cyanide and halide ions. Thiols form sufficiently stable complexes to permit the oxidation of the electrode.

$$2RSH + Hg \longrightarrow (RS)_2Hg + 2H^+ + 2e^- \tag{5.1}$$

Lee[31] and Mairesse-Ducarmois *et al.*[32,33] using fast linear sweep and differential pulse polarography at mercury electrodes, respectively, investigated the electro-chemistry of cystine/cysteine and concluded that at 0 V mercurous cysteinate was formed by Reaction 5.2.

$$RSH + Hg \longleftrightarrow RSHg + H^+ + e^- \tag{5.2}$$

Stricks and Kolthoff[34] using DC polarography found that between pH 1 and pH 10.5 and at *ca* -0.25 V *vs* SCE, Reaction 5.2 also applied to GSH. Mairesse-Ducarmois *et al.*[33] showed that following the oxidation of GSH, the product, GSHg, was adsorbed onto the mercury surface. Similar results have been observed with silver electrodes.[35] Gold amalgam electrodes can also be used (see below).

Thiols can be oxidised at a variety of solid electrodes, such as noble metals, carbon and carbon with chemically modified surfaces. Mefford and Adams[36] found that relatively high voltages, greater than $+1$ V, were required to oxidise GSH and cysteine at glassy carbon electrodes. Chemically modified surfaces can reduce the oxidation potentials required and hopefully increase specificity and sensitivity. Halbert and Baldwin[37] used cyclic voltammetry to study the electrochemistry of cysteine, homocysteine, *N*-acetylcysteine and GSH at both unmodified and cobalt phthalocyanine-modified carbon paste electrodes. This non-chromatographic technique was used to measure the relatively high concentrations of GSH in whole blood.[38] The electrochemistry of thiols has been reviewed.[4,39]

5 Electrochemical Detection of Thiols

5.1 Detection at Mercury Electrodes

Since the oxidation of thiols at mercury electrodes is facile and specific, it should be adaptable to HPLC detection. Rabenstein and Saetre[40] first reported this and so gave rise to a significant improvement in the assay of thiols in complex mixtures. Cysteine and other amino-thiols were separated on a cation-exchange column and detected at a mercury-pool EC detector ($+0.05$ V *vs* SCE) with a sensitivity of <1 nmol for most of the thiols tested. Sample preparation consisted of little more than diluting a protein-free extract of a biofluid. These authors subsequently detailed many applications of this assay, *e.g.* D-penicillamine in blood and urine,[18,41,42] cysteine and homocysteine in plasma and urine[43] and cysteine and glutathione in fruit.[44] However, the detector was not commercially available and few workers copied their system.

5.2 Detection at Gold Amalgam Electrodes

Bioanalytical Systems were the first to manufacture metal electrodes that would interchange with their standard GCE blocks in their amperometric detector cell. Thiols could be detected by forming an amalgam on a gold electrode. The static electrode had increased sensitivity relative to the mercury-pool electrode requiring similar low potentials but importantly was easier to operate. Nobel metal electrodes work at potentials intermediate between those of mercury-based electrodes and those required by a GCE (Figure 5.1). With the development of these metal electrodes the availability and the applicability of the methods described above were extended.

These new gold-mercury electrode configurations facilitated study of the pharmacokinetics of many thiol-containing drugs. Wiesner *et al.*[45] measured plasma D-penicillamine following both i.v. and oral dosage. The same method was used to characterise the protein binding and pharmacokinetics of D-penicillamine following cation-exchange chromatography.[12,46] Others used Au/Hg electrodes to measure free D-penicillamine[20]; sensitivities of less than 100 fmol injected were achieved. Applications of Au/Hg electrodes to other biological thiols and drugs have been reported. Perrett and Drury[23] detected captopril in plasma following RPLC (Figure 5.2) (see also chapter 6, Section 14). Drummer *et al.*[20] and Stenken *et al.*[47] measured N-acetylcysteine following IP-RPLC, whilst Lunte and Kissinger[48] and Demaster *et al.*[49] measured GSH. The radioprotective agents WR2721 and WR1065 were also measured in blood and biological samples in this way,[50,51] although more recently coulometric detection has been employed[52] (Chapter 6, Section 18).

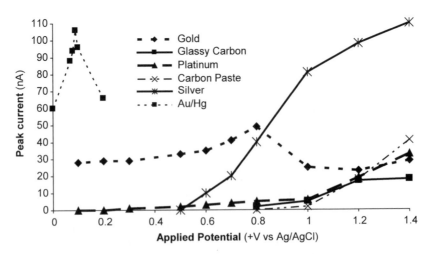

Figure 5.1 *Hydrodynamic voltammograms for captopril (100 nmol injected) at various electrodes Eluent: 100 mmol L⁻¹ potassium dihydrogen orthophosphate, pH 2; Flow-rate: 1 mL min⁻¹.*

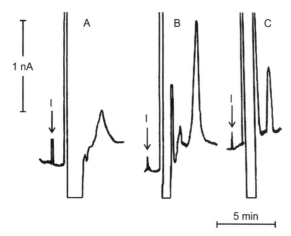

Figure 5.2 *HPLC of captopril in plasma and urine (re-drawn from Perrett and Drury[23]).*
Column: 5 μm Hypersil ODS; Eluent: pH 2, 100 mmol L^{-1} potassium
dihydrogen orthophosphate; Flow-rate: 1 mL min^{-1}; Detection: Hg/Au
electrode, +0.05 V vs Ag/AgCl; A. Patient plasma blank, B. Plasma blank
(1 mL) with 25 pmol captopril added, C. Urine from patient 1 h after receiving
captopril (6 mg, oral).

5.2.1 Preparation of Gold Amalgam Electrodes

Au/Hg electrodes are easily formed. Firstly, any existing amalgam should be removed from the gold electrode with a drop of concentrated nitric acid, which is then washed off with water. If necessary the electrode should be polished according to the manufacturer's instructions – the better the preparation of the gold surface the higher the sensitivity of the final electrode. With the electrode block standing flat in a tray a drop of triple distilled mercury is placed on the metal so that the surface is completely covered. This is left for a few minutes for the gold to adsorb a thin layer of mercury and the excess mercury carefully removed by wiping the surface of the electrode with a piece of card. The electrode surface should be shiny without any gold showing. A poorly made or fouled electrode usually has a dull surface and the gold may show through. The electrode housing should be reassembled using a double thickness gasket according to the manufacturer's instructions. On applying an appropriate potential a large current will be observed that will slowly decay and, depending on the desired operating conditions, a stable background should be reached after about one hour. The lifetime of the amalgam can range from a few injections to 1–2 weeks; 1 week is usual. Loss of surface is indicated by either by a gradual decline in the response to standard injections or rapid loss of response. The electrode can also lose sensitivity *via* an increase in background noise. Such losses of sensitivity necessitate both frequent injections of standards and frequent re-making of the electrode. Nevertheless, with due care, the Au/Hg electrode is precise, extremely sensitive, and thiol specific.

Gold amalgam electrodes have been successfully used with CZE for certain applications. Jin and Wang[53] showed that a gold amalgam electrode used without eluent deoxygenation could be used to detect cysteine in standard solutions. The electrode was based on a 100 μm gold wire coated with mercury. Later they measured glutathione in single human erythrocytes[54] and in individual mouse peritoneal macrophages.[55] Subsequently, they bundled carbon fibre and Au/Hg dual electrodes together in order to measure glutathione and neurochemicals such as tryptophan simultaneously in standard mixtures, hepatocyte extracts and single hepatocytes. In single hepatocytes the amount of GSH found was approximately 11 fmol.[56]

5.3 Detection at Other Solid Electrodes

Carbon electrodes were used for thiol analysis by Mefford and Adams at potentials >+1 V *vs* Ag/AgCl.[36] As little as 5 pmol of cysteine and GSH could be measured in acid extracts of animal tissue. The ready availability of GCE electrodes led others[21,57] to use them to detect D-penicillamine. At GCEs the maximal response for simple thiols occurs at relatively high potentials (+1 to +1.5 V *vs* Ag/AgCl) as compared to 0 V at Hg electrodes (see Figure 5.1). Such high working potentials mean that many other electroactive compounds in plasma and urine also respond, so the selectivity of the extraction and HPLC column must be correspondingly greater. In such circumstances the higher resolution of CZE can be advantageous – Wang *et al.*[58] (2000) have reported use of a carbon fibre electrode with CZE to detect L-cysteine, carbimazole, methimazole, GSH and 6-thiopurine.

Carro-Ciampi *et al.*[59] used dual PGEs to detect GSH and GSSG. The upstream electrode was at +0.7 V, presumably to reduce background interferences, and analytes were detected at the downstream electrode held at +0.9 V *vs* Pd. Maximum sensitivity was 1.5 ng GSH injected, but even using a clean-up electrode sensitivity was lost over 100 analyses. Stulik and Pacakova[60] measured standard solutions of thiouracil at +1.4 V at a carbon paste electrode.

Perrett[61] used flow injection analysis (FIA) to investigate various electrode materials for the measurement of electroactive compounds, since it is only in such a system that the effect of pH on the electrochemistry can be studied independently of the chromatography. All thiols could be detected at working potentials above +1 V on both metal and carbon surfaces. Many thiols reacted in a semi-specific manner with some metal electrodes (see Figure 5.1). Some, including thiomalate and D-penicillamine, also exhibited a second maximum at *ca* +0.75 V *vs* Ag/AgCl with a gold electrode. If, as has been suggested,[57] the peak at above +1 V is due to the direct oxidation of thiols to disulfides, the reason for these thiols giving two peaks becomes unclear. Possibly the lower voltage is associated with an intermediate oxidation product or the formation of a surface metal complex. With physiological fluids, gold electrodes offer less selectivity than Au/Hg electrodes, but more selectivity than carbon electrodes. Gold electrodes are however more robust than Au/Hg electrodes when physiological samples are analysed.

These observations at un-modified metal surfaces led to electrodes of this type being used in amperometric detectors. Kreuzig and Frank[62] oxidised

D-penicillamine at a gold electrode at +0.8 V *vs* Ag/AgCl following cation-exchange chromatography, and found comparable sensitivity to the Au/Hg electrode. Rudge *et al.*[24] used a gold electrode and RPLC to measure pmol L^{-1} concentrations of thiomalate in plasma and urine of rheumatoid arthritis patients receiving sodium aurothiomalate injections. The same system was applied to captopril[63] and D-penicillamine (Figure 5.3). Gold and silver electrodes lose sensitivity with use, but the rate is less than with Au/Hg electrodes. Silver electrodes rapidly lose sensitivity, probably due to the thiol blackening of the surface. It is possible that pulsed electrode procedures could be useful in prolonging the life of gold and platinum electrodes when detecting thiols.

In the FIA studies mentioned earlier platinum electrodes gave a secondary peak only with cysteamine. In a subsequent assay[64] cysteamine was separated by iso-cratic SCX-HPLC and detected at +0.45 V *vs* Ag/AgCl. The analysis time was less than 3 min.

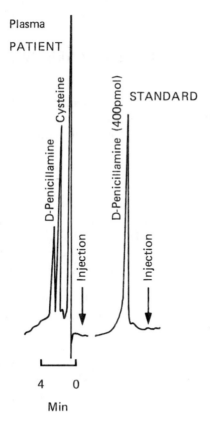

Figure 5.3 *Measurement of D-penicillamine in plasma using a gold electrode. Column: 5 μm Hypersil ODS; Eluent: pH 2, 50 mmol L^{-1} potassium dihydrogen orthophosphate; Flow-rate: 1 mL min^{-1}; Detection: Gold electrode, +0.65 V vs Ag/AgCl; A. Standard injection of D-penicillamine (400 pmol), B. Plasma from patient receiving 1000 mg d^{-1} D-penicillamine.*

5.4 Formation of Electroactive Derivatives of Thiols

In HPLC most derivatising reagents are used to increase sensitivity by forming chromophores and fluorophores. There would seem little point in forming electrophores of already electroactive compounds, such as thiols. However, Shimada et al.[65] used N-(4-anilinophenyl)maleimide to confer not only electroactivity, but also stability to N-acetylcysteine, cysteine, GSH and D-penicillamine. The reaction was complete in 10 min at 0 °C and the derivatives were separated in 12 min using isocratic RPLC. They also used derivatisation with a related compound, N-(4-dimethylaminophenyl)-maleimide, to measure captopril in plasma.[66]

At alkaline pH (9.2) Mairesse-Ducarmois et al.[32,33] found that cystine was reduced at −0.91 V vs SCE. As a preliminary to the development of a dual electrode detector for disulfides, Allison and Shoup[25] performed cyclic voltammetry of GSSG at pH 3 at an Au/Hg electrode and found three reductive peaks. Two peaks were attributed to the reduction of adsorbed species, but that at −1.1 V resulted from the direct reduction of GSSG in solution. After an exhaustive reduction of GSSG they were able to show that the only product was GSH.

6 The Electrochemistry of Disulfides and Related Compounds

At mercury electrodes Lee[31] reported that cystine dissolved in perchloric acid was reduced to cysteine (Reaction 5.3) at a potential of −0.43 V, but that there was also evidence of the transient formation of mercurous cysteinate as a reaction intermediate:

$$RSSR' + 2e^- + 2H^+ \longrightarrow RSH + R'SH \tag{5.3}$$

Injection of excessive amounts of DTT also rapidly degrades electrodes, particularly Au/Hg electrodes. Newer reductants based on phosphine chemistry are claimed to be more efficient. Tri-n-butylphosphine[67] is efficient, but can be explosive in excess! Triphenylphosphine[68] is also an efficient reductant, but tris-(2-carboxyethyl)-phosphine[69] solves most problems. Dithionite (sodium hydrosulfite) has also been used.

Since disulfides can be completely reduced to thiols at voltages below −1 V, off-line EC reduction can be useful. Saetre and Rabenstein[18,43] reduced D-penicillamine-containing disulfides in plasma and blood after protein precipitation at a mercury-pool electrode held at −6 V. This procedure was also adopted by Bergstrom et al.[46] A disadvantage of this approach is the assumption that only one disulfide is present and that it is totally reduced. No chromatographic identification of the disulfides is possible since only the reduced thiols are analysed.

7 The Electrochemical Detection of Disulfides plus Thiols

Comprehensive reviews of the measurement of disulfides have been published.[5,7]

7.1 Direct Detection of Disulfides

Disulfides have been directly detected at GCEs held at positive potentials (+1 V *vs* Ag/AgCl or thereabouts). Hence, under appropriate conditions, analytical systems that detect thiols may also be used to detect disulfides. Abounassif and Jefferies[21] found that cystine and cysteine-D-penicillamine disulfide gave a maximum response at +0.9 to +1.1 V and D-penicillamine disulfide a maximum response at +1.2 V, but the sensitivity was 1000-fold less than for the corresponding thiols. The reason for this was not discussed. Alternatively, disulfides are readily reduced at noble metal electrodes. Figure 5.4 shows the measurement of GSSG at a gold

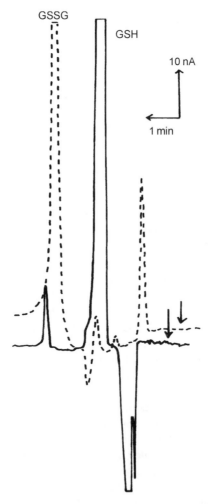

Figure 5.4 *Measurement of GSSG at a gold electrode in the reductive mode (−1.1 V vs Ag/ AgCl) and GSH in oxidative mode (+0.8 V vs Ag/AgCl). The same gold working electrode was used for both analytes. Eluent: 0.1 mol L⁻¹ mono-chloroacetic acid pH 3-methanol (98 + 2); Injection: 2 nmol each compound.*

electrode. Note that there is some sensitivity towards GSSG in the oxidative mode at the gold electrode, although the sensitivity is less than 5% of that obtained in the reductive mode.

7.2 Indirect Detection of Disulfides

Disulfides can be reduced chemically, electrochemically or enzymatically. Chemical reduction is often used prior to derivatisation of thiols to fluorophores, but it can also be used with HPLC-ED. Common reducing agents are mercapto-ethanol, DTT, dithioerythritol, cyanide and sodium borohydride, but their efficiency varies towards different disulfides. Cyanide or borohydride rapidly reduce cystine at room temperature, but require over 24 h to reduce cysteine-D-penicillamine disulfide completely, whilst the sterically hindered D-penicillamine disulfide is not reduced at all.[70]

A yield of 89% for D-penicillamine from its homodisulfide using a 10-fold excess of DTT at 60 °C (4 h) has been claimed.[21] Comparative studies with cysteine-D-penicillamine disulfide or protein-bound D-penicillamine were not reported. Kelly *et al.*[64] reduced cysteamine disulfides and protein-cysteamine with DTT prior to HPLC-ED. However, excess DTT can produce analytical problems since both DTT and its oxidised form are readily separated by RPLC and can be detected electrochemically.

7.3 In-Series Detection of Disulfides

To overcome the disadvantages in the off-line reduction of disulfides, Eggli and Asper[70] introduced a dual electrode flow-through HPLC detector for the measurement of both thiols and disulfides. The first electrode (reduction cell) consisted of a short column of amalgamated silver grains through which the eluate containing the thiols and disulfides flowed. The disulfides were reduced and immediately entered a mercury-pool electrode (detection cell). Using the eluate from an SCX-modified silica column (pH 2.6, 90 mmol L^{-1} phosphate/citrate buffer) at a flow-rate of 0.3 mL min^{-1} complete reduction of cystine was achieved when the reduction electrode was held at below −1.1 V vs SCE. Simultaneous separation and detection of cysteine, cystine and D-penicillamine was demonstrated. Eggli and Asper's design, although electrochemically efficient, generates significant post-column band broadening. Allison *et al.*[71] adapted this approach to a dual electrode cell using two Au/Hg electrodes in one cell block separated by only 1–2 mm as the reducing and detecting electrodes. Figure 5.5 shows the separation and detection of GSSG using this dual electrode approach.

The efficiency of the flow-through reducing electrode is limited by a number of factors, including diffusion coefficient, eluent pH, applied potential and very importantly the chemical structure of the disulfide. Figure 5.5 shows that the sensitivity towards GSSG is less than *ca* 2% that of GSH. As only amperometric electrodes were used it should be expected that <5% of the disulfide might be reduced at the first electrode and then <5% of the thiol(s) produced would be detected at the second. Since two thiol molecules are formed from one disulfide it is to be expected that the detector sensitivity to disulfides would be <10% that of

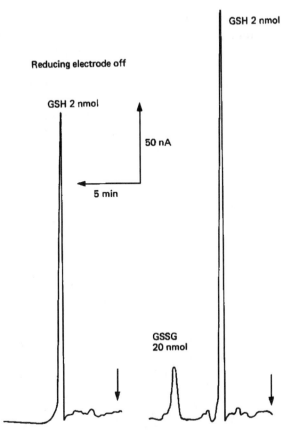

Figure 5.5   *Use of dual electrodes in series to measure GSH and GSSG in a single sample. Column: 3 μm ODS-Hypersil 100 × 4.6 mm; Eluent: 100 mmol L^{-1} monochloroacetic acid (pH 3)-methanol (98 + 2); Flow-rate: 1 mL min^{-1}; Detection electrode Au, +0.7 V vs. Ag/AgCl; Reducing electrode Hg/Au, −0.9 V vs Ag/AgCl.*

the thiol. Such a reduction/detection ratio is possible, but difficult to achieve routinely. Amongst the precautions necessary is the need to thoroughly deoxygenate the system. All plastic tubing must be replaced with metal tubing and the eluent should be constantly refluxed under nitrogen or helium. In addition it is usually also necessary to de-oxygenate the sample *in vacuo* or by purging with helium. As with the assay of thiols alone, the instability of the gold amalgam at the detection electrode is also an important factor. Some disulfides are more difficult to detect than others, *e.g.* D-penicillamine disulfide reduction is sterically inhibited by its two methyl groups and its detection sensitivity is therefore much less than that of cystine. Cysteine-D-penicillamine disulfide would be expected to have an intermediate level of sensitivity.

Many people now use commercially available dual cells, which employ independent controllers, for disulfide measurement. The reductive electrode does not necessarily require such a controller and Stein *et al.*[72] employed a simple voltage divider circuit and a battery to generate the necessary reductive potential. We have also used a simple mains powered, but stable, d.c. voltage source for the reduction. The dual electrode approach has also been employed in CZE to measure cysteine, glutathione and cystine.[73]

Sun *et al.*[74] have described an amalgam cell which contained three gold electrodes and which was employed to monitor the thiol and disulfide composition of complex peptide mixtures when some peptides contained both moieties. The gold electrodes were mounted in an L-shape, such that one electrode only measured thiols whilst the other two electrodes formed a reduction/detection pair for the measurement of disulfides.

Kleinman and Richie[75] used the in-series approach with gold amalgam electrodes to measure 12 thiols and disulfides, including GSH, GSSG, cysteine, cystine, homocysteine, homocystine and several mixed disulfides in rat tissues. The first (upstream) electrode was maintained at −1.0 V and the second (downstream)

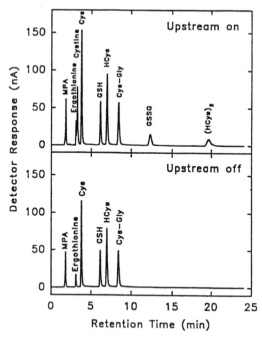

Figure 5.6 *Upper: HPLC-ED of thiols and disulfides detected at Hg/Au electrode (+0.15 V vs Ag/AgCl) after reduction of the disulfides at the first electrode (Hg/Au, −1.0 V vs Ag/AgCl); Lower: The disulfides are not detected when the upstream electrode is turned off. Standard mixture: 25 μmol L^{-1} each compound; Injection: 50 μL. MPA = metaphosphoric acid.*
(With permission from Kleinman and Richie[75])

Table 5.3 *Published papers on the measurement of endogenous thiols/disulfides by HPLC-ED*

Sulfide	Sample	Column	Sensitivity	References
Cystine	Standards	SCX	0.1 ng	70
	Urine	IP-RPLC		76
	Blood	IP-HPLC	5 pmol	77
GSSG	Citrus leaves Whole blood	RPLC	5.7 pmol	25
	Microsomes	IP-RPLC		48
	Liver	IP-RPLC		19
	Liver, kidney	IP-RPLC	12 pmol	72
	Blood	IP-RPLC	5 pmol	77
GSH/GSSG	Hepatocytes cardiomyocytes	IP-RPLC	40 pg (GSH), 300 pg (GSSG)	78
Cysteine-D-penicillamine	Urine	IP-RPLC		76
D-Penicillamine disulfide	Plasma	IP-RPLC	600 pmol	79
Homocystine	Plasma	IP-RPLC	<100 pmol	80
Homocysteine	Standards	RPLC	1 pmol	81
N-Acetylcysteine disulfide	Urine	IP-RPLC	1 pmol	47
Cysteinyl peptides	Standards	RPLC	1 pmol	74,82
Twelve thiols and disulfides inc. GSH/GSSG	Blood and tissues	IR-RPLC	0.02–1 pmol	75

at $+0.15$ V *vs* Ag/AgCl. The analysis was performed using an ODS-modified silica column with methanol-*N,N*-dimethylformamide-aq. monochloroacetic acid (0.1 mol L^{-1}) $(5+1.75+93.25)$ containing heptanesulfonic acid $(2.25 \text{ mmol L}^{-1})$ as eluent. Although ergothioneine was not completely resolved from cystine, it could be easily differentiated by turning the upstream electrode off (Figure 5.6). LoDs were between 0.02 (cysteine) and 1.0 pmol (homocystine).

Since its introduction, the on-line reduction-detection system has been used with varying success. Published applications are summarised in Table 5.3.

8 Other Electrochemical Methods for Thiols and/or Disulfides

Although the methods described so far are generally successful for thiols and disulfides, novel methods continue to be published. Qian *et al.*[83] used an HPLC flow-through differential-pulse polarograph at negative potentials in an ammonium chloride cobalt electrolyte to measure thiols and disulfide compounds, but the sensitivity was only moderate. Kuninori and Nishiyama[35] separated thiols by IP-RPLC and, following post-column addition of silver nitrate in ammonium nitrate at pH 10.5, they measured the amount of silver ions consumed by the thiol with a platinum disc electrode held at -0.1 V *vs* Ag/AgCl. For the measurement of disulfides, sodium sulfite (10–100 mmol L^{-1}) was included in the post-column reagent. Sensitivity to thiols was *ca* 1 pmol and the method was claimed to be simpler and more stable than using Au/Hg electrodes. Nishiyama and Kuninori[84] (1992) also showed that NEM-thiol adducts could be quantitatively electro-reduced prior to thiol measurement.

Another post-column reaction system, based on the on-line electrochemical generation of bromine from bromide, was claimed to be as sensitive as the dual electrode system for thiols, disulfides and thioethers.[85] Following a reaction coil, the change in bromine concentration was measured at an amperometric Pt electrode. A negative peak was observed when sulfur-containing compounds eluted. The fact that the system detected all sulfur compounds relatively specifically was employed in a study of the changes in the intracellular concentrations of GSH, GSSG and paracetamol-GSH conjugates.[86] This bromine generating reaction was also employed by Holland and Lunte[87] to detect cysteine, GSH, methionine and *N*-acetylcysteine after CZE. Three different configurations for the dual electrode configuration were investigated and it was concluded that a novel platinum wire-wire on-capillary electrode system was best.

9 Analysis of Selenium-Containing Compounds

Considering the close chemical similarity of thiols and selenols it is not surprising that selenols, diselenides and selenenyl sulfides are electrochemically active. Killa and Rabenstein[88] used dual Au/Hg amperometric detection to measure these compounds following RPLC in the presence of OSA. The optimum potential was found to be $+0.15$ V vs Ag/AgCl. With correct choice of chromatographic conditions, selenols could also be resolved from the thiols occurring in biological matrices.

10 Conclusions

Although many procedures are now available for thiol assay, difficulties still exist in measuring these compounds in physiological fluids. Most HPLC methods are capable of good resolution and ED usually offers sufficient sensitivity. The major problems, however, are linked to sample preparation since many thiols, particularly the aliphatic thiols, are unstable in solution unless precautions are taken to prevent oxidation. This problem is considerably greater in physiological fluids. Binding to proteins both *in vivo* and *in vitro* is so rapid that some thiols can be completely bound within a few minutes. Steps to prevent this must therefore be taken if an analysis is to be meaningful.

For pharmacokinetic and metabolic studies it is essential that thiols and disulfides are not measured in isolation. Pre-column reduction procedures partially overcome this problem, but qualitative data on the disulfides are lacking. Post-column reduction can be restrictive since sensitivity depends on the ability of the reactor to reduce disulfides; in some cases this is impossible. HPLC-ED is proving in some cases to be the method of choice for this dual application.

Where sample numbers are large and assays can be performed within a few days of sample collection HPLC-ED using metal electrodes offers both speed and sensitivity. Good column maintenance is essential to avoid poor chromatography, broad late-eluting peaks and also 'ghost' peaks.

References

1 S. Patai (ed), *The Chemistry of the Sulfur-containing Functional Groups*, Wiley, Chichester, 1993.

2 M.R.F. Ashworth, *The determination of sulfur-containing groups*, 3rd edn, Academic Press, London, 1977.

3 A. Russo and E.A. Bump, Detection and quantitation of biological sulfhydryls in *Methods of Biochemical Analysis* D. Glick (ed) 1988, **33**, 165–241.

4 P.C. White, N.S. Lawrence, J. Davis and R.G. Compton, Electrochemical determination of thiols: A perspective, *Electroanalysis*, 2002, **14**, 89–98.

5 D. Perrett and S.R. Rudge, The determination of thiols and related compounds using high-performance liquid chromatography, *J. Pharm. Biomed. Anal.*, 1985, **3**, 3–27.

6 E. Camera and M. Picardo, Analytical methods to investigate glutathione and related compounds in biological and pathological processes, *J. Chromatogr. B*, 2002, **781**, 181–206.

7 J. Lock and J. Davis, The determination of disulfide species within physiological fluids, *Tr. Anal. Chem.*, 2002, **21**, 807–815.

8 P.C. Jocelyn, *The Biochemistry of the SH Group*, Academic Press, New York, 1972.

9 K. Kuriyama, R. Huxtable and H. Iwata (eds), *Sulfur Amino Acids: Biochemical and Clinical Aspects*, Alan R. Liss, New York, 1983.

10 A. Meister, Selective modification of glutathione metabolism, *Science*, 1983, **220**, 472–477.

11 O. Nekrassova, N.S. Lawrence and R.G. Crompton, Analytical determination of homocysteine: a review, *Talanta*, 2003, **60**, 1085–1095.

12 R.F. Bergstrom, D.R. Kay and J.G. Wagner, The *in vitro* loss of penicillamine in plasma, albumin solutions, and whole blood: implications for pharma-cokinetic studies of penicillamine, *Life Sci.*, 1980, **27**, 189–198.

13 S.R. Rudge, D. Perrett and A.J. Swannell, Free thiomalate in plasma and urine of patients receiving sodium aurothiomalate, *Ann. Rheum. Dis.*, 1984, **43**, 66–69.

14 B.J. Mills, J.P. Richie Jr., and C.A. Lang, Sample processing alters glutathione and cysteine values in blood, *Anal. Biochem.*, 1990, **184**, 263–267.

15 J. Reglinski, W.E. Smith, M. Brzeski, M. Marabani R.D. Sturrock, Clinical analysis by 1H spin-echo NMR, 2. Oxidation of intracellular glutathione as a consequence of penicillamine therapy in rheumatoid arthritis, *J. Med. Chem.*, 1992, **35**, 2134–2137.

16 M.P. Brigham, W.H. Stein and S. Moore, The concentration of cysteine and cystine in human blood plasma, *J. Clin. Invest.*, 1960, **39**, 1633–1638.

17 S. Roston, Determination of cysteine, *Anal. Biochem.*, 1963, **6**, 486–490.

18 R. Saetre and D.L. Rabenstein, Determination of penicillamine in blood and urine by high performance liquid chromatography, *Anal. Chem.*, 1978, **50**, 278–280.

19 S.M. Lunte and P.T. Kissinger, Detection of thiols and disulfides in liver samples using liquid-chromatography electrochemistry, *J. Liq. Chromatogr.* 1985, **8**, 691–706.

20 O.H. Drummer, N. Christophidis, J.D. Horowitz and W.J. Louis, Measurement of penicillamine and *N*-acetylcysteine in human blood by high-performance liquid chromatography and electrochemical detection, *J. Chromatogr.*, 1986, **374**, 251–257.

21 M.A. Abounassif and T.M. Jefferies, The determination of *D*-penicillamine and its disulfide in plasma by reversed-phase ion-pair high-performance liquid chromatography, *J. Pharm. Biomed. Anal.*, 1983, **1**, 65–72.

22 S. Perlman and J. Kirschbaum, High-performance liquid chromatographic analyses of the antihypertensive drug captopril, *J. Chromatogr.*, 1981, **206**, 311–317.

23 D. Perrett and P.L. Drury, The determination of captopril in physiological fluids using high performance liquid chromatography with electrochemical detection, *J. Liq. Chromatogr.*, 1982, **5**, 97–110.

24 S.R. Rudge, D. Perrett, P.L. Drury and A.J. Swannell, The determination of thiomalate in physiological fluids by high-performance liquid chromatography and electrochemical detection, *J. Pharm. Biomed. Anal.*, 1983, **1**, 205–210.

25 L.A. Allison and R.E. Shoup, Dual electrode liquid chromatography detector for thiols and disulfides, *Anal. Chem.*, 1983, **55**, 8–12.

26 D. Dupuy and S. Szabo, Measurement of tissue sulfhydryls and disulfides in tissue protein and non-protein fractions by high-performance liquid-chromatography using electrochemical detection, *J. Liq. Chromatogr.*, 1987, **10**, 107–119.

27 D. Perrett and S.R. Rudge, Problems associated with the high-performance liquid chromatography of thiols. *J Chromatogr. A*, 1984, **294**, 380–384.

28 R.E. Jonkers, B. Oosterhuis, R.J. ten Berge and C.J. van Boxtel, Analysis of 6-mercaptopurine in human plasma with a high-performance liquid chromatographic method including post-column derivatization and fluorimetric detection. *J. Chromatogr.*, 1982, **233**, 249–255.

29 M. Dixon and J.H. Quastel, A new type of reduction-oxidation system, Part I. Cysteine and glutathione, *J. Chem. Soc.* 1923, **123**, 2943–2953.

30 I.M. Kolthoff and C. Barnum, The anodic reaction and waves of cysteine at the dropping mercury electrode and at the platinum micro wire electrode, *J. Am. Chem. Soc.*, 1940, **62**, 3061–3065.

31 W. Lee, Rapid-sweep polarography of cystine, *Biochem. J.*, 1971 **121**, 563–565.

32 C.A. Mairesse-Ducarmois, G.J. Patriarche and J.L. Vandenbalck, Contribution a l'électrochimie des thiols et disulfures: Partie I. Cystéine et cystine, *Anal. Chim. Acta*, 1974, **71**, 165–174.

33 C.A. Mairesse-Ducarmois, G.J. Patriarche and J.L. Vandenbalck, Contribution à l'électrochimie des thiols et disulfures: Partie II. Polarographie d.c., a.c. et impulsionnelle différentielle du glutathion, *Anal. Chim. Acta*, 1975, **76**, 299–308.

34 W. Stricks and I.M. Kolthoff, Polarography of glutathione, *J. Am. Chem. Soc.*, 1952, **74**, 4646–4653.

35 T. Kuninori and J. Nishiyama, Measurement of biological thiols and disulfides by high performance liquid chromatography and electrochemical detection of silver mercaptide formation, *Anal. Biochem.*, 1991, **197**, 19–24.

36 I. Mefford and R.N. Adams, Determination of reduced glutathione in guinea pig and rat tissue by HPLC with electrochemical detection, *Life Sci.*, 1978, **23**, 1167–1173.

37 M.K. Halbert and R.P. Baldwin, Determination of cysteine and glutathione in plasma and blood by liquid chromatography with electrochemical detection using a chemically modified electrode containing cobalt phthalocyanide, *J. Chromatogr.*, 1985, **345**, 43–49.

38 S.A. Wring, J.P. Hart and B.J. Birch, Development of an improved carbon electrode chemically modified with cobalt phthalocyanine as a re-usable sensor for glutathione, *Analyst*, 1989, **114**, 1563–1570.

39 J.P. Hart, *Electroanalysis of Biologically Important Compounds*, Ellis Horwood, Chichester, 1990.

40 D.L. Rabenstein and R. Saetre, Mercury-based electrochemical detector of liquid chromatography for the detection of glutathione and other sulfur-containing compounds, *Anal. Chem.*, 1977, **49**, 1036–1039.

41 A.S. Russell, R. Saetre, P. Davis and D.L. Rabenstein, A rapid, sensitive technique to assay penicillamine levels in blood and urine, *J. Rheumatol.*, 1979, **6**, 15–19.

42 M. Butler, G. Carruthers, M. Harth, D. Freeman, J. Percy and D. Rabenstein, Pharmacokinetics of reduced D-penicillamine in patients with rheumatoid arthritis, *Arthritis Rheum.* 1982, **25**, 111–116.

43 R. Saetre and D.L. Rabenstein, Determination of cysteine in plasma and urine and homocysteine in plasma by high-pressure liquid chromatography, *Anal. Biochem.* 1978, **90**, 684–692.

44 R. Saetre and D.L. Rabenstein, Determination of cysteine and glutathione in fruit by high-performance liquid chromatography, *J. Agric. Food Chem.*, 1978, **26**, 982–983.

45 R.H. Wiesner, E.R. Dickson, G.L. Carlson, L.W. McPhaul and V.L.W. Go, The pharmacokinetics of D-penicillamine in man, *J. Rheumatol.*, 1981, **8 (Suppl 7)**, 51–55.

46 R.F. Bergstrom, D.R. Kay and J.G. Wagner, High-performance liquid chromatographic determination of penicillamine in whole blood, plasma, and urine, *J. Chromatogr.*, 1981, **222**, 445–452.

47 J.A. Stenken, D.L. Puckett, S.M. Lunte and C.E. Lunte, Detection of N-acetylcysteine, cysteine and their disulfides in urine by liquid chromatography with a dual-electrode amperometric detector, *J. Pharm. Biomed. Anal.*, 1990, **8**, 85–89.

48 S.M. Lunte and P.T. Kissinger, The use of liquid chromatography with dual-electrode electrochemical detection in the investigation of glutathione oxidation during benzene metabolism, *J. Chromatogr.*, 1984, **317**, 579–688.

49 E.G. Demaster, F.N. Shirota, B. Redfern, D.J. Goon and H.T. Nagasawa, Analysis of hepatic reduced glutathione, cysteine and homocysteine by cation-exchange high-performance liquid chromatography with electrochemical detection, *J. Chromatogr.*, 1984, **308**, 83–91.

50 L.M. Shaw, H. Bonner, A. Turrisi, A.L. Norfleet and D.J. Glover, A liquid-chromatographic electrochemical assay for S-2-(3-aminopropylamino)ethyl-phosphorothioate (WR2721) in human-plasma, *J. Liq. Chromatogr.*, 1984, **7**, 2447–2465.

51 L.M. Shaw, H. Bonner, A. Turrisi, A.L. Norfleet and M. Kligerman, Measurement of S-2-(3-aminopropyl-amino)ethanethiol (WR1065) in blood and urine, *J. Liq. Chromatogr.*, 1986, **9**, 845–859.

52 F. Bai, M.N. Kirstein, S.K. Hanna and C.F. Stewart, New liquid chromatographic assay with electrochemical detection for the measurement of amifostine and WR1065, *J. Chromatogr. B*, 2002, **772**, 257–265.

53 W.R. Jin and Y. Wang, Determination of cysteine by capillary zone electrophoresis with end-column amperometric detection at a gold/mercury amalgam microelectrode without deoxygenation, *J. Chromatogr. A*, 1997, **769**, 307–314.

54 W.R. Jin, W. Li and Q. Xu, Quantitative determination of glutathione in single human erythrocytes by capillary zone electrophoresis with electrochemical detection, *Electrophoresis*, 2000, **21**, 774–779.

55 W.R. Jin, Q. Dong, X.Y. Ye and D.Q. Yu, Assay of glutathione in individual mouse peritoneal macrophages by capillary zone electrophoresis with electrochemical detection, *Anal. Biochem.*, 2000, **285**, 255–259.

56 W. Jin, L. Xiujun and N. Gao, Simultaneous determination of tryptophan and glutathione in individual rat hepatocytes by capillary zone electrophoresis with electrochemical detection at a carbon fiber bundle-Au/Hg dual electrode, *Anal. Chem.*, 2003, **75**, 3859–3864.

57 I.C. Shaw, A.E. McLean and C.H. Boult, Analysis of D-penicillamine by high-performance liquid chromatography with glassy carbon electrochemical detection, *J. Chromatogr.*, 1983, **275**, 206–210.

58 A. Wang, L. Zhang, S. Zhang and Y. Fang, Determination of thiols following their separation by CZE with amperometric detection at a carbon electrode, *J. Pharm. Biomed. Anal.*, 2000, **23**, 429–436.

59 G. Carro-Ciampi, P.G. Hunt, C.J. Turner and P.G. Wells, A high-performance liquid chromatographic assay for reduced and oxidised glutathione in embryonic, neonatal, and adult tissues using a porous graphite electrochemical detector, *J. Pharmacol. Methods*, 1988, **19**, 75–83.

60 K. Stulik and V. Pacakova, High-performance liquid chromatography of biologically important pyrimidine derivatives with ultraviolet-voltammetric-polarographic detection, *J. Chromatogr.*, 1983, **273**, 77–86.

61 D. Perrett, Comparative studies on different electrochemical detector systems in: *Drug Determination in Therapeutic and Forensic Contexts, Methodological Surveys in Biochemistry* 14, E. Reid, and I.D. Wilson (eds) Plenum, New York, 1984, 111–121.

62 F. Kreuzig and J. Frank, Rapid automated determination of D-penicillamine in plasma and urine by ion-exchange high-performance liquid chromatography with electrochemical detection using a gold electrode, *J. Chromatogr.*, 1981, **218**, 615–620.

63 D. Perrett, S.R. Rudge and P.L. Drury, Determination of captopril by an improved high-performance liquid-chromatography electrochemical assay, *Biochem. Soc. Trans.*, 1984, **12**, 1059–1060.

64 M.J. Kelly, D. Perrett and S.R. Rudge, The determination of cysteamine in physiological fluids by HPLC with electrochemical detection, *Biomed. Chromatogr.*, 1988, **2**, 216–220.

65 K. Shimada, M. Tanaka and T. Nambara, Sensitive derivatization reagents for thiol compounds in high-performance liquid chromatography with electrochemical detection, *Anal. Chim. Acta*, 1983, **147**, 375–380.

66 K. Shimada, M. Tanaka, T. Nambara, Y. Imai, K. Abe and K. Yoshinaga, Determination of captopril in human blood by high-performance liquid chromatography with electrochemical detection, *J. Chromatogr. B*, 1982, **227**, 445–451.

67 D. Tang, L-S. Wen and P.H. Santschi, Analysis of biogenic thiols in natural water samples by high-performance liquid chromatographic separation and fluorescence detection with ammonium 7-fluorobenzo-2-oxa-1,3-diazole-4-sulfonate (SBD-F), *Anal. Chim. Acta*, 2000, **408**, 299–307.

68 A.R. Ivanov, I.V. Nazimov and L.A. Baratova, Qualitative and quantitative determination of biologically active low-molecular-mass thiols in human blood by reversed-phase high-performance liquid chromatography with photometry and fluorescence detection, *J. Chromatogr. A*, 2000 **870**, 433–442.

69 E. Causse, C. Issac, P. Malatray, C. Bayle, P. Valdiguie, R. Salvayre, *et al.*, Assays for total homocysteine and other thiols by capillary electrophoresis-laser-induced fluorescence detection, I. Preanalytical condition studies, *J. Chromatogr. A*, 2000, **895**, 173–178.

70 R. Eggli and R. Asper, Electrochemical flow-through detector for the determination of cystine and related compounds, *Anal. Chim. Acta*, 1978, **101**, 253–259.

71 L.A. Allison, J. Keddington and R.E. Shoup, Liquid-chromatographic behavior of biological thiols and the corresponding disulfides, *J. Liq. Chromatogr.*, 1983, **6**, 1785–1798.

72 A.F. Stein, R.L. Dills and C.D. Klaassen, High-performance liquid chromatographic analysis of glutathione and its thiol and disulfide degradation products, *J. Chromatogr. B*, 1986, **381**, 259–270.

73 B.L. Lin, L.A. Colon and R.N. Zare, Dual electrochemical detection of cysteine and cystine in capillary zone electrophoresis, *J. Chromatogr. A*, 1994, **680**, 263–270.

74 Y.P. Sun, D.L. Smith and R.E. Shoup, Simultaneous detection of thiol-containing and disulfide-containing peptides by electrochemical high-performance liquid-chromatography with identification by mass-spectrometry, *Anal. Biochem.*, 1991, **197**, 69–76.

75 W.A. Kleinman and J.P. Richie, Determination of thiols and disulfides using high-performance liquid chromatography with electrochemical detection, *J. Chromatogr. B*, 1995, **672**, 73–80.

76 D.C. Sampson, P.M. Stewart and J.W. Hammond, Measurement of urinary cystine and cysteinyl-penicillamine in patients with cystinuria, *Biomed. Chromatogr.*, 1986, **1**, 21–26.

77 J.P. Richie and C.A. Lang, The determination of glutathione, cyst(e)ine, and other thiols and disulfides in biological samples using high-performance liquid chromatography with dual electrochemical detection, *Anal. Biochem.*, 1987, **163**, 9–15.

78 F. Remião, H. Carmo, F. Carvalho and M.L. Bastos, Simultaneous determination of reduced and oxidized glutathione in freshly isolated rat hepatocytes and cardiomyocytes by HPLC with electrochemical detection, *Biomed. Chromatogr.*, 2000, **14**, 468–473.

79 G.T. Yamashita and D.L. Rabenstein, Determination of penicillamine, penicillamine disulfide and penicillamine-glutathione mixed disulfide by high-performance liquid-chromatography with electrochemical detection, *J. Chromatogr. B*, 1989, **491**, 341–354.

80 D.L. Rabenstein and G.T. Yamashita, Determination of homocysteine, penicillamine, and their symmetrical and mixed disulfides by liquid-chromatography with electrochemical detection, *Anal. Biochem.*, 1989, **180**, 259–263.

81 O. Chailapakul, W. Siangproh, B.V. Sarada, C. Terashima, T.N. Rao, D.A. Tryk and A. Fujishima, The electrochemical oxidation of homocysteine at boron-doped diamond electrodes with application to HPLC amperometric detection, *Analyst*, 2002, **127**, 1164–1168.

82 C.T. Garvie, K.M. Straub and R.K. Lynn, Quantitative liquid chromatographic determination of disulfide-containing peptide analogues of vasopressin with dual Hg/Au electrochemical detection, *J. Chromatogr.*, 1987, **413**, 43–52.

83 X.-X. Qian, K. Nagashima, T. Hobo, Y.-Y. Guo and C. Yamaguchi, High-performance liquid chromatography of thiols with differential pulse polarographic detection of the catalytic hydrogen evolution current, *J. Chromatogr. A*, 1990, **515**, 257–264.

84 J. Nishiyama and T. Kuninori, Assay of thiols and disufides based on the reversibility of *N*-ethylmaleimide alkylation of thiols combined with electrolysis, *Anal. Biochem.*, 1992, **200**, 230–234.

85 W.T. Kok, U.A.T. Brinkman R.W. Frei, On-line electrochemical reagent production for detection in liquid chromatography and continuous flow systems, *Anal. Chim. Acta*, 1984, **162**, 19–32.

86 A.J.J. Debets, R. Van De Straat, W.H. Voogt, H. Vos, N.P.E. Vermeulen and R.W. Frei, Simultaneous determination of glutathione, glutathione disulfide, paracetamol and its sulfur containing metabolites by HPLC with electro-chemical detection with on-line generated bromine, *J. Pharm. Biomed. Anal.*, 1988, **6**, 329–336.

87 L.A. Holland and S.M. Lunte, Postcolumn reaction detection with dual-electrode capillary electrophoresis – electrochemistry and electrogenerated bromine, *Anal. Chem.*, 1999, **71**, 407–412.

88 H.M.A. Killa and D.L. Rabenstein, Determination of selenols, diselenides, and selenenyl sulfides by reversed-phase liquid-chromatography with electro-chemical detection, *Anal. Chem.*, 1988, **60**, 2283–2287.

CHAPTER 6

Applications of HPLC-ED in the Analysis of Drugs

The aim of this section is to discuss applications of HPLC-ED in the analysis of drugs and their metabolites in biological specimens, usually plasma. Analytes (International Nonproprietary Names, INNs are used when possible) are listed mainly by pharmacological category. Chemical structures are given for most analytes and internal standards, as are ED conditions, column, eluent, sample size and limit of detection (LoD) and low limit of quantitation (LLoQ). Brief details of sample preparation procedures are also included in many instances.

The last few years have seen a marked increase in the use of PGEs in the ED of drugs and poisons with the aim of increasing sensitivity/selectivity. The use of serial GCEs is also under investigation with this aim in view. These trends look set to continue. However, use of ED in conjunction with mini-bore (2 mm i.d.) or narrower columns provides a means of enhancing sensitivity with existing equipment.

The use of EC detection either in the form of PGE arrays or technically simpler systems in the qualitative analysis of basic drugs in urine, possibly as an adjunct to diode array or scanning UV detection, is a further topic worthy of continued investigation. An additional area where developments might be expected is in the application of post-column photolysis to generate electroactive species.

1 Anaesthetics

Halbert and Baldwin[1] measured lignocaine (lidocaine) (Figure 6.1) and its principal metabolites, monoethylglycinexylidide and glycinexylidide, in serum following i.v. lignocaine (200 mg). The analysis was performed on an unmodified silica column (10 μm) with acetonitrile-aq. ammonium phosphate (40 mmol L^{-1}, pH 7.8) (1 + 3) as eluent. Sample preparation was by SPE using ODS-modified silica cartridges. Bupivacaine (Figure 6.1) was the internal standard. The compounds were detected at glassy carbon or carbon paste electrodes (+1.2 V *vs* Ag/AgCl). HPLC using alkyl-modified silica columns with low wavelength (195–215 nm) UV detection has proved popular in the analysis of other local anaesthetics.[2] However, Jane *et al.*[3] demonstrated that many similar compounds can be successfully analysed

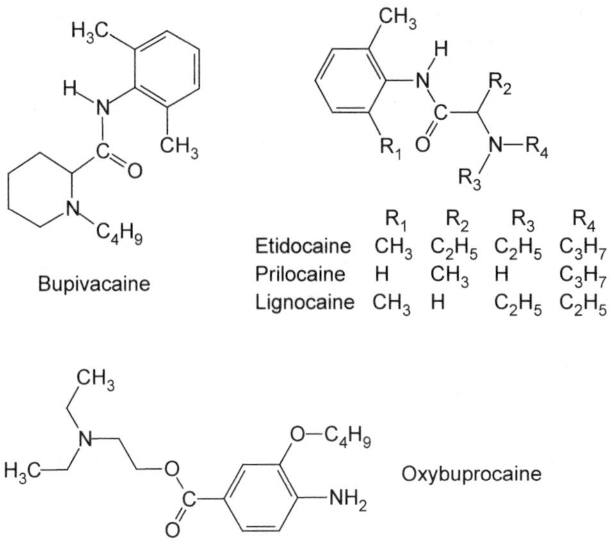

	R₁	R₂	R₃	R₄
Etidocaine	CH_3	C_2H_5	C_2H_5	C_3H_7
Prilocaine	H	CH_3	H	C_3H_7
Lignocaine	CH_3	H	C_2H_5	C_2H_5

Figure 6.1 *Some local anaesthetics.*

on unmodified silica using methanolic ammonium perchlorate ($10\,mmol\,L^{-1}$, pH 6.7) as eluent and that ED often offered greater sensitivity than UV detection (254 nm). To improve the retention of lignocaine, a strong cation-exchange column (Spherisorb S5SCX) has been used.[4] The method was applied to lignocaine measurement in fetal and placental tissue samples. Specimens were digested using the proteolytic enzyme Subtilisin A, after which lignocaine and the internal standard (bupivacaine) were extracted into methyl *t*-butyl ether (MTBE). Detection was at a GCE ($+1.0\,V$ *vs* Ag/AgCl). The LoD was better than $0.1\,mg\,kg^{-1}$ using 100 mg tissue (RSD 6.3%).

Plasma prilocaine (Figure 6.1) concentrations following i.v. and intraligamentary (dental) injections have been measured by HPLC-ED with PGEs ($+0.9\,V$ *vs* Pd). HPLC on a cyanopropyl-modified silica analytical column using acetonitrile-aq. phosphoric acid ($10\,mmol\,L^{-1}$) (3 + 7) as eluent gave acceptable peak shapes.[5] Prilocaine and the internal standard, etidocaine (Figure 6.1), were extracted into toluene under alkaline conditions and the analytes adsorbed onto acid-prepared diol AASP (Varian) cartridges. Using the AASP, which resulted in a high proportion of the drug in the sample being injected, together with HPLC-ED gave a LoD of $5\,\mu g\,L^{-1}$ for a relatively small sample volume (0.2 mL). Although the AASP is no longer manufactured, the method illustrates how SPE can be used to concentrate samples after liquid-liquid extraction. Oxybuprocaine (Figure 6.1) has been measured together with diclofenac by HPLC-ED in human aqueous humour (Section 16).

A number of HPLC-ED methods have been described for the general anaesthetic propofol (2,6-di-isopropylphenol) in plasma. Mazzi and Schinella[6] used a phenyl-propyl-modified silica column and a GCE ($+0.8\,V$ *vs* Ag/AgCl). The eluent was methanol-aq. phosphate buffer (25 mmol L^{-1}, pH 6.0) (1 + 3). Sample preparation was by SPE and thymol (5-methyl-2-(1-methylethyl)phenol) was used as the internal standard. Uebel *et al.*[7] also used ED (PGEs, E$_1$: +0.20, E$_2$: +0.65 V *vs* Pd). Sample preparation was by LLE into ethyl acetate and chlorprothixene (Figure 6.36) was the internal standard. Dowrie *et al.*[8] measured propofol in human plasma and in rat plasma and tissues using a 3 µm phenyl-modified silica column with methanol-aq. phosphate buffer (50 mmol L^{-1}, pH 2.8) (3 + 2) as eluent and detection at PGEs (E$_1$: +0.3, E$_2$: +0.8 V *vs* Pd). The internal standard was 2,6-*t*-butylmethylphenol. Skin and fat samples were homogenised in sodium deoxycholate (5 mmol L^{-1}) and steam distilled, the distillate being extracted with pentane. Plasma and other tissues (homogenised in acetonitrile-aq. phosphate buffer (50 mmol L^{-1}, pH 7.0) (1 + 1)) were extracted with pentane.

Trocewicz *et al.*[9] analysed propofol using a 5 µm LiChrosorb RP-18 column. The eluent was methanol-aq. phosphate buffer (50 mmol L^{-1}, pH 3.8) (3 + 1), although in a figure legend the eluent is described as methanol-aq. ammonium phosphate, pH 7 (7 + 4). The applied potentials (PGEs) were guard: +0.2 V and E$_1$: +0.45 V *vs* Pd. Propofol was extracted from urine using a supported liquid membrane – porous PTFE impregnated with undecane. Sample and sulfuric acid (50 mmol L^{-1}) were pumped through a mixing coil and over the membrane. Propofol was collected in a 'stagnant' acceptor solution maintained at an alkaline pH (*e.g.* 0.1 mol L^{-1} sodium hydroxide). Thymol was again used as the internal standard. The LoD was 10 ng L^{-1}. HPLC-ED was also one of several methods for propofol and its hydroxylated and conjugated metabolites reported by Favetta *et al.*[10] HPLC was performed using a 5 µm Spherisorb ODS2 column with acetonitrile-aq. disodium hydrogen orthophosphate buffer (25 mmol L^{-1}, pH 3.8) (3 + 2) as eluent and detection at PGEs (E$_1$: +0.40 V, E$_2$: +0.65 or +0.85 V *vs* Pd). Propofol and metabolites were extracted from urine by SPE with ODS cartridges.

2 β$_2$-Adrenergic Agonists

The major use for β-adrenergic agonists is in the treatment of asthma. Non-selective agonists such as isoprenaline or the β$_1$-selective, dobutamine, may be used for their inotropic properties (Section 14). Some β$_2$-selective agonists may be used to relax uterine smooth muscle, whilst other β$_2$-agonists are used as illegal growth promoters in animals and illegal performance enhancers in sport, hence there is considerable interest in their analysis in biofluids and tissues. Because β-agonists are generally catechol-containing compounds, such as catecholamines, they are usually amenable to HPLC-ED with enhanced sensitivity. The electroactive moiety is invariably a phenolic hydroxyl or primary aromatic amine group.

One of the first HPLC-ED methods reported was that for isoetharine (Figure 6.2) using an ODS-modified silica column with methanol-aq. sodium sulfate/phosphate (0.1 mol L^{-1}, pH 3) (12.5 + 87.5) as eluent.[11] Detection was at a GCE (+0.6 V *vs* Ag/AgCl). The LoD was 1 µg L^{-1} (150 µL sample). Colterol (Figure 6.2), the active metabolite of bitolterol, was the internal standard. Causon *et al.*[12] used an

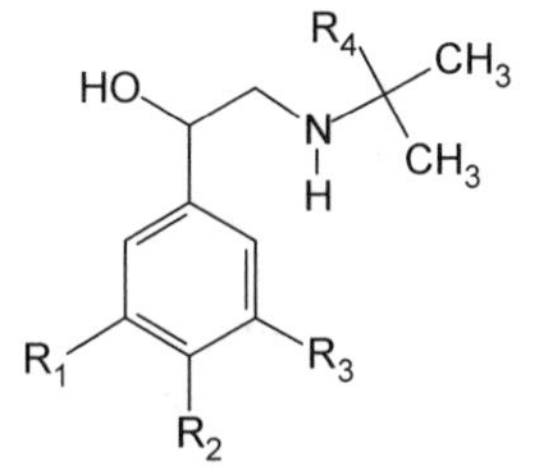

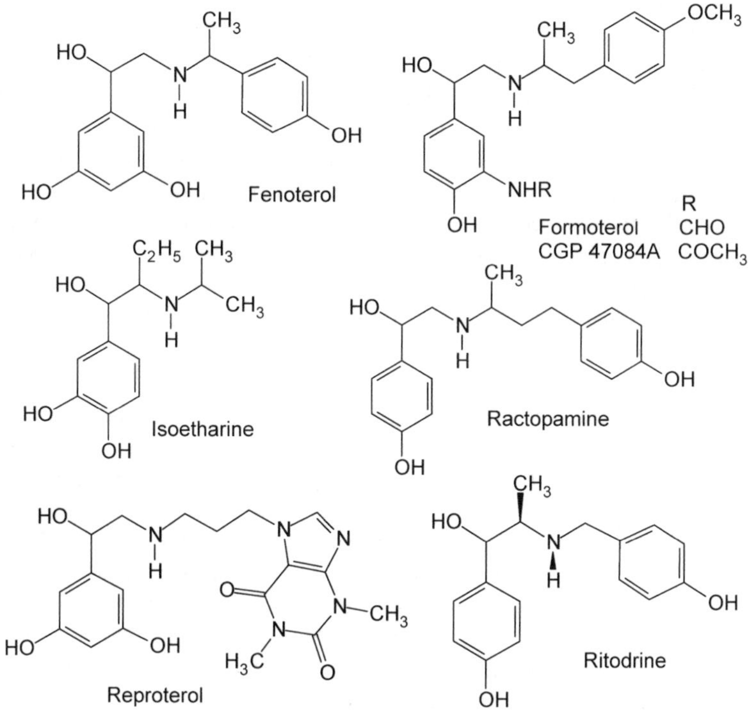

Figure 6.2 *Some β-adrenergic agonists.*

	R_1	R_2	R_3	R_4
Clenbuterol	Cl	Cl	NH_2	CH_3
Colterol	OH	OH	H	CH_3
Isoprenaline	OH	OH	H	H
Mabuterol	Cl	NH_2	CF_3	CH_3
Metaproterenol	OH	H	OH	H
Salbutamol	CH_2OH	OH	H	CH_3
Terbutaline	OH	H	OH	CH_3

octyl-modified silica column eluted with methanol-aq. citrate/phosphate buffer (pH 6.0) (3 + 97) to measure isoprenaline (Figure 6.2) in plasma and in urine. Isoprenaline 3-*O*-sulfate was measured as isoprenaline after pre-column hydrolysis. *N*-Methyldopamine was the internal standard. The compounds were detected at GCEs (+0.5 V *vs* Ag/AgCl). The LoD was 0.5 µg L^{-1} using a 2 mL sample.

CE with end-column amperometric detection (33 μm diameter carbon fibre electrode, +0.65 V *vs* Ag/AgCl) has also been used to detect isoprenaline enantiomers in microdialysates of rat plasma.[13] Analyses were performed at 20 kV in fused-silica capillaries (65 cm × 50 mm i.d.). The analysis buffer consisted of aq. lithium acetate (100 mmol L^{-1}, pH 4.75), methyl-*O*-β-cyclodextrin (100 g L^{-1}) and disodium EDTA (0.5 mmol L^{-1}). Samples were introduced by electrokinetic injection (18 kV, 3 s) or by pH-mediated stacking (electrokinetic injection, 18 kV 15 s, followed by electrokinetic injection of 0.1 mol L^{-1} aq. hydrochloric acid, 18 kV, 20 s). Inclusion of DHBA as the internal standard was claimed to increase the precision of the analysis from 3.2 to 1.4% RSD. The LoD was 600 ng L^{-1} (S/N = 3, 1 μL sample).

The β$_2$-adrenoceptor agonists salbutamol[14] (Figure 6.2) and reproterol[15] (Figure 6.2) have been measured in plasma using alkyl-modified silica analytical columns and ED using rotating disc carbon paste electrodes. Both methods used on-line extract enrichment, the former with an SCX-modified silica column (50 × 4.6 mm i.d.), the latter with an ODS-modified column (20 × 4.6 mm i.d.). The potentials used were + 0.95 *vs* SCE and +0.90 V *vs* Ag/AgCl, respectively, giving sensitivities of 0.5–1 μg L^{-1} (1 mL samples). Using a GCE (+0.8 V *vs* Ag/AgCl), Tan and Soldin[16] claimed a LoD of 0.4 μg L^{-1} for salbutamol extracted from 0.5 mL plasma, although the RSD at 2.0 μg L^{-1} was 7.3%. Salbutamol and the internal standard (fenoterol, Figure 6.2) were extracted by SPE (ODS-modified silica cartridge), eluted with methanol and the dry residue subjected to further purification steps, including ion-pair extraction. The final extract was analysed on an ODS-modified silica column using methanol-aq. sodium phosphate (70 mmol L^{-1}, pH 6.8) containing sodium 1-heptanesulfonate (2 mmol L^{-1}) (25 + 75) as eluent. Even with a LoD of 0.1 μg L^{-1} they concluded that it would not be possible to measure fenoterol after single oral doses (2.5 mg) in man.

Salbutamol enantiomers have been measured in urine using an α$_1$-acid glycoprotein column with 0.1% (v/v) TEA in aqueous citrate buffer (5.3 mmol L^{-1}, pH 7.2) as eluent and ED (+0.75 V *vs* Ag/AgCl).[17] Sample preparation was by SPE (octyl-modified silica column). The LoD was 0.25 mg L^{-1} (1 mL sample). (±)-Terbutaline (Figure 6.2) was the internal standard. Jarvie *et al.*[18] assayed salbutamol and terbutaline in plasma using an ODS-modified silica analytical column with methanol-aq. phosphate buffer (67 mmol L^{-1}, pH 6)-aq. SDS (40 g L^{-1})-DEA (90 + 210 + 1.5 + 0.06) as eluent and ED (GCE, +0.90 V *vs* Ag/AgCl). Sample preparation was by SPE using ODS-modified silica cartridges and an extensive extract purification procedure. Pholedrine (4-(2-methylaminopropyl)phenol) was the internal standard. The LoD was 3 μg L^{-1} for each compound (1 mL sample).

Sagar *et al.*[19] also used ED to measure plasma terbutaline, but employed a carbon fibre working electrode (14 μm diameter), an Ag/Ag$_3$PO$_4$ reference electrode and a stainless steel counter electrode. The applied potential was +1.3 V. Sample preparaion was by column switching and HPLC was performed using an ODS-modified silica column with methanol-aq. phosphate buffer (67 mmol L^{-1}, pH 5)-aq. SDS (40 g L^{-1})-DEA (55 + 45 + 0.5 + 0.02) as eluent (compare with Jarvie *et al.*[18]).

No internal standard was used. A better LoD, using a 1 mL sample (0.8 as against $2 \mu g L^{-1}$), was claimed using the carbon fibre microelectrode than was attainable using a conventional GCE ($+1.3$ V *vs* Ag/Ag_3PO_4) but in reality this difference in sensitivity is negligible. Salbutamol can be measured using this same system (LoD $1 \mu g L^{-1}$, carbon fibre electrode).[20] As salbutamol and terbutaline possess native fluorescence (excitation wavelength 200 nm), exploitation of this property can give better baseline stability and hence greater sensitivity than attainable with ED.[21] However, nowadays fluorescence detectors fitted with deuterium lamps are not easily available.

Terbutaline has been measured after extraction from plasma using ion-exchange chromatography, concentration on a pre-column and then analysis on an ODS-modified silica column with methanol-aq. citrate/phosphate buffer (pH 6.0) containing sodium perchlorate ($50 \, mmol \, L^{-1}$) (6 + 44) as eluent.[22] ED was at a GCE ($+0.9$ V *vs* Ag/AgCl). The *N*-(1,1-dimethylpropyl) analogue of terbutaline was the internal standard. The LoD was approximately $1 \mu g L^{-1}$ (2 mL sample). An analogous procedure was used to assay terbutaline after administration of bambuterol (a dicarbamate derivative of terbutaline) except that 1. an esterase inhibitor was added to blood samples to prevent *in vitro* hydrolysis of bambuterol and other metabolites to terbutaline, 2. SPE on ODS-modified silica was used instead of ion-exchange chromatography and 3. the *N*-(1-methylpropyl) analogue of terbutaline was the internal standard.[23] Herring and Johnson[24] based their extraction of terbutaline on that of Kennedy *et al.*[23] but chose metaproterenol (Figure 6.2) as the internal standard. Analyses were performed using a $5 \mu m$ ODS-modified silica column (150×4.6 mm i.d.) with methanol-aq. phosphate buffer ($25 \, mmol \, L^{-1}$, pH 7.4) (23 + 77) with $2 \, mmol \, L^{-1}$ octanesulfonic acid as eluent. A Model 5011 cell (ESA) was used (PGEs, E_1: $+0.45$ V E_2: $+0.7$ V *vs* Pd). Although the detector response to terbutaline peaked at approximately $+1$ V, running at the lower voltage reduced the frequency of fouling of the electrode and the need for rinsing with $0.9 \, mol \, L^{-1}$ acetic acid. The inter-batch variation at $1 \mu g \, L^{-1}$ was less than 6.5%.

Formoterol, a potent long-acting β_2-agonist developed for treating asthma, gives sub-$\mu g \, L^{-1}$ concentrations in plasma when administered to man. Van den Berg *et al.*[25] reported a LoD of $20 \, ng \, L^{-1}$ (S/N = 3) using a 2 mL sample. Plasma was extracted using propylsulfonic acid SPE columns prior to HPLC on a $3 \mu m$ octyl-modified silica column (100×4.6 mm i.d.) with methanol-aq. citrate buffer ($12 \, mmol \, L^{-1}$, pH 6.0) (38.5 + 61.5) as eluent. Detection was at a single GCE ($+0.63$ V *vs* Ag/AgCl). Bromoformoterol (structure not given in the paper) was used as the internal standard. Campestrini *et al.*[26] based their method on that given above, but increased the sensitivity some 10-fold, obtaining a LLoQ of $3 \, ng \, L^{-1}$ with an intra-assay RSD of 14%. The analysis was performed on $5 \mu m$ ODS-modified silica in glass columns ($2 \times 100 \, mm \times 3$ mm i.d.) using acetonitrile-methanol-aq. phosphate buffer ($35 \, mmol \, L^{-1}$, pH $6.0 + 400 \, mg \, L^{-1}$ EDTA) (5 + 25 + 70) as eluent. The detector (GCE, $+0.63$ V *vs* Ag/AgCl) and column were maintained at 33 °C. The internal standard was CGP 47086A (Figure 6.2).

Ritodrine (Figure 6.2), a selective β_2-adrenoceptor agonist used primarily in obstetrics, has been assayed following LLE into ethyl acetate at pH 9.4.

The analytical column was ODS-modified silica with methanol-aq. phosphate buffer (pH 4.5) containing sodium octanesulfonate ($0.3\,mmol\,L^{-1}$) ($25 + 75$) as eluent. Detection was at a GCE ($+0.90\,V$ *vs* Ag/AgCl). The LoD was $0.2\,ng$ on column.[27]

Clenbuterol (Figure 6.2) differs from the other adrenoceptor bronchodilators as it contains an aromatic primary amine moiety rather than a phenolic hydroxyl. Because of the low plasma concentrations encountered it may not be possible to quantify clenbuterol in human plasma by HPLC-ED, but plasma and urine concentrations in animals have been measured. Diquet *et al.*[28] used a GCE ($+1.15\,V$ *vs* Ag/AgCl), in conjunction with an octyl-modified silica column and a methanol-aq. ammonium phosphate ($1\,mmol\,L^{-1}$, pH 6.0) ($9 + 1$) eluent, to measure clenbuterol in mouse plasma. The alkaloid yohimbine was the internal standard. Using LLE into chloroform a recovery of approximately 45% was obtained and the LoD was $3\,\mu g\,L^{-1}$ ($0.5\,mL$ sample). However, the lowest calibrator was $10\,\mu g\,L^{-1}$ and the RSD at $40\,\mu g\,L^{-1}$ was 4.4%. Hooijerink *et al.*[29] used serial UV (244 nm) and ED (GCE, $+1.25\,V$ *vs* Ag/AgCl) in the measurement of clenbuterol in calf urine. Sample preparation was by SPE on two columns in succession (Extrelut-3 (Merck) and Bondelut C2 (Analytichem)), and the LoD was $0.5\,\mu g\,L^{-1}$ ($4\,mL$ sample). Qureshi and Eriksson[30] used a PGE ($+0.75\,V$ *vs* Pd) to measure clenbuterol and mabuterol (Figure 6.2) in equine plasma. Sample preparation was by LLE under alkaline conditions into diethyl ether-(2-butanol) ($9 + 1$) and HPLC was performed on an ODS-modified silica column eluted with acetonitrile-aq. phosphate buffer ($0.1\,mol\,L^{-1}$, pH 4.0) ($23 + 77$). The LoDs were 0.5 and $2\,\mu g\,L^{-1}$ for clenbuterol and mabuterol, respectively ($1\,mL$ sample).

More recent developments in sample preparation for β_2-agonists include the use of immunoaffinity columns (IAC), as exemplified by Stevenson's group which evaluated columns containing anti-clenbuterol antibodies.[31] Koole *et al.*[32] included the use of an IAC, in which the antibody was raised against salbutamol, in their sample preparation of human and calf urine containing clenbuterol, bromobuterol, mabuterol and mapenterol. Alkalinised urine was analysed using Extrelut columns eluted with MTBE-hexane ($1 + 4$) and the dried residues applied to either the IAC or a mixed mode SPE column (LiChrolut TSC). HPLC was with a 'base-deactivated' ODS-modified silica column and acetonitrile-water containing phosphate buffer (pH 6.5, $25\,mmol\,L^{-1}$), OSA ($2.5\,mmol\,L^{-1}$), sodium chloride ($2.5\,mmol\,L^{-1}$) and EDTA ($50\,mg\,L^{-1}$) ($3 + 7$) as eluent. Detection was at a GCE ($+0.8\,V$ *vs* Ag/AgCl).

Ractopamine (Figure 6.2) is a β-adrenergic agonist developed as a feed additive since it reduces fat and increases muscle mass. To analyse this compound, plasma was extracted using a weak ion-exchange (–COOH) SPE column and the ractopamine was eluted with 2% (v/v) ammonium hydroxide in ethyl acetate.[33] The organic phase was shaken with borate buffer (pH 10.3) before evaporation to dryness. An ODS-modified silica HPLC column was used with acetonitrile-ammonium phosphate buffer ($50\,mmol\,L^{-1}$, pH 4.5) ($21.5 + 79$) as eluent. Detection was at PGEs (E_1: $+0.4\,V$, E_2: $+0.7\,V$ *vs* Pd). The LoD was $0.5\,pg\,L^{-1}$ (S/N = 3) and the LLoQ $2\,pg\,L^{-1}$ (S/N = 10) for a $0.5\,mL$ sample.

3 Anticancer Drugs

3.1 Alkylating Agents

Cisplatin (cis-dichlorodiamineplatinum(II)) and related complexes have been studied by HPLC-ED with varying degrees of success. Dropping mercury and hanging mercury drop electrodes (employing differential pulse amperometry), thin-layer gold/mercury electrodes, glassy carbon and a halide-catalysed oxidation platinum electrode have been used.[34–40] In general, LoDs are dependent on the ligands involved in forming the platinum complex and are higher (typically in the micromolar range) than those normally associated with ED in HPLC. However, most of the complexes studied lack significant UV absorption and thus ED is especially valuable. O'Dea *et al.*[41] have compared the utility of a HMDE and GCE in HPLC-ED systems for the measurement of cisplatin. The LoDs were approximately $5 \, mg \, L^{-1}$ using the HMDE ($+0.05 \, V$ *vs* Ag/AgCl), $0.37 \, mg \, L^{-1}$ using a single GCE ($+1.2 \, V$ *vs* Ag/AgCl), and $25 \, \mu g \, L^{-1}$ using dual GCEs in parallel (differential operation, oxidation potentials not stated). HPLC was on an ODS-modified silica column with aq. sodium acetate ($10 \, mmol \, L^{-1}$, pH 4.6) containing OSA ($5 \, mmol \, L^{-1}$) as eluent. The analysis of biological samples was not reported.

The alkylating agent melphalan (Figure 6.3) has been measured in plasma using a phenyl-modified silica column with acetonitrile-aq. sodium phosphate buffer ($0.1 \, mol \, L^{-1}$, pH 4.5) ($6 + 44$) as eluent and ED (PGEs, E_1: $+0.1 \, V$, E_2: $+0.6 \, V$, *vs* Pd).[42] Sample preparation involved perchloric acid treatment and LLE into chloroform to remove interferences. No internal standard was used. The LoD was $2 \, \mu g \, L^{-1}$ (1 mL sample).

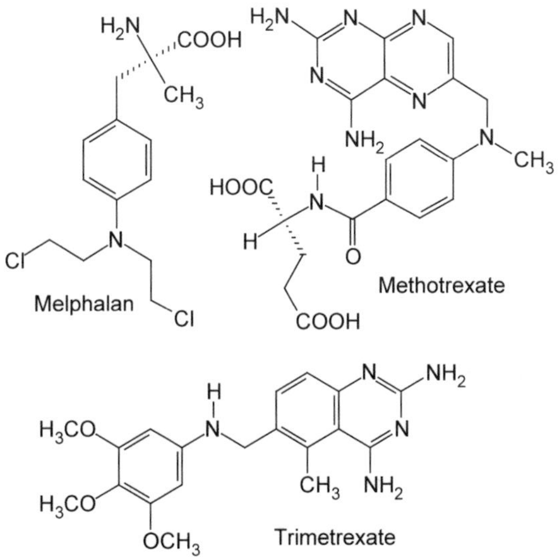

Figure 6.3 *Melphalan, methotrexate and trimetrexate.*

3.2 Antimetabolites

Separation methods for methotrexate (Figure 6.3) and its analogues have been extensively reviewed.[43] Few of the methods used electrochemical detection. However, methotrexate has been measured in serum by HPLC-ED after SPE using an ODS-modified silica analytical column (3 µm) with acetonitrile-methanol-aq. acetate buffer $(0.05 \, \text{mol L}^{-1}, \text{pH} \, 6.5) \, (13 + 13 + 174)$ as eluent.[44] ED was at a GCE $(+0.95 \, \text{V} \, vs \, \text{Ag/AgCl})$. An internal standard was not used. The LoD was $1 \, \mu\text{g L}^{-1}$. A coulometric method for methotrexate using dual PGEs in the screen mode $(\text{E}_1: +0.5 \, \text{V}, \, \text{E}_2: +1.1 \, \text{V}, \, vs \, \text{Pd})$ demonstrated the value of coulometric detection in removing interfering substances.[45] An octyl-modified silica HPLC column was used with methanol-aq. phosphate buffer $(0.1 \, \text{mol L}^{-1}, \text{pH} \, 6.9) \, (20 + 80)$ as eluent. Serum (0.2 mL) was extracted with ethyl acetate-2-propanol $(10 + 1)$. The methotrexate concentration of the lowest calibrant was $10 \, \mu\text{g L}^{-1}$. 4-Aminoacetophenone was used as the internal standard.

Lin *et al.*[46] based their HPLC separation of trimetrexate (Figure 6.3) on that of Weir *et al.*,[47] who demonstrated the increased sensitivity afforded by EC as compared to UV detection for this analyte. The eluent was prepared from acetonitrile-water-acetic acid $(170 + 822 + 8)$ containing sodium dihydrogen orthophosphate $(50 \, \text{mmol L}^{-1})$. The final pH was adjusted to 6.5 with triethyl-amine. An ODS-modified silica HPLC column was used. Detection was at a GCE $(+0.6 \, \text{V} \, vs \, \text{Ag/AgCl})$. Plasma (1 mL) and faeces were pretreated by SPE (ODS-modified silica), but urine was injected directly after appropriate dilution. The LODs were reported as $18.5 \, \mu\text{g L}^{-1}$ (0.05 µmol/L) for plasma and urine, and $7.4 \, \mu\text{g L}^{-1}$ (0.02 µmol/L) for faeces. An internal standard was used, but not described.

Lometrexol (Figure 6.4) has been measured in plasma by HPLC-ED using a phenyl-modified silica analytical column with acetonitrile-aq. sodium acetate $(50 \, \text{mmol L}^{-1}, \text{pH} \, 4.0)$ as eluent and detection at PGEs in the screen mode $(\text{E}_1: +0.4 \, \text{V}, \, \text{E}_2: +0.9 \, \text{V}, \, vs \, \text{Pd})$.[48] Plasma (1 mL) plus internal standard (LY277413, Figure 6.4) was extracted by SPE (aromatic sulfonic acid-modified silica column). The LoD was $5 \, \mu\text{g L}^{-1}$.

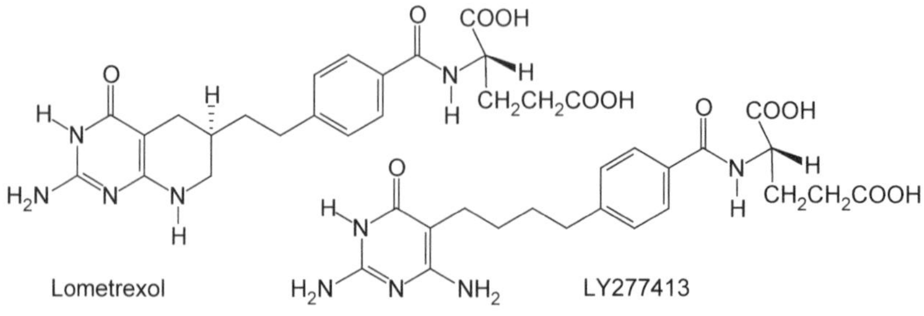

Figure 6.4 *Lometrexol and internal standard (LY277413).*

3.3 Cytotoxic Antibiotics

Many anticancer drugs are quinones or hydroquinones and are thus amenable to ED. Indeed, the anthracycline antibiotics, doxorubicin (adriamycin) and daunorubicin (daunomycin) (Figure 6.5), can be detected by either oxidative or reductive ED due to the presence of both a quinone and a hydroquinone moiety.[49] Akpofure *et al.*[50] measured daunorubicin and its metabolites in plasma using HPLC-ED (GCE, +0.65 V *vs* Ag/AgCl). Sensitivity was comparable to that obtained using fluorescence detection (2 ng on-column). Kotake *et al.*[51] used a phenylpropyl-modified silica column and a GCE (+0.7 V *vs* Ag/AgCl) to measure doxorubicin. The eluent was acetonitrile-glacial acetic acid-water (27 + 1 + 72) adjusted to pH 4.3 using aqueous sodium acetate (20% w/v). Sample preparation was by SPE (phenylpropyl-modified silica). Daunorubicin was the internal standard. Riley *et al.*[52,53] and Riley and Runyan[54] have given details of additional methods for doxorubicin, daunorubicin and related compounds, which include the use of fluorescence detection (excitation 480 nm, emission 560 nm) as well as ED.

Mou *et al.*[55] used SPE (ODS-modified silica) to extract doxorubicin and prochlorperazine (Figure 6.35) from plasma. However, although prochlorperazine was detected amperometrically, doxorubicin and daunorubicin (internal standard) were measured by fluorimetry. Epirubicin (an analogue of doxorubicin differing only in the position of the C-4 hydroxyl group of the sugar moiety), doxorubicin and their principal (13-*S*-dihydro) metabolites have been analysed on an ODS-modifed silica column with acetonitrile-water containing disodium hydrogen orthophosphate (50 mmol L^{-1}) and 0.05% (v/v) triethylamine (final pH adjusted to 4.6 with citric acid) as eluent.[56] Plasma samples (200 μL) were prepared by SPE on a polymeric column (Oasis HLB, Waters). The analytes were detected with a Model 5014 high performance cell (ESA) with an amperometric 'enhanced response' GCE coupled with a coulometric PGE and operated in the redox mode (E$_1$: +0.4 V, E$_2$: −0.3 V, *vs* Pd). The LoD was 1 μg L^{-1} (S/N = 3) for all compounds.

Mitoxantrone (Figure 6.6) has been measured in plasma using an ODS-modified silica column with ED (GCE, +0.75 V *vs* Ag/AgCl).[57] The eluent was acetonitrile-aq.

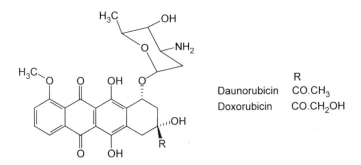

	R
Daunorubicin	CO.CH$_3$
Doxorubicin	CO.CH$_2$OH

Figure 6.5 *Daunorubicin and doxorubicin.*

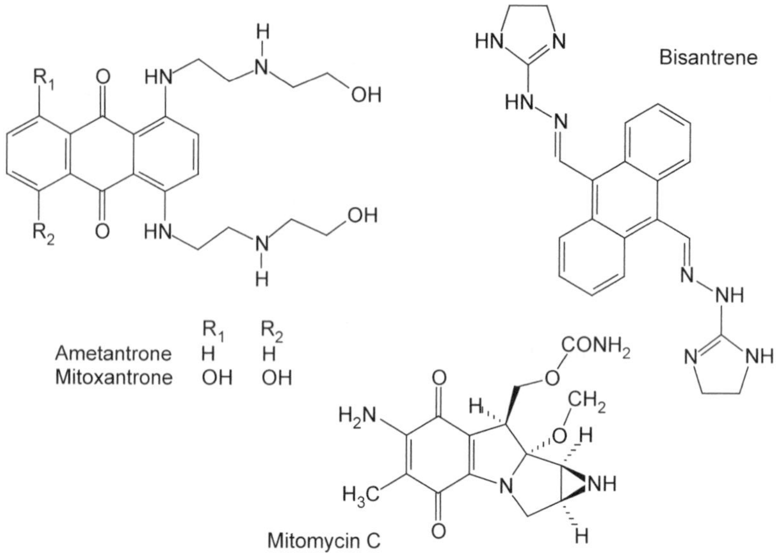

Figure 6.6 *Ametantrone, bisantrene, mitomycin C and mitoxantrone.*

sodium formate (80 mmol L^{-1}, pH 3.0) (28 + 72). Sample preparation was by LLE using dichloromethane. Bisantrene (Figure 6.6) was the internal standard. The LoD was 0.1 µg L^{-1} (1 mL sample). Similar methodology has been applied to measuring mitoxantrone in bone marrow extracts using ametantrone (Figure 6.6) as the internal standard.[58]

Mitomycin C (Figure 6.6) has been measured using a HMDE (−0.6 V *vs* Ag/AgCl) *via* reduction of the quinone moiety.[59] However, UV absorption (360 nm) was equally sensitive and was considered more practical because of the need to remove oxygen from the mobile phase when reduction was used.

3.4 Photosensitisers

Photodynamic therapy is the use of photosensitisers which, when irradiated with light of an appropriate wavelength, produce cytotoxic species. The photosensitiser temoporfin (5,10,15,20-tetra(*m*-hydroxyphenyl)chlorin, mTHPC, Figure 6.7) has been analysed on an ODS-modified silica column using acetonitrile-phosphate buffer (10 mmol L^{-1}, pH 2.5) (45 + 55) as eluent.[60] The extraction of tissue and plasma was based on the method of Wang *et al.*[61] The more electroactive *p*-isomer was used as the internal standard and the analytes detected at a PGE (+0.8 V vs Pd). A LoD of 5 µg L^{-1} (0.5 mL plasma) was claimed. Photodynamic agents are usually highly fluorescent and have high extinction coefficients. Temoporfin absorbs strongly at 415 nm ($\varepsilon = 146,000$ L^{-1} mol^{-1} cm^{-1}, methanol) and UV detection of this compound can be as sensitive as fluorescence detection.[62]

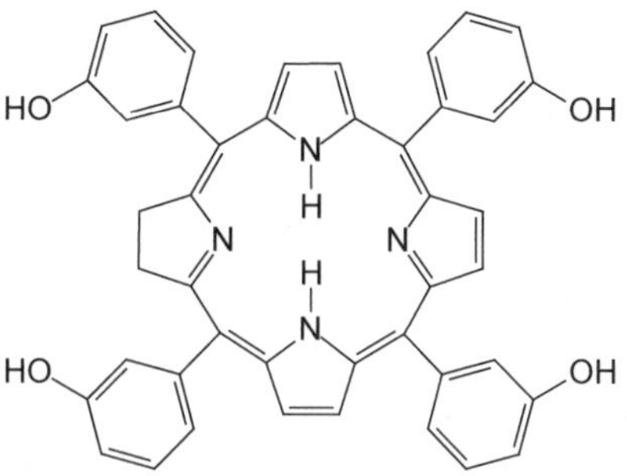

Figure 6.7 *Temoporfin.*

3.5 Podophyllotoxins

Etoposide and teniposide (Figure 6.8) and their metabolites have been measured in plasma, urine and CSF after LLE into ethyl acetate.[63] The analytical column was phenylpropyl-modified silica and the eluent was acetonitrile-water-acetic acid $(30 + 68 + 2$ for teniposide and $25 + 74 + 1$ for etoposide). Teniposide was detected at $+0.75$ V and etoposide at $+0.8$ V *vs* Ag/AgCl. The lowest assay calibrator used was $50 \, \mu g \, L^{-1}$. Etoposide or teniposide were used as the internal standard when measuring the other compound. A similar approach was used by Canal *et al.*[64] to measure plasma teniposide. An ODS-modified silica column was used with methanol-ammonium acetate $(250 \, mmol \, L^{-1})$-glacial acetic acid $(54 + 45 + 1)$ as eluent. Detection was at a GCE $(+0.9$ V *vs* Ag/AgCl). Three extraction techniques were investigated: LLE into chloroform, LLE into ethyl acetate and SPE using ODS-modified silica cartridges. LLE with chloroform was the preferred option. The LoD was 200 pg (S/N = 5), equivalent to a concentration of $10 \, \mu g \, L^{-1}$. However, the RSD at $1 \, mg \, L^{-1}$ was 6.2%. To assay etoposide in

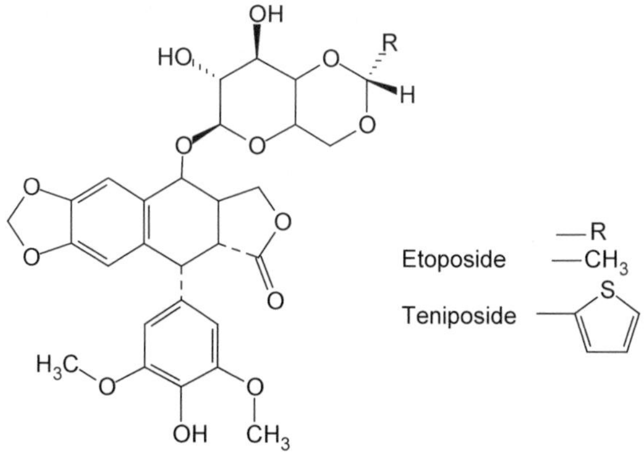

Figure 6.8 *Etoposide and teniposide.*

plasma El-Yazigi and Martin[65] used an ODS-modified silica column with methanol-aq. phosphate buffer (10 mmol L^{-1}, pH 3.0) (493 + 507) as eluent. Teniposide was the internal standard. Samples (1 mL) were extracted with chloroform. Detection was at +0.8 V (Waters Model 656 with Metrohm potentiostat). The LoD was 10 µg L^{-1} (not 10 ng L^{-1} as claimed in the paper); the inter-assay RSD at 25 µg L^{-1} was 7%.

Etoposide and teniposide have both been measured in plasma by HPLC with column switching and either UV or ED.[66,67] The analytical column was either ODS- or phenylpropyl-modified silica and the final eluent was (ODS column) aqueous phosphate buffer (10 mmol L^{-1}, pH 7.0) containing SDS (40 mmol L^{-1}) and 1-propanol (0.5 mol L^{-1}) or (phenyl column) methanol-aq. phosphate buffer (10 mmol L^{-1}, pH 7.0) (55 + 45). ED was at a GCE (+0.50 V *vs* Ag/AgCl). A LoD of 20 µg L^{-1} teniposide was claimed. Etoposide concentrations in plasma (protein bound and free) and leukaemic cells have been measured using dual PGEs in the screen mode (E$_1$: +0.2 V; E$_2$: +0.5 V *vs* Pd).[68] A phenyl-modified silica pre-column was used and it is assumed that the analytical column (7 µm Nucleosil) was packed with the same material. The eluent was acetonitrile-methanol-water-acetic acid (4 + 43 + 52 + 1). Plasma samples (0.5 mL) were extracted with chloroform (2 mL) and 25 µL of the extract reconstituted in 1 mL water:methanol (1 + 1) was injected; ultrafiltrates of plasma (50 µL) were injected directly. The LLoQs were 100 µg L^{-1} for whole plasma and 10 µg L^{-1} for ultrafiltrates. Teniposide was used as the internal standard.

The cytotoxic catechol metabolite formed by demethylation of etoposide has been measured by HPLC-ED.[69] An ODS-modified silica column was used with methanol-disodium hydrogen orthophosphate (10 mmol L^{-1}, final pH 6.0) (43 + 57) as eluent. The analyte was stabilised in plasma with ascorbic acid and plasma proteins precipitated by addition of two volumes of acetonitrile. Detection was at PGEs (E$_1$: +0.1 V; E$_2$: +0.45 V *vs* Pd). The LoD was 10 µg L^{-1} (150 µL sample). No internal standard was used. The catechol metabolite and parent etoposide have been measured simultaneously using a phenyl-modified silica column with acetonitrile-aq. citric acid (25 mmol L^{-1})-aq. sodium phosphate (50 mmol L^{-1}, pH 2.4) as eluent.[70] Sample preparation was by LLE into chloroform. Detection was at PGEs (E$_1$: +0.01 V; E$_2$: +0.37 V *vs* Pd). The LoDs were 0.7 and 0.1 µg L^{-1} for etoposide and desmethyletoposide, respectively (200 µL sample).

The podophyllotoxin derivative TOP-53 and its glucuronide metabolite (Figure 6.9) have been measured by HPLC-ED (GCE +0.7 V *vs* Ag/AgCl).[71] Samples were extracted into dichloromethane and back-extracted into phosphate buffer (100 mmol L^{-1}, pH 3.0), which then washed with ethyl acetate. The conjugate was quantified as the aglycone after hydrolysis with β-glucuronidase. HPLC was performed using an ODS-modified silica column with methanol-phosphate buffer (100 mmol L^{-1}, pH 7.0) (43 + 57) as eluent. A diethylamino podophyllotoxin homologue was used as the internal standard (Figure 6.9). LLoQs were 2 and 10 µg L^{-1} for each compound in rat plasma and urine, respectively.

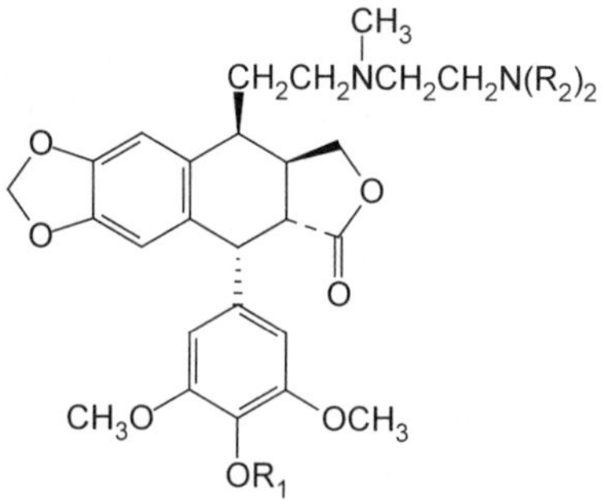

	R_1	R_2
TOP-53	H	CH_3
TOP-53 D-glucuronide	D-glucuronate	CH_3
Internal standard	H	C_2H_5

Figure 6.9 *TOP-53 and associated compounds.*

3.6 Vinca Alkaloids

Vendrig *et al.*[72] evaluated the use of diol columns for the SPE of vincristine (22-oxovinblastine) and vinblastine (Figure 6.10) from plasma and urine. HPLC-ED (GCE +0.85 V *vs* Ag/AgCl) used an ODS-modified column with methanol-aq. phosphate buffer (10 mmol L^{-1}, pH 7.0) (65 + 35) as eluent. The LoDs (S/N = 3) were 100 pg and 250 pg for vinblastine and vincristine, respectively. One alkaloid was used as the internal standard when measuring the other and *vice versa*. In a later paper,[73] cyanopropyl SPE columns were used for vinblastine, vincristine, vindesine (desacetylvinblastine amide) and desacetylvinblastine, giving a LoD of 100 pg for each compound. The RSDs for plasma assays (10 μg L^{-1}, 1 mL) were <6% in each case.

Vincristine has been measured using an ODS-modified silica analytical column with acetonitrile-methanol-aq. phosphate buffer (25 mmol L^{-1}, pH 7.0) (20 + 48 + 32) as eluent and ED (GCEs, +0.75 V or +0.83 V, *vs* Ag/AgCl).[74]

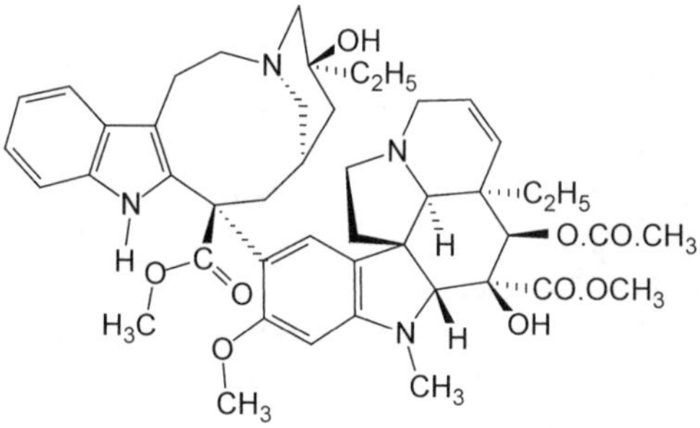

Figure 6.10 *Vinblastine.*

Sample preparation was by column switching using a pellicular ODS-modified silica preconcentration column. Vinblastine was the internal standard. The LoD was $0.3\,\mu g\,L^{-1}$ (1.2 mL sample). Koopmans *et al.*[75] developed an automated method for vincristine using column switching. Both the preparation and analytical columns were based on ODS-modified silica. The eluent was acetonitrile-methanol-aq. phosphate buffer (20 mmol L^{-1}, pH 7.0) (15 + 50 + 35). The GCE was operated at +0.83 V *vs* Ag/AgCl. The LLoQ (0.3 mL sample) was $0.48\,\mu g\,L^{-1}$ (RSD = 10.3%). Vinblastine was the internal standard. When the method was applied to the assay of vincristine in mononuclear cells from children with acute lymphoblastic leukaemia the LLoQ was $1.17\,\mu g\,L^{-1}$ (RSD = 5.7%).[76]

Vintriptol (4-deacetyl-3-L-*O*-ethyltryptophan-vinblastine) has been measured in plasma by HPLC-ED (GCE, +0.70 V *vs* Ag/AgCl).[77] An ODS-modified silica column was used with acetonitrile-aq. phosphate buffer (50 mmol L^{-1}, pH 6.0) containing sodium chloride (0.5 g L^{-1}) and EDTA (20 mg L^{-1}) (62 + 38) as eluent. Sample preparation was by LLE and vinblastine was the internal standard. The LoD was $2\,\mu g\,L^{-1}$ (0.5 mL sample). A coulometric method (PGEs, E_1: +0.4 V, E_2: +0.7 V, *vs* Pd) for vinorelbine (nor-5'-anhydrovinblastine) in rabbit plasma (0.4 mL), using teniposide as the internal standard, has also been described.[78] Sample preparation was by LLE into diethyl ether. HPLC used an ODS-modified silica column with acetonitrile-methanol-aq. sodium dihydrogen orthophosphate (20 g L^{-1}) containing heptane sulfonate (0.8 g L^{-1}), pH 3.0 (3 + 2 + 5) as eluent. A LoD of $1\,\mu g\,L^{-1}$ (S/N = 3) was claimed, although the inter-assay RSD at $5\,\mu g\,L^{-1}$ was more than 20%.

3.7 Miscellaneous Anticancer Drugs

Elliptinium, the N_2-methyl derivative of 9-hydroxyellipticine (Figure 6.11) has been measured in plasma using an ODS-modified silica column and methanol-water (60 + 40) containing ammonium acetate (100 mmol L^{-1}, final pH 6.0) as eluent.[79] Sample preparation was by ion-pair extraction into ethyl acetate containing sodium tetraphenylborate. The N_2-propyl homologue was used as the internal standard. Detection was at a GCE (0.6 V *vs* Ag/AgCl). The LoD was 250 pg (S/N = 2), equivalent to $25\,\mu g\,L^{-1}$ for a 200 μL sample.

Amperometric, UV and fluorescence detection for narciclasine (Figure 6.11) in mouse serum have been compared.[80] HPLC was performed using an ODS-modified silica column with methanol-aq. potassium dihydrogen orthophosphate buffer (25 mmol L^{-1}, pH 6.5) (1 + 1) as eluent. Sample preparation was by protein precipitation and direct analysis of the supernatant. No internal standard was used. Detection with carbon fibre electrodes (+1.8 V *vs* Ag/AgCl) gave a LoD of $15.4\,\mu g\,L^{-1}$ (S/N = 2). Fluorescence detection (excitation 360 nm, emission 480 nm) gave a LoD of $32\,\mu g\,L^{-1}$, but post-column eluent pH adjustment was needed. UV detection (254 nm) gave a LoD of $65\,\mu g\,L^{-1}$ and interfering peaks were visible on the chromatograms.

Procarbazine (Figure 6.11) has been measured using a secondary amino/cyano-modified silica column (Partisil PAC, Whatman) and a carbon paste

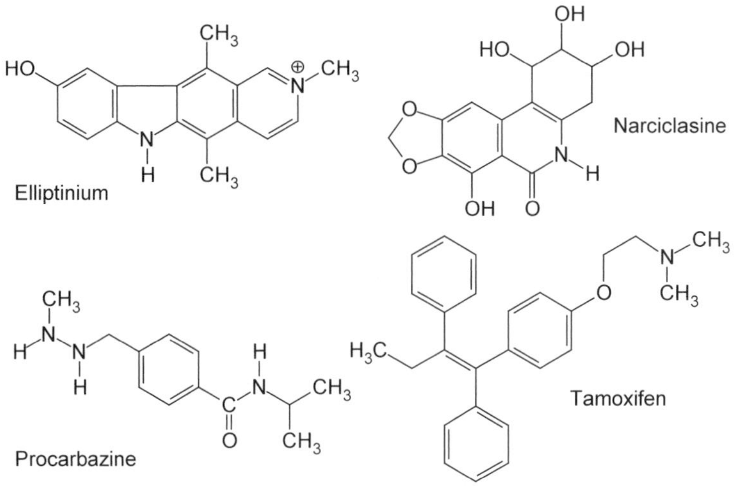

Figure 6.11 *Elliptinium, narciclasine, procarbazine and tamoxifen.*

electrode ($+0.75$ V *vs* Ag/AgCl).[81] The eluent was methanol-aq. ammonium phosphate (0.1 mol L^{-1}, pH 7) $(1 + 9)$. The LoD was 2 ng procarbazine on-column. No internal standard was used and the analysis of patient samples was not reported.

Tamoxifen (Figure 6.11), 2-methyltamoxifen and 4-hydroxy-2-methyltamoxifen have been measured in cell culture media using a cyanopropyl-modified silica column with ED (GCE, $+1.1$ V *vs* Ag/AgCl).[82] The eluent was acetonitrile-methanol-aq. sodium dihydrogen phosphate (50 mmol L^{-1}) $(19.4 + 11.6 + 69)$ containing EDTA (0.11 mmol L^{-1}). Sample preparation was by SPE (phenylpropyl-modified silica). An internal standard was not used.

Tirapazamine (3-amino-1,2,4-benzotriazine-2,4-di-*N*-oxide, SR4233) and its plasma metabolites (Figure 6.12) have been measured in microdialysates using dual GCEs in series operated at -0.45 and $+0.40$ V *vs* Ag/AgCl.[83] This arrangement allowed the compounds of interest to be quantified at the second

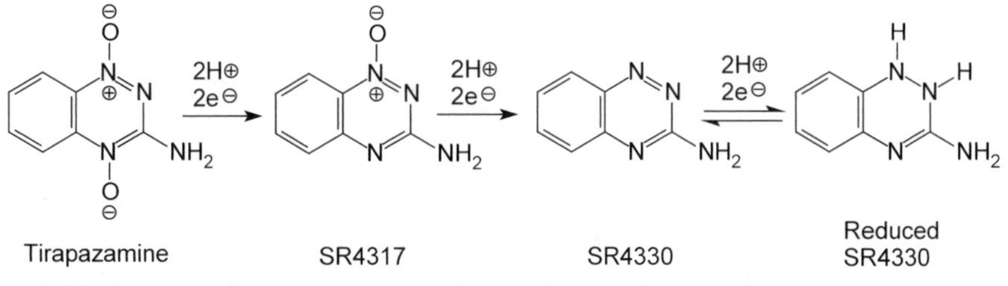

Figure 6.12 *Mechanism for the electrochemical reduction of tirapazamine and its plasma metabolites SR4317 and SR 4330.*[83]

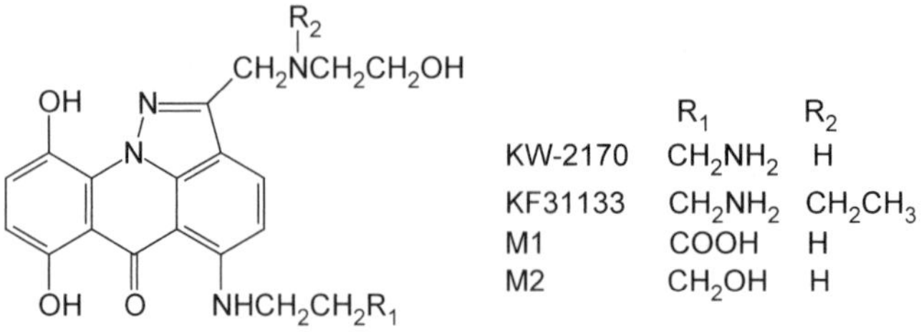

Figure 6.13 *KW-2170 and associated compounds.*[84]

electrode without interference from oxygen in the sample. A microbore (100 mm × 1 mm i.d.) column packed with 3 μm ODS-modified silica was used. The eluent was acetonitrile-phosphate buffer (50 mmol L^{-1}, pH 2.5). The LoD (5 μL sample) was 12.5 μg L^{-1} (70 nmol L^{-1}) for tirapazamine and 8.9 μg L^{-1} (50 nmol L^{-1}) for the two metabolites (S/N = 3).

A new pyazoloacridone derivative (KW-2170), which is being developed as an anti-neoplastic agent, has been measured in human and dog plasma together with its hydroxylated and carboxylic acid metabolites by HPLC-ED[84] (Figure 6.13). Sample preparation was by SPE using weak cation exchange columns. An analogue of the parent compound (KF31133, Figure 6.13) was used as the internal standard. The eluent was acetonitrile-McIlvaine's buffer, pH 4.0, containing disodium EDTA (1 mmol L^{-1}) and OSA (2.5 mmol L^{-1}) (22 + 78). The analytical column was a YMC-Pack pro C_{18} column. Detection was at a GCE (WE-GC, Eicom) (+0.35 V, reference electrode not stated). The LLoQ was 0.1 μg L^{-1} (intra-batch RSD 6.6% for human plasma, 16.6% for dog plasma).

4 Anticholinergics

Leroy and Nicolas[85] reported that atropine extracted from belladonna powder could be assayed by HPLC-ED, but the only antimuscarinic drug that appears to have been assayed in biological samples by HPLC-ED is hyoscine (scopolamine, Figure 6.14).[86] An unmodified silica column was used with acetonitrile-aq. ammonium nitrate (0.1 mol L^{-1}, pH 8.9) (9 + 1) as eluent. Hyoscine, apohyoscine (a hyoscine metabolite) and *N*-ethylnorhyoscine (Figure 6.14), the internal standard, were detected at PGEs (+0.9 V *vs* Pd). The method was applied to urine, which was extracted with toluene under alkaline conditions. The analytes were concentrated by adsorption onto acid-prepared cyanopropyl-modified silica SPE columns and eluted with HPLC eluent. The LoD was 1 μg L^{-1} (5 mL sample).

Despite their structural similarity, atropine (± hyoscyamine) and homatropine cannot be assayed in the same way as hyoscine (6,7-epoxyatropine) as the former compounds are very strongly retained on unmodified silica. Reversed-phase methods usually suffer from analyte peak tailing and, although some success with base deactivated columns has been achieved, it has not proved possible to develop

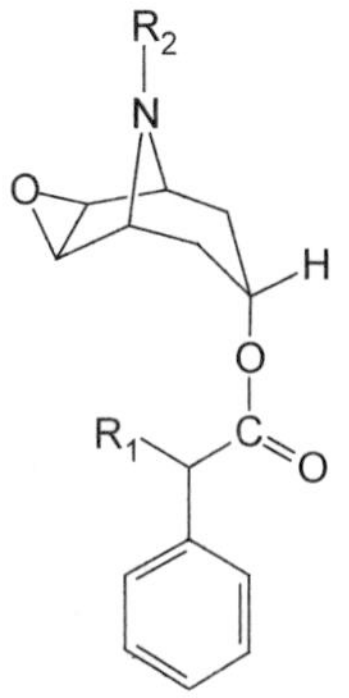

	—R$_1$	R$_2$
Apohyoscine	=CH$_2$	CH$_3$
N-Ethylnorhyoscine	—CH$_2$OH	C$_2$H$_5$
Hyoscine	—CH$_2$OH	CH$_3$

Figure 6.14 *Hyoscine, apohyoscine and N-ethylnorhyoscine.*

a sufficiently sensitive assay using such systems as yet (Whelpton and Hurst, unpublished).

5 Anticholinesterases

The anticholinesterase physostigmine and its derivatives (Figure 6.15) possess native fluorescence and although ED tends to be more sensitive, fluorescence detection offers an alternative approach that may be sensitive enough for some applications.[87-89] The first published HPLC-ED method for physostigmine in plasma was that of Whelpton.[90] An unmodified silica column was used with methanol-aq. ammonium nitrate (1 mol L^{-1}, pH 8.6) (90 + 10) as eluent. Sample preparation was by LLE into diethyl ether, which had been freshly distilled to remove antioxidants. Using a GCE (+0.8 V *vs* Ag/AgCl), the LoD was 0.5 µg L^{-1} (3 mL sample), which was barely sufficient to measure the drug after a sub-cutaneous injection of 1 mg physostigmine salicylate. Neostigmine (10 mg L^{-1}), added to blood samples to prevent enzymatic hydrolysis by plasma esterases, is a quaternary ammonium compound and is not oxidised at the applied potential used.

Several modifications, including the use of dual electrode coulometric detection, a high speed reciprocating pump and the *N,N*-dimethyl analogue of physostigmine as the internal standard, decreased the LoD to 25 ng L^{-1} (RSD=19.6%, 4 mL

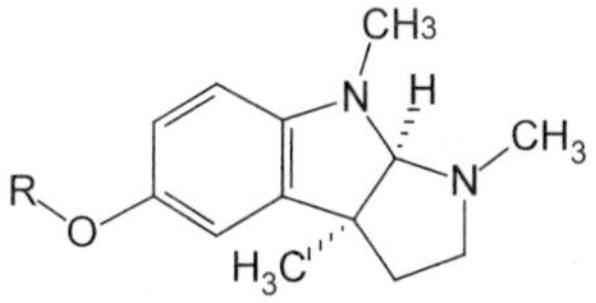

	R
Eseroline	H
Heptylphysostigmine	CO.NH.(CH$_2$)$_6$.CH$_3$
Physostigmine	CO.NH.CH$_3$

Figure 6.15 *Eseroline, heptylphysostigmine and physostigmine.*

sample).[91] The method was further improved by replacing LLE with SPE and using a three working electrode system so that physostigmine was electrochemically hydrolysed and the product subjected to redox at the analytical electrodes (Figure 3.13). As little as 50 pg physostigmine in 0.5 mL plasma or blood could be detected (RSD = 8.1%).[92] A similar electrode arrangement was employed by Knapp *et al.*[93] but sample preparation was by LLE into diethyl ether after protein precipitation with perchloric acid. The LoD was not given, but the RSD (n = 4) at 75 ng L^{-1} was 13.2%. Lawrence and Yatim[94] also employed PGEs (+0.8 and −0.2 V, respectively, *vs* Pd) for the ED of this drug.

Isaksson and Kissinger[95,96] measured physostigmine using dual GCEs in series after chromatography on an octyl-modified silica column using acetonitrile-aq. phosphate buffer (0.1 mol L^{-1}, pH 3.0) (40 + 60) containing SDS (0.5% w/v) as eluent. Physostigmine was oxidised at the first electrode (+1.0 V *vs* Ag/AgCl) and the reaction product reduced at the second (+0.1 V). Unni *et al.*[97] also used dual GCEs (+0.25 and +0.95 V *vs* Ag/AgCl, respectively) to measure plasma physostigmine. However, they employed an unmodified silica column with methanol-aq. sodium acetate (10 mmol L^{-1}, pH 4.6) (10 + 1) as eluent. Sample preparation was by LLE. An internal standard was not employed. The LoD was 0.5 µg L^{-1} (2 mL sample).

Plasma heptylphysostigmine (Figure 6.15) has been measured using an unmodified silica column with acetonitrile-THF-water-glacial acetic acid (92 + 1 + 5 + 2) containing sodium acetate (10 mmol L^{-1}, apparent pH 4.6) and ED (serial dual GCEs, E$_1$: +0.24 V, E$_2$: +0.95 V, *vs* Ag/AgCl).[98] Sample preparation was *via* LLE into heptane with back-extraction. The heptylphysostigmine metabolite eseroline (Figure 6.15) was resolved on this system, but no internal standard was used. The LoD was 125 ng L^{-1} (2 mL sample). Zecca *et al.*[99] have also described the assay of heptylphysostigmine on an unmodified silica column. However, acetonitrile-methanol-aq. ammonium nitrate (80 mmol L^{-1}, pH 8.9) (5 + 4 + 1) was used as eluent and detection was at a single GCE (+0.75 V *vs* Ag/AgCl). Sample preparation was by one-step LLE into heptane. Physostigmine was the internal standard. The LoD was 50 ng L^{-1} (1 mL sample). Pyridostigmine was added in advance to all sample tubes to inhibit esterase activity.

The microsomal oxidation of the anticholinesterase tacrine (9-amino-1,2,3,4-tetrahydroacridine) to 9-hydroxylamine-1,2,3,4-tetrahydroacridine has been studied using HPLC-ED.[100] An interesting use of electrochemistry was the production of the hydroxylamine reference compound for the metabolism study. Tacrine was chemically oxidised to the nitro compound and then electrochemically reduced to the hydroxylamine (Figure 6.16). Tacrine and its possible metabolites were analysed on a Hamilton RPR-1 column using acetonitrile-aq. ammonium phosphate buffer (0.5 mol L^{-1}, pH 11) (1 + 4) as eluent and ED was by dual GCEs in series (+0.8 and −0.4 V, respectively, *vs* Ag/AgCl). Tacrine and its ring-hydroxylated metabolites may be measured by HPLC with fluorescence detection.[101]

Post-column hydrolysis of carbamate esters, such as carbaryl, was discussed in Chapter 4, Section 2. Further consideration of anticholinesterase insecticides can be found in Chapter 7, Section 4.

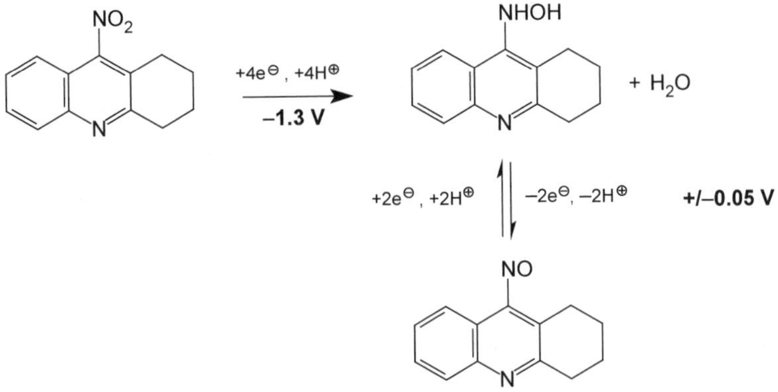

Figure 6.16 *Electrochemical reduction of 9-nitro-1,2,3,4-tetrahydroacridine and redox of 9-hydroxylamino/9-nitroso-1,2,3,4-tetrahydroacridine.*[100]

6 Antidepressants

Antidepressants can be crudely divided into three classes: the monoamine oxidase inhibitors (MAOIs), the tricyclic antidepressants (TCAs) and related compounds, and the selective serotonin reuptake inhibitors (SSRIs). HPLC-ED did not feature in a review of methods for five leading SSRIs: citalopram, fluoxetine, fluvoxamine, paroxetine and sertraline.[102] More recently, Clement *et al.*[103] measured venlafaxine (Figure 6.17) and *O*-desmethylvenlafaxine by HPLC with coulometric

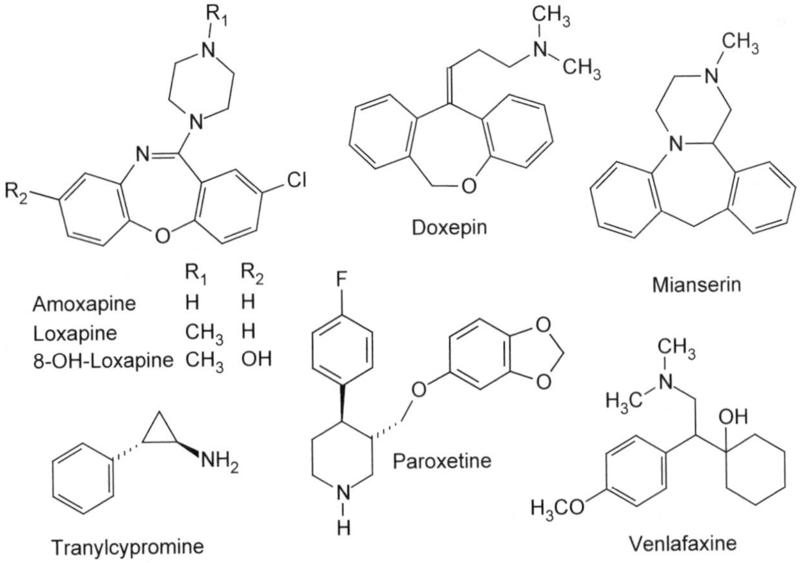

Figure 6.17 *Amoxapine, doxepin, 8-hydroxyloxapine, loxapine, mianserin, paroxetine, tranylcypromine and venlafaxine.*

detection (PGEs; E_1: +0.65 V, E_2: +0.95 V *vs* Pd). Sample preparation was by SPE (carboxymethyl-modified silica column) and the compounds were separated on a mixed mode ODS/CN column eluted with methanol-phosphate buffer (50 mmol L^{-1}, pH 4.8) (70 + 30). Paroxetine (Figure 6.17) was used as the internal standard giving inter-assay RSDs of 11.8% and 11.1%, for 15 µg L^{-1}, for the drug and metabolite, respectively. However, UV detection still appears to be preferred for this group of compounds.[104,105]

The only MAOI that appears to have been analysed by HPLC-ED is tranylcypromine (Figure 6. 17). Jane *et al.*[3] showed that tranylcypromine has an excellent EC response at a GCE at +0.8 V *vs* Ag/AgCl at pH 6.7 (EC$_{1.2 \text{ v}}$/UV$_{254 \text{ nm}}$ ratio of 1200). Plasma tranylcypromine has been measured using an ODS-modified silica column with acetonitrile-aq. phosphate buffer (0.1 mol L^{-1}, pH 3), containing sodium chloride (0.3 g L^{-1}) and sodium octanesulfonate (0.76 g L^{-1}) (20 + 80), as eluent with detection at a GCE (+1.0 V *vs* Ag/AgCl).[106] Sample preparation was by LLE involving back-extraction into dilute phosphoric acid. Apomorphine (Figure 6.34) was the internal standard. The LoD was 5 µg L^{-1} (1 mL sample).

That there are considerably fewer EC methods for the TCAs as compared with those for the phenothiazine antipsychotics probably reflects the generally higher potentials that are required and the poor EC response of some of the class of drugs.[3] This is particularly true of compounds, such as amitriptyline and nortriptyline (Figure 6.18), that possess no alicyclic nitrogen. The fact that imipramine *N*-oxide produced a response similar to that of imipramine (Figure 6.18) shows that the alicyclic nitrogen is amenable to oxidation.

Imipramine, desipramine (Figure 6.18) and their 2-hydroxy metabolites have been measured in plasma after LLE into diethyl ether.[107] 8-Hydroxyclomipramine was the internal standard. The analysis was performed on an ODS-modified silica column with acetonitrile-aq. acetate buffer (0.1 mol L^{-1}, pH 4.2) (40 + 60) containing sodium heptanesulfonate (5 mmol L^{-1}) as eluent and ED (GCE, +1.05 V *vs* Ag/AgCl). The LoDs were approximately 5 µg L^{-1} for each compound (1 mL sample). Koyama *et al.*[108] have also reported the assay of imipramine and its metabolites (including imipramine *N*-oxide) by HPLC-ED

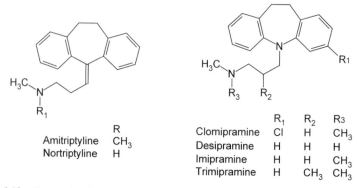

Figure 6.18 *Some tricyclic antidepressants.*

(GCE, $+0.85$ V *vs* Ag/AgCl). Sample preparation was by LLE into diethyl ether-2-propanol $(9+1)$ followed by back-extraction. Pericyazine (propericiazine, Figure 6.35) was the internal standard. The LoDs (0.5 mL plasma, S/N > 3) were 20 ng L^{-1} for the hydroxylated metabolites and 0.3 μg L^{-1} for desipramine, 0.5 μg L^{-1} for imipramine and 1.0 μg L^{-1} for imipramine *N*-oxide.

A similar approach was used by Chen *et al.*[109] to again measure imipramine, desipramine and their 2- and 10-hydroxy metabolites. The eluent was acetonitrile-aq. dipotassium hydrogen orthophosphate (100 mmol L^{-1}) $(3+7)$, final pH adjusted to pH 6.0 with orthophosphoric acid. The column was ODS-modified silica (300 mm $\times$ 3.9 mm i.d., 10 μm). Detection was at $+1.0$ V (EDT Chromajet, no further details given). LoDs (S/N > 5) were 3 μg L^{-1} for imipramine and desipramine, and 1.5 μg L^{-1} for the hydroxylated metabolites. The inter-assay RSDs (n $= 5$) at 50 μg L^{-1} were in the range 6.2–6.8%. A coulometric method (PGEs E$_1$: $+0.2$ V, E$_2$: $+0.68$ V *vs* Pd) for imipramine, desipramine and their 2-hydroxy metabolites has also been described.[110] Plasma samples (1 mL) to which 8-hydroxyclomipramine (internal standard) was added, were extracted with heptane-ethyl acetate $(80+10)$ and back-extracted into potassium dihydrogen orthophosphate (25 mmol L^{-1}, pH 2.4). The HPLC column was ODS-modified silica used with acetonitrile-aq. phosphate buffer (10 mmol L^{-1}, pH 2.4) containing tetramethylammonium chloride (5 mmol L^{-1}) as eluent. Calibration was between 0.5 and 20 μg L^{-1}; the inter-assay RSDs (%) at the lowest concentration being 4.9, 7.6, 2.5 and 2.3 for imipramine, desipramine, 2-hydroxyimipramine and 2-hydroxydesipramine, respectively.

Lofepramine (*N*-(4-chlorobenzylmethyl)desipramine) and its metabolites, desipramine, desmethyldesipramine and 2-hydroxydesipramine have been quantified by oxidation at a GCE ($+1.0$ V *vs* Ag/AgCl) after analysis on a cyanopropyl-modified silica column eluted with acetonitrile-methanol-aq. phosphate buffer (0.02 mol L^{-1}, pH 6.8) $(55+5+40)$ as eluent.[111] Lofepramine, which is readily hydrolysed to desipramine, was reduced with sodium borohydride to a stable amino alcohol derivative (Figure 6.19) before LLE into MTBE. The LoD for lofepramine and for desipramine was 0.5 μg L^{-1} (0.4 mL plasma sample).

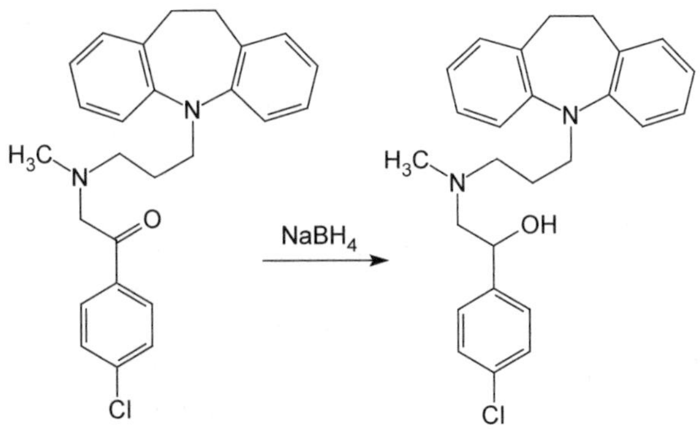

Figure 6.19 *Reduction of lofepramine with sodium borohydride.*

For trimipramine (Figure 6.18) and metabolites ethyl acetate-heptane $(20 + 80)$ has been used as the extraction solvent.[112] A methyl-modified silica column was employed with acetonitrile-aq. potassium dihydrogen orthophosphate $(0.05 \, \text{mol} \, L^{-1})$ $(35 + 65)$ containing butylamine $(1.2 \, \text{mL} \, L^{-1})$ and orthophosphoric acid $(1.0 \, \text{mL} \, L^{-1})$ as eluent. Detection was at a GCE $(+1.1 \, V \, vs \, \text{Ag/AgCl})$. Gulaid *et al.*[113] used PGEs $(E_1: +0.3 \, V, E_2: +0.85 \, V \, vs \, \text{Pd})$ in the analysis of these compounds.

Clomipramine (Figure 6.18) and its hydroxylated and *N*-desmethyl metabolites have been measured in plasma and in urine using an ODS-modified silica column together with ED (PGEs, $E_1: +0.40$, $E_2: +0.73 \, V \, vs \, \text{Pd}$).[114] The eluent was acetonitrile-aq. potassium phosphate $(10 \, \text{mmol} \, L^{-1}$, pH 2.4) containing tetra-methylammonium chloride $(5 \, \text{mmol} \, L^{-1})$ $(43 + 57)$. Sample preparation was by LLE and involved back-extraction into aqueous acid. Desipramine was the internal standard. The absolute sensitivity claimed was 15 pg on-column for all the compounds studied.

Suckow *et al.*[115] used an almost identical system to that used for imipramine to measure mianserin (Figure 6.17), normianserin and 8-hydroxymianserin in plasma, except that the eluent acetonitrile content was reduced to 33% and MTBE was used in the extraction. Desipramine (Figure 6.18) was the internal standard. The sensitivity and reproducibility of the mianserin assay were similar to those reported for imipramine and its metabolites. Brown *et al.*[116] reported the analysis of mianserin alone using an ODS-modified silica analytical column with acetonitrile-methanol-aq. phosphate buffer $(70 \, \text{mmol} \, L^{-1}$, pH approximately 6.2) $(380 + 185 + 436)$ as eluent and ED in the screen mode (PGEs, $E_1: +0.4 \, V$, $E_2: +0.7 \, V$ *vs* Pd). Sample preparation was by SPE (octyl-modified silica column) and clomipramine (Figure 6.18) was the internal standard. The LoD was $2 \, \mu g \, L^{-1}$ (0.5 mL sample). A methyl-modified silica column was also used in the analysis of amoxapine (Figure 6.17) and its 7- and 8-hydroxy metabolites, but with a modified eluent – the proportion of acetonitrile was reduced to 25% and heptanesulfonic acid $(5 \, \text{mmol} \, L^{-1})$ was added.[116] MTBE was the extraction solvent and 8-hydroxylox-apine (Figure 6.17) the internal standard. Detection at PGEs (analytical: $+0.75 \, V$ *vs* Pd) was said to be more sensitive than use of a GCE in a thin-layer configuration.

HPLC-ED detection of doxepin (Figure 6.17) and nordoxepin has been reported,[117] although the paper is mainly concerned with the assay of carbamazepine in the presence of imipramine and desimpramine. The analytes were extracted (LLE) into hexane-(2-methylbutanol) $(98 + 2)$. HPLC was performed using an ODS-modified silica column with acetonitrile-methanol-phosphate buffer $(20 \, \text{mmol} \, L^{-1}$, pH 6.7) $(45 + 45 + 10)$ as eluent. Detection was at a GCE $(+0.9 \, V \, vs \, \text{Ag/AgCl})$. The sensitivity was described as 'at ng and μg level for the tricyclics studied and for carbamazepine, respectively'.

Trazodone and its metabolite 1-(3-chlorophenyl)piperazine (*m*-CPP) (Figure 6.20) have been measured by HPLC-ED.[118] A trimethylsilyl-modified silica column was used. The eluent was acetonitrile-phosphate buffer $(50 \, \text{mmol} \, L^{-1}$, pH 3.0) $(9 + 1)$ containing sodium heptane sulfonate $(5 \, \text{mmol} \, L^{-1})$ and nonylamine $(5 \, \text{mmol} \, L^{-1})$. Sample preparation was by LLE with MTBE. Etoperidone (trazodone in which the pyridine ring has been replaced by ethyl moities) and 1-(4-tolyl)piperazine were used as internal standards. Detection was at a GCE $(+1.1 \, V \, vs \, \text{Ag/AgCl})$.

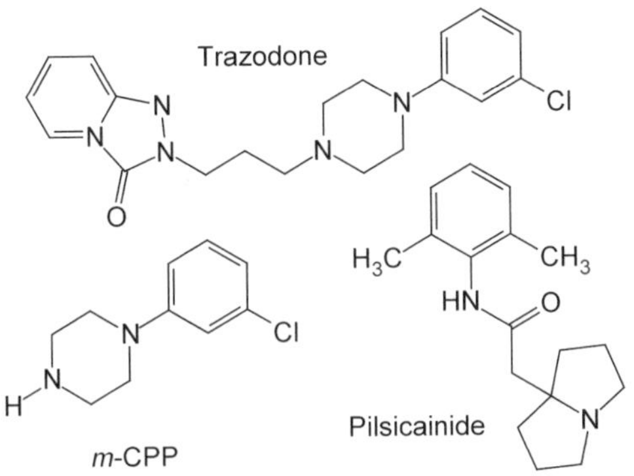

Figure 6.20 *Trazodone, 1-(3-chlorophenyl)piperazine (m-CPP) and pilsicainide.*

Intra-assay precision for trazodone ($100 \mu g \, L^{-1}$) and *m*-CPP ($5 \mu g \, L^{-1}$) was 1.5 and 2.8%, respectively. Ohkubo *et al.*[119] used UV detection (254 nm) for trazodone and coulometric detection (PGE, $+0.84$ V *vs* Pd) for *m*-CPP. The analysis was performed using a octyl-modified silica column with acetonitrile-aq. potassium dihydrogen orthophosphate ($5 \, g \, L^{-1}$) ($23 + 77$), final pH adjusted to pH 2.5 with orthophosphoric acid, as eluent. Plasma (1 mL) was extracted by SPE on nitrile-modified silica cartridges. Pilsicainide (Figure 6.20) was used as the internal standard. The LoDs for trazodone and *m*-CPP were 50 and $2.5 \mu g \, L^{-1}$, respectively. Note that at the electrometer sensitivity required to determine *m*-CPP, the EC response to the parent compound would have been beyond the linear range of the coulometric detector.

7 Antihistamines

Phenothiazine antihistamines are amenable to HPLC-ED and the methods reported are similar to those described for the structurally related phenothiazine anti-psychotics (Section 11). Wallace *et al.*[120] established the use of cyanopropyl-modified silica analytical columns and detection with a GCE at $+0.9$ V *vs* Ag/AgCl. The eluent was acetonitrile-aq. potassium dihydrogen orthophosphate ($20 \, mmol \, L^{-1}$, no pH adjustment) ($55 + 45$). Using a 2 mL plasma sample, 2,4-dichloropromazine as the internal standard and LLE into hexane, they achieved a LoD of $0.2 \mu g \, L^{-1}$ (RSD $= 15.4\%$). To reduce fouling of the electrode surface, Leelavathi *et al.*[121] included column switching to remove early-eluting contaminants and used ammonium rather than potassium orthophosphate in the eluent. Promethazine (Figure 6.35) has been measured in serum (1 mL) using cyanopropyl-modified silica columns.[122] Detection was at PGEs (E_1: $+0.5$ V; E_2: $+0.70$ V *vs* Pd). Chlorpromazine was used as the internal standard. The LLoQ was $0.2 \mu g \, L^{-1}$ (RSD 7.5%). Trimeprazine (Figure 6.35) has been assayed using similar methods to those described above[123,124] as have the antipsychotic phenothiazines – see Section 11. The major site of EC oxidation of these compounds is not the ring

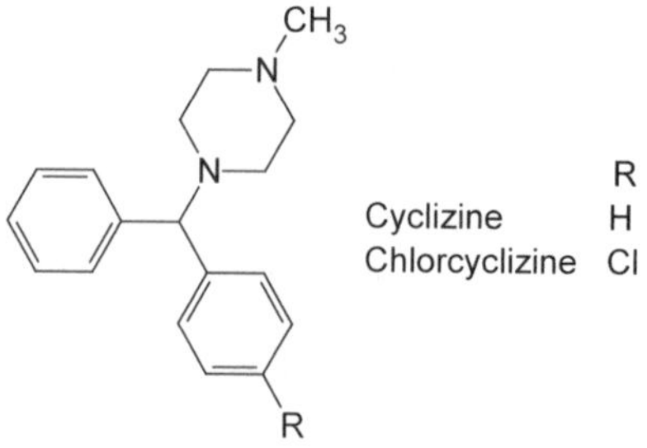

Figure 6.21 *Cyclizine and chlorcyclizine.*

nitrogen, as claimed by Radzik and Lunte,[125] but the phenothiazine sulfur, as shown by the observation that sulfoxide metabolites are not readily oxidised whilst *N*-oxide and other metabolites show similar EC characteristics to the parent compounds. Once the sulfoxides have been reduced they can be detected in the same way as the parent sulfides[126] (Figure 3.7). However, in their assay of promethazine, monodesmethylpromethazine, and the corresponding sulfoxides, Vanapalli *et al.*[127] chose to measure the sulfoxides by UV detection (236 nm). Plasma (0.2 mL) made alkaline with aq. sodium hydroxide (1 mol L^{-1}, 0.8 mL) was extracted with chloroform containing the internal standard, trifluoperazine (Figure 6.35). Separations were achieved with a cyanopropyl-modified column using methanol-ammonium acetate (150 mmol L^{-1}, pH 6.0)-water (38 + 50 + 12) as eluent. The sulfides were detected at a GCE (+0.8 V *vs* Ag/AgCl). LoDs for the four compounds were between 1 and 2 µg L^{-1}.

For the measurement of non-phenothiazine antihistamines in plasma using HPLC-ED, the EC$_{(+1.2 V)}$/UV$_{(254 nm)}$ response ratios reported by Jane *et al.*[3] suggest that many commonly used antihistamines can be assayed in this way. The high EC/UV ratios of compounds such as bromdiphenhydramine (420) and chlorcyclizine (410) are partly due to their relatively low absorption at 254 nm. Cinnarazine (22) has a relatively high absorbance ($\varepsilon = 21,600$ L mol^{-1} cm^{-1}) at this wavelength. Cyclizine, however, has one of the highest EC/UV ratios (950) of the group. Cyclizine (Figure 6.21) and *N*-desmethylcyclizine (norcyclizine) have been measured by HPLC-ED using an ODS-modified silica column with acetonitrile-phosphate buffer (50 mmol L^{-1}, pH 3) (3 + 7) at 30 °C as eluent.[128] Detection was at dual PGEs in the screen mode (E$_1$: +0.55 V; E$_2$: 0.9 V *vs* Pd). Extracts of serum (1 mL) were prepared by SPE on ODS-modified silica cartridges. Chlorcyclizine was used as the internal standard. The LLoQ was 1 µg L^{-1} for both compounds (RSDs < 10%).

8 Anti-infectives

This category includes antibiotics, antifungals, antivirals and antimalarials. Chromatographic methods for selected antibiotics were reviewed[129] in a special volume of the Journal of Chromatography devoted to separation methods for antibiotics and other antibacterial agents.[130]

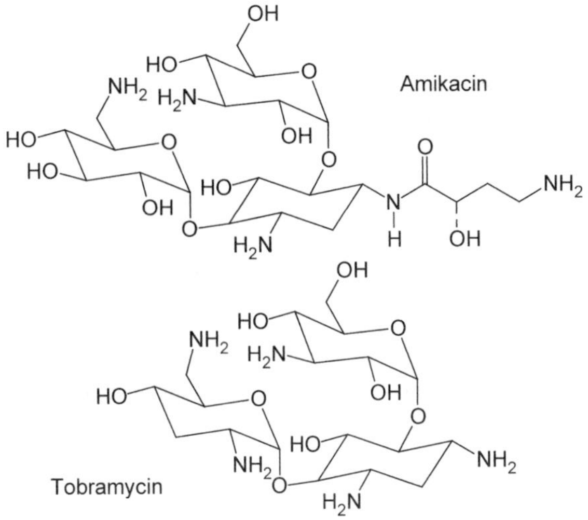

Figure 6.22 *Amikacin and tobramycin.*

8.1 Aminoglycoside Antibiotics

Aminoglycosides are derivatives of amino sugars and have been assayed using PAD with gold electrodes. Most published work concerns their measurement in tablets. Thus, tobramycin (Figure 6.22) has been measured in pharmaceutical preparations by HPLC-PAD (Dionex 4500i, gold electrode) using a Dionex CarboPac PA1 polymeric anion-exchange column.[131] The minimum detectable quantity was 2 ng tobramycin on-column. A similar approach was adopted by Szúnyog *et al.*[132] Indeed this group applied PAD to several aminoglycosides, including neomycin, kanamycin, gentamicin, netilmicin and amikacin (Figure 6.22).[133–137]

McLaughlin and Henion[138] described the use of HPLC-PAD and HPLC-MS for the determination of four aminoglycosides (streptomycin, spectinomycin, hygomycin B, dihydrostreptomycin) in spiked bovine muscle homogenates. An ODS-modified silica column was used and mobile phases based on acetonitrile-aq. PFPA or aq. HFBA were investigated. The compounds were detected at a gold electrode (+0.1 V, reference electrode not given) with a pulse duration of 480 ms and a current sampling period of 200 ms). Potentials for cleaning the electrodes were +0.6 V and −0.8 V, with pulse durations of 120 and 300 ms, respectively. The pH of the eluent was made alkaline by the post-column addition of 300 mmol L^{-1} aq. sodium hydroxide. LoDs were in the range 4–14 ng injected (S/N = 4).

8.2 Anti-Folate Drugs

Inhibitors of bacterial folate synthesis are considered here. Folate antagonists used as antineoplastic agents, *i.e.* methotrexate and lometrexol, are discussed in Section 3.

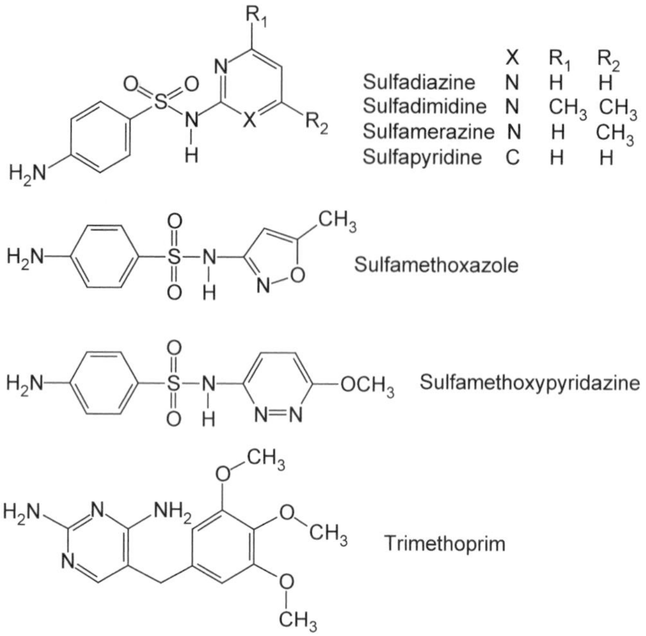

Figure 6.23 *Some sulfonamide antimicrobials and trimethoprim.*

Trimethoprim (Figure 6.23) and sulfonamide antimicrobials can be detected at +1.1 to +1.2 V *vs* Ag/AgCl with generally better sensitivity than with UV detection.[125,139] Nordholm and Dalgaard[140] measured trimethoprim in plasma after LLE into ethyl acetate. An ODS-modified silica column was used with methanol-aq. phosphate buffer (70 mmol L^{-1}, pH 4.75) as eluent. However, as the sulfonamides especially have good UV absorption (254 nm) and are generally present in blood in relatively high concentration, they can be measured in small volumes (10 µL) of whole blood.[141] ED thus offers little, if any, practical advantage over HPLC-UV. Nevertheless, HPLC-ED methods for sulfonamides have been reported. Alawi and Rüssel[142] measured sulfadimidine (sulfamethazine) and sulfadiazine (Figure 6.23) in milk (GCE, +1.1 V *vs* Ag/AgCl). Samples were prepared by LLE into chloroform. Mallett *et al.*[143] measured sulfadiazine, sulfamerazine and sulfapyridine (Figure 6.23) in plasma by HPLC-ED (GCE, +1.0 V *vs* Ag/AgCl). von Baer *et al.*[144] measured sulfamethoxazole, sulfamethoxypyridazine (Figure 6.23), and their the N_4-acetyl metabolites by increasing the oxidation potential to +1.25 V *vs* Ag/AgCl.

8.3 Antifungals

The antifungal ketoconazole (Figure 6.24) has been measured in plasma using an ODS-modified silica column with methanol-aq. formic acid (0.15 mol L^{-1}) containing dibutylamine (10 mmol L^{-1}, pH 3.0) (1 + 1) and ED (GCE, +1.0 V *vs* Ag/AgCl).[145] Sample preparation was by LLE into hexane-ethyl acetate (1 + 1)

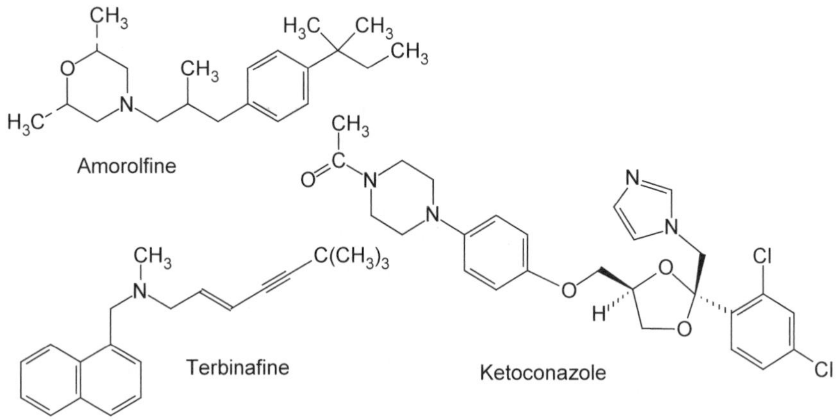

Figure 6.24 *Amorolfine, ketoconazole and terbinafine.*

and perphenazine (Figure 6.35) was the internal standard. The LoD was approximately 5 μg L^{-1} (1 mL sample). Amorolfine (Figure 6.24) has been measured in pharmaceutical preparations (nail lacquer and creams) by HLPC-ED (GCE, +0.8 V *vs* Ag/AgCl) after samples had been diluted and prepared by column switching.[146] However, this approach is unlikely to be of use in measuring percutaneous absorption.

Terbinafine (Figure 6.24) and desmethylterbinafine have been measured in human plasma and milk by HPLC-ED.[147] The HPLC conditions were: column, octyl-modified silica; eluent, methanol-aq. phosphate (10 mmol L^{-1}, pH 7.0) (77 + 23); and detection, GCE (+1.1 V *vs* Ag/AgCl). Samples (0.1 mL) were prepared by precipitating the proteins and injecting a sample of the supernatant layer. The LODs for both compounds were 50 mg L^{-1} and 150 mg L^{-1} for plasma and milk, respectively. No internal standard was used. HPLC-UV was used for urine samples.

8.4 Antimalarials

The antimalarial artemisinin (qinghaosu, Figure 6.25) is derived from the Chinese herb quing hao (*Artemisia annua*).[148] However, its bioavailability is low and esters,

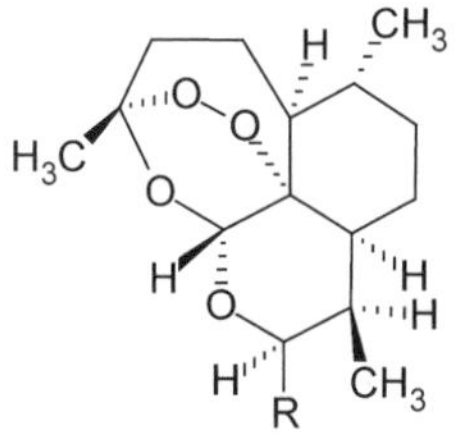

	R
Artemisinin	=O
Dihydroartemisinin	—OH
Artemether	$—OCH_3$
Arteether	$—OC_2H_5$
Artesunate	$—OCO(CH_2)_2COOH$

Figure 6.25 *Artemisinin, dihydroartemisinin and some derivatives (full stereochemical structures not shown).*

such as the methyl (artemether), ethyl (arteether) and hemisuccinate (artesunate, artesunic acid) of dihydroartemisinin (DHA), are used therapeutically. The active metabolite, DHA, exists as diastereoisomers, α-DHA and β-DHA, and although these can be separated by HPLC, it is usual for sample extracts to be allowed to stand for several hours to maximise the proportion (*ca* 4:1) of the α-tautomer. Artemisinin derivatives have no useful UV absorption, hence methods relying on UV detection require analyte conversion to a suitable chromophore under acid or alkaline conditions. Thus, the tautomers have to be separated after pre-column conversion or post-column reaction has to be employed.[149–151] However, the peroxide bond (which is necessary for antimalarial activity) can be reduced at GCEs (−1.0 V *vs* Ag/AgCl). This low reduction potential necessitates precautions to exclude oxygen, *viz.* heating the eluent, sparging with helium or argon and using stainless steel or PEEK tubing throughout.

DHA has been measured as a metabolite of artesunate[152] using reductive ED (Au/ Hg working electrode, −0.8 V *vs* Ag/AgCl). HPLC was performed using phenyl-modified silica columns with acetonitrile-aq. ammonium acetate (0.1 mol L^{-1}, pH 7.1) (18 + 82) as eluent. Sample preparation was by single stage LLE into MTBE. The method was validated down to 4 mg L^{-1} for artesunate and 0.2 mg L^{-1} for DHA. Arteether (Figure 6.25) and DHA have been measured in plasma using a cyanopropyl-modified silica column with acetonitrile-aq. acetate buffer (pH 6.0) (25 + 75, 35 + 65 or 40 + 60 depending on the separation required) as eluent and ED (parallel dual GCEs, −1.0 V *vs* Ag/AgCl).[153] Different eluent buffer concentrations and acetonitrile-water proportions and different internal standards were used depending on the assay variant. Eluents were deoxygenated by bubbling with argon (2 h, 35 °C). Sample preparation was by LLE into 1-chlorobutane-ethyl acetate (90 + 10). The LoDs (different assay procedures) were 5 and 25 µg L^{-1} for arteether and DHA, respectively (1 mL samples). A number of HPLC-ED methods for DHA and its derivatives have appeared subsequently. They have all used GCEs (−1.0 V *vs* Ag/AgCl) and are basically modifications of the method given above (Table 6.1).

Primaquine (Figure 6.26) and its metabolite carboxyprimaquine have been measured in human plasma and urine by HPLC-ED.[154] Plasma was partially deproteinated with acetonitrile and analysed directly using a nitrile-modified silica column with acetonitrile-citrate buffer (80 mmol L^{-1}, pH 6.0) (77 + 23) as eluent. Detection was at PGEs (E$_1$: +0.3 V; E$_2$: +0.5 V *vs* Pd) giving a LoD of 2 ng on column (S/N > 3). Dean *et al.*[155] adopted a similar approach to measure these analytes, precipitating plasma samples (0.5 mL) with acetonitrile after addition of the internal standard (WR6026, Figure 6.26). The analysis was performed using an octyl-modified silica column with acetonitrile-methanol-aq. chloroacetic acid (7 mmol L^{-1}) containing decanesulfonic acid (0.5 mmol L^{-1}) (24 + 20 + 56) as eluent. Detection was at +0.82 V (presumably, GCE, *vs* Ag/AgCl). LLoQs of 5 and 20 µg L^{-1} for primaquine and its metabolite, respectively, were claimed.

Pamaquine, primaquine and carboxyprimaquine (Figure 6.26) have been measured in calf plasma using an ODS-modified polymer gel column with acetonitrile-aq. phosphate buffer (70 mmol L^{-1}, pH 6.8) (23 + 77) containing EDTA (10 mg L^{-1}) as eluent and ED (GCE, +0.75 V *vs* Ag/AgCl).[156] Sample

Table 6.1 *Measurement of artemisinin and its derivatives at GCEs* (+ *1.0 V vs Ag/AgCl*)

Analytes	Extraction	HPLC Column	Eluent	LoD (S/N) (μg L^{-1})	Reference
Arteether DHA DHA propyl ether (IS)	Chlorobutane-ethyl acetate	Nitrile	Acetonitrile-aq. sodium acetate (50 mmol L^{-1}, pH 6.0) (15+85)		153
Artemether DHA Artemisinin (IS)	Isooctane-chlorobutane	Nitrile	Acetonitrile-aq. sodium acetate (50 mmol L^{-1}, pH 6.0) (15+85)	2.5 (3) 1.25 (3)	157
Artemisinin DHA	MTBE	Nitrile	Acetonitrile-aq. ammonium acetate (100 mmol L^{-1}, pH 6.5) (25+75)	12.5 (5)	158
Artemether α-DHA Artemisinin (IS)	Dichloromethane-MTBE *or* Chlorobutane-ethyl acetate	Nitrile	Acetonitrile-aq. sodium acetate (100 mmol L^{-1}, pH 6.0) (20+80)	5 (3) 3 (3)	159
Artesunic acid DHA Artemisinin (IS)	SPE (ODS-modified silica)	Octyl	Acetonitrile-aq. sodium acetate (50 mmol L^{-1}, pH 6.0) (42+58)	4 (3) 4 (3)	160
Artemether DHA Artemisinin (IS)	Isooctane-chlorobutane	Nitrile	Acetonitrile-aq. sodium acetate (100 mmol L^{-1}, pH 6.0) (15+85)	10.9 (13%)* 11.2 (16%)*	161
Artemether α-DHA β-DHA Arteether Artemisinin (IS)	Chlorobutane-ethyl acetate	Nitrile	Acetonitrile-aq. sodium acetate (100 mmol L^{-1}, pH 6.0) (40+60)	5 (5.9%)*	162
Artesunic acid DHA Artemisinin (IS)	Dichloromethane-MTBE	Octa-decyl	Acetonitrile-aq. sodium acetate (100 mmol L^{-1}, pH 4.8) (45+55)	10 (16.7%)* 10 (7.8%)*	163

* LLoQ (RSD%)

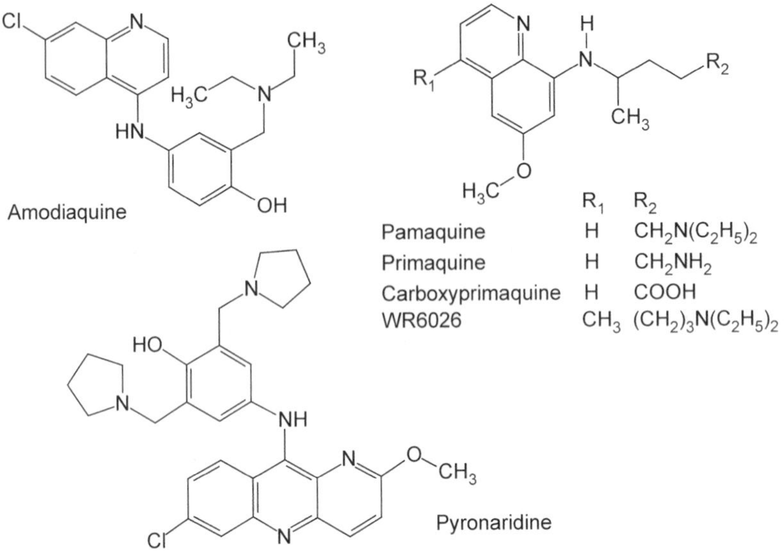

Figure 6.26 *Amodiaquine, pamaquine, primaquine, carboxyprimaquine, WR6026 and pyronaridine.*

preparation was by protein precipitation with acetonitrile followed by filtration. 2-Methoxy-5-methylaniline was the internal standard. The LoD was $10\,\mu g\,L^{-1}$ for each analyte (0.5 mL sample).

The antimalarial pyronaridine (Figure 6.26) has been measured in blood and urine using a styrene-divinylbenzene co-polymer column with methanol-water $(34 + 66)$ containing TEA (1% v/v) as eluent.[164] A GCE (+0.75 V *vs* Ag/AgCl) was used. An isobutyl analogue of desethylamodiaquine was the internal standard. The LoD was $20\,\mu g\,L^{-1}$ (0.5 mL sample). In addition, amodiaquine (Figure 6.26) has been measured using ED (GCE, +0.8 V *vs* Ag/AgCl) after gradient elution from an ODS-modified silica column.[165] However, the sensitivity attained was no greater than when using UV detection (340 nm). An automated method for pyronaridine in whole blood (1 mL) has been described, in which blood was treated with a haemolysis reagent (water (90 mL)-sodium azide (10 mg)-Triton X-100 (10 mL)) before application to CBA SPE columns in an ASPEC XL.[166] The analysis was performed on an ODS-modified silica column with acetonitrile-phosphate buffer (10 mmol L^{-1}, pH 2.5)-sodium perchlorate (1.0 mol L^{-1}) as eluent. UV detection (278 nm) proved superior to ED (GCE, +0.75 V *vs* Ag/AgCl), the within-batch RSD and bias values, respectively, at 39 $\mu g\,L^{-1}$ (75 nmol L^{-1}) being 4.3% and −0.5% (UV) and 19.0% and −35.6% (ED). Amodiaquine was evaluated as the internal standard, but the precision was better when external standardisation was used.

Although dapsone (Figure 6.27) is important in treating leprosy, it is also used in the prophylaxis of malaria and it is in this context that a method for measuring the drug in serum and saliva has been described.[167] Sample preparation was by protein

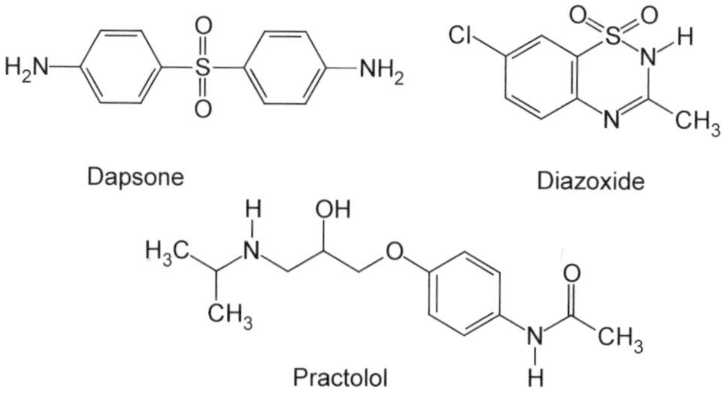

Figure 6.27 *Dapsone and internal standards.*[167]

precipitation with perchloric acid (PCA) (serum, UV detection (295 nm), internal standard: diazoxide (Figure 6.27)) or manual column switching (saliva, ED, internal standard: practolol (Figure 6.27)). The analytes were separated on an ODS-modified silica column using acetonitrile-methanol-ammonium dihydrogen ortho-phosphate (50 mmol L^{-1}, pH 4.6) (9 + 4 + 87) as eluent when ED was used (PGEs, E$_1$: +0.5 V; E$_2$: +0.7, V *vs* Pd). The ED method when appled to saliva samples was approximately an order of magnitude more sensitive than that using UV as judged by the inter-assay RSDs; 7.5% at 5 µg L^{-1} compared to 6.9% at 50 µg L^{-1}, for EC and UV detection, respectively.

8.5 Antiviral Drugs

Chromatographic methods for aciclovir (acyclovir) (Figure 6.28) and related antiviral drugs have been reviewed,[168] and although ganciclovir (Figure 6.28) has been measured in pharmaceutical preparations by anodic differential pulse voltammetry,[169] HPLC-ED has not been used for biological samples. However, Zhang *et al.*[170] have applied CE with UV and ED to the analysis of aciclovir in urine. The LoD was 0.15 mg L^{-1} using either detector. A further antiviral agent, enviradene (Figure 6.28) has been measured using an ODS-modified silica column with GCE (+0.9 V *vs* Ag/AgCl).[171] The eluent was methanol-aq. sodium acetate (0.2 mol L^{-1}) (75 + 25) containing 3 mg L^{-1} EDTA. The LoD was 2 µg L^{-1} (1 mL sample). The reverse protease inhibitor indinavir has been measured in cells and cultivation media using a 'base-deactivated' ODS-modified silica column and detection at PGEs (E$_1$: +0.4 V; E$_2$: +0.75, V *vs* Pd).[172] The eluent was acetonitrile-sodium dihydrogen orthophosphate (10 mmol L^{-1}, pH 6.3) (35 + 65). The LoD was 2 µg L^{-1} (S/N = 3) and the LLoQ 2 µg L^{-1} (RSD = 12%).

The antiviral agent foscarnet (trisodium phosphonoformate) has been measured in plasma by HPLC using an ODS-modified silica column with ED. Pettersson *et al.*[173] used PGEs (E$_1$: +0.75, E$_2$: +0.90 V *vs* Pd). The eluent was methanol-aq. phosphate buffer (43 mmol L^{-1}, pH 6.8) (25 + 75) containing tetrahexylammonium hydrogen sulfate (1 mmol L^{-1}). Sample preparation involved ultrafiltration and

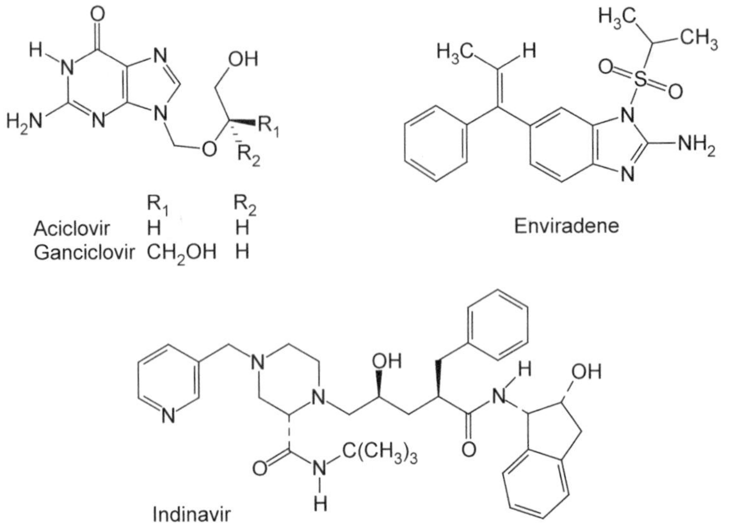

Figure 6.28   *Aciclovir, ganciclovir, enviradine and indinavir.*

dilution. Hassanzadeh *et al.*[174] used a similar procedure save that sample preparation consisted of ultrafiltration followed by heat- and ethanol-treatment to inactivate HIV. Ba *et al.*[175] reported an automated SPE method for foscarnet using SAX extraction columns. HPLC was with ODS-modified silica packed in a PEEK column to reduce the interaction of the drug with metal ions. The eluent was methanol-phosphate buffer (40 mmol L^{-1}, pH 7.6) containing tetrahexylammonium hydrogen sulfate (0.25 mmol L^{-1}) (25 + 75) with detection at a GCE (+1.15 V *vs* Ag/AgCl). The LLoQ was 1.8 mg L^{-1} (15 μmol L^{-1}) (RSD = 7.2%). Alternative assays for foscarnet and cidofovir (another phosphorus-containing antiviral agent) have been reviewed.[176]

8.6 Inhibitors of Bacterial Cell Wall Synthesis

β-Lactam antibiotics can be measured by HPLC-ED after post-column photolysis, HPLC-*hv*-ED.[177] Aspoxicillin, for example, has been measured in this way in broncho-alveolar lavage fluid (GCE, +0.80 V *vs* Ag/AgCl) after addition of a reactive internal standard, amoxicillin (Figure 6.29), and ultrafiltration.[178] The LoD was 1 μg L^{-1} (1 mL sample). HPLC-*hv*-EC has been applied to the detection of five penicillins (penicillin G (Figure 6.29), penicillin V, oxacillin, cloxacillin and dicloxacillin) in bovine muscle.[179] Samples were prepared by LLE and on-line SPE. HPLC was on an ODS-modified silica column with acetonitrile-aq. phosphate buffer (0.2 mol L^{-1}, pH 3.0) (1 + 9) containing disodium EDTA (2 mmol L^{-1}) as eluent, and detection was at a GCE (+0.65 V *vs* Ag/AgCl). LoDs were in the range 1.2–4.6 ng. Amoxicillin has also been measured in plasma using an ODS-modified silica column and fluorimetric detection after post-column on-line EC oxidation (PGE, +0.78 V *vs* Pd).[180] The eluent was acetonitrile-aq. methanesulfonic

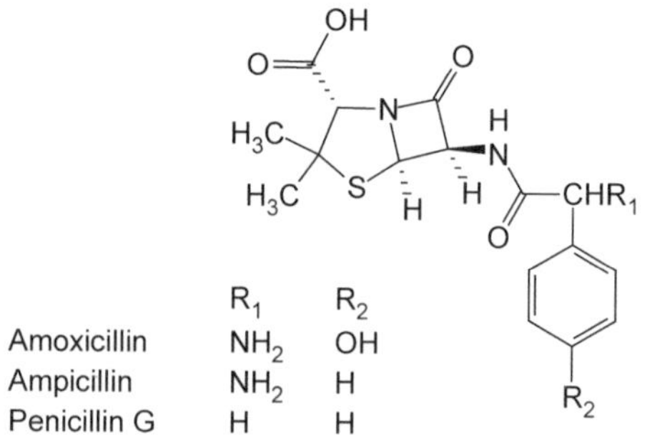

	R_1	R_2
Amoxicillin	NH$_2$	OH
Ampicillin	NH$_2$	H
Penicillin G	H	H

Figure 6.29   *Structural formulae of amoxicillin, ampicillin and penicillin G.*

acid (20 mmol L^{-1}) (7.5 + 92.5) and sample preparation consisted of protein precipitation by adding PCA. An internal standard was not used. The LoD was 0.1 mg L^{-1}.

A simple method for measuring plasma vancomycin has been described.[181] Plasma (0.1 mL) was deproteinised with methanol (0.1 mL) and further diluted (80-fold) with eluent (acetonitrile-water (12 + 88) containing disodium hydrogen phosphate (25 nmol L^{-1}), final pH 7.0) before HPLC using an ODS-modified silica column. Detection was at PGEs in the screen mode (E$_1$: +0.4 V; E$_2$: +0.7 V vs Pd). The high dilution gave good chromatograms with stable baselines. The LLoQ was 1 mg L^{-1} (RSD < 15%) or 0.25 mg L^{-1} if the dilution was reduced to 1:20.

8.7 Macrolide Antibiotics

Methods for the analysis of macrolide antibiotics, in fermentation media, pharmaceutical preparations, blood fractions, urine and various tissues, have been reviewed.[182] Although the review was not exclusively about HPLC-ED, more than 20 of the 140 references cite HPLC-ED methods for erythromycin, azithromycin, clarithromycin, roxithromycin and oleandomycin (see Figure 6.30 for structures of some compounds). Generally, these drugs have been detected at GCEs at potentials of +0.9 to +1.5 V vs Ag/AgCl, or at dual PGEs in the screen mode (E$_1$: +0.5 to + 0.7 V; E$_2$: +0.75 to +0.9 V, vs Pd). HPLC has tended to be with hydrophobic packings used with eluent buffers in the range pH 5–8 for silica based materials, up to pH 10.5 with polymeric supports, and even pH 11.0 when alumina columns were used.[183] Various proportions of acetonitrile and methanol, either individually or in admixture have been used as organic modifiers. Macrolides are lipophilic and LLE with MTBE has been a popular mode of sample preparation.

Kees *et al.*[184] introduced the use of nitrile-bonded silica column packings in their methods for roxithromycin, clarithromycin and azithromycin by HPLC-ED. The eluent composition, acetonitrile-methanol-phosphate buffer (50 mmol L^{-1}, pH 6.6–7.5), was varied according to the compounds being assayed and detection was at PGEs (E$_1$: +0.5 to +0.6 V; E$_2$: +0.8 to +0.9 V vs Pd). Samples were prepared by LLE into MTBE. LLoQs (RSD, 0.5 mL plasma) were clarithromycin, 30 μg L^{-1}

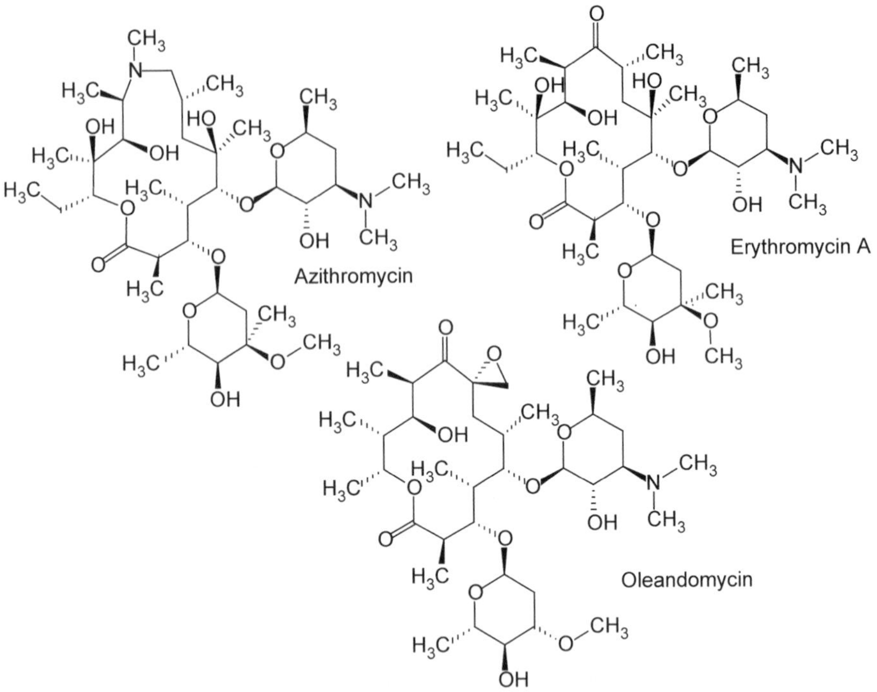

Figure 6.30 *Some macrolide antibiotics.*

(4.1%), roxithromycin, $50 \mu g \, L^{-1}$ (3.1%) and azithromycin, $10 \mu g \, L^{-1}$ (11.7%). Raines *et al.*[185] also described the HPLC-ED analysis of azithromycin and 2 metabolites, 9-α-*N*-desmethylazithromycin and *N*-desmethylazithromycin, in human plasma and tears. LLE with MTBE was used to extract plasma (0.5 mL) or tears (0.05 mL) and HPLC was with an ODS-modified silica column. The eluent was acetonitrile-methanol-aq. buffer (disodium hydrogen orthophosphate, $35 \, mmol \, L^{-1}$; tetrabutylammonium phosphate, $5 \, mmol \, L^{-1}$; sodium perchlorate, $5 \, mmol \, L^{-1}$; pH 7.0) (456 + 216 + 1800). Detection was at a PGE (+0.85 V *vs* Pd). A conditioning cell (PGE, +0.7 V *vs* Pd) preceded the analytical cell. The LoDs were $0.1 \, mg \, L^{-1}$ for azithromycin and $0.2 \, mg \, L^{-1}$ for the azithromycin metabolites.

Taninaka *et al.*[186] measured erythromycin, roxithromycin and azithromycin in rat plasma using clarithromycin as the internal standard, or clarithromycin using roxithyromycin as the internal standard. Plasma (0.15 mL) was extracted with MTBE and HPLC used an ODS-modified silica column. In contrast to the method described by Kees *et al.*,[184] a single eluent (acetonitrile-aq. phosphate buffer, $50 \, mmol \, L^{-1}$, pH 7.2 (43 + 57)) was used. Detection was at a GCE (+0.95 V *vs* Ag/AgCl). The LLoQs were similar to those reported by others, *viz.* $0.03 \, mg \, L^{-1}$ for clarithromycin or azithromycin and $0.1 \, mg \, L^{-1}$ for erythromycin and roxithromycin. A rapid method for the measurement of clarithromycin in human plasma[187] used a nitrile-modified silica column with acetonitrile-methanol-aq. phosphate buffer ($40 \, mmol \, L^{-1}$, pH 6.9) (532 + 9 + 39) as eluent. Plasma (0.5 mL) was extracted with ethyl acetate-hexane (1 + 1) after addition of roxithromycin as the internal standard. Detection was with a Waters PAD (+1.0 V presumably *vs* Ag/AgCl). The LLoQ was 0.1 mg L^{-1} and the total run time was 11 min. Pappa-Louisi *et al.*[188] measured clarithromycin using roxithromycin as the internal standard, and *vice versa*, in plasma. In a systematic study of the effects of eluent pH, the proportion of organic modifier and column temperature on peak area, peak shape, assay reproducibility and general robustness, they obtained maximum sensitivity with methanol-aq. phosphate buffer (0.2 mol L^{-1}, pH 7.0) (80 + 20) at 40 °C when the LoD was 0.1 mg L^{-1} (S/N = 3.4 for clarithromycin and 2.4 for roxithromycin). An ODS-modified silica column was used and detection was at a GCE (1.0 V *vs* Ag/AgCl). Of a number of SPE cartridges evaluated, Waters Oasis HLB was pronounced the best.

Although oleandomycin has been used as an internal standard in assays of other macrolides with tertiary amine groups (*e.g.* Laakso *et al.*,[189]) it does not appear itself to have been analysed by HPLC-ED.

Erythromycin, oleandomycin and a further macrolide, spiramycin, have been separated by CE (fused-silica capillary, 80×50 μm i.d) in phosphate buffer (0.2 mmol L^{-1}, pH 7.5) at an applied potential of 12.5 kV.[190] Reductive fast voltammetric detection was used at an Hg-film electrode (6 μm film on 25 μm diameter Au disk) scanned between −1.0 and −1.8 V *vs* SCE. A scan rate of 50 V s^{-1} was used as it gave the highest signal-to-noise values. Samples were introduced hydrostatically (10 cm, 15 s). The reproducibility (RSD) of migration time and detector response was <1 and <5%, respectively (n = 5 in each case), at 5 μmol L^{-1} (*ca* 4 mg L^{-1}). The LoDs were 75 pmol L^{-1} (64 ng L^{-1}) for spiramycin and 300 p mol L^{-1} (*ca* 220 μg L^{-1}) for erythromycin and oleandomycin. The method was not applied to biological samples.

8.8 Miscellaneous Antibiotics

In common with other compounds containing an aromatic nitro moiety, chloramphenicol (Figure 6.31) can be assayed in blood by EC reduction using an Hg film electrode.[125] Abou-Khalil *et al.*[191] used both reductive and oxidative ED in series with UV detection in the HPLC of chloramphenicol and four analogues.

Although highly fluorescent, tetracycline antibiotics also have several electro-active moieties. The EC behavior of tetracycline, oxytetracycline, chlortetracycline

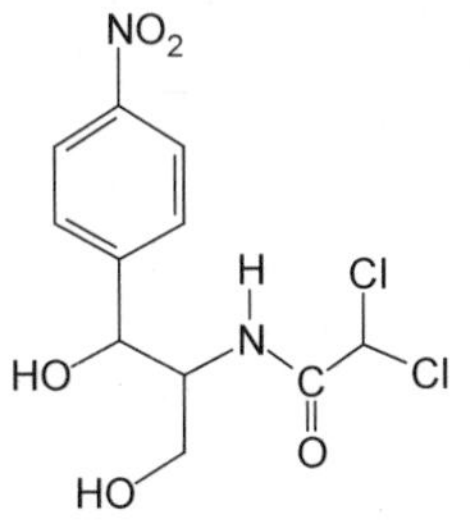

Figure 6.31 *Chloramphenicol.*

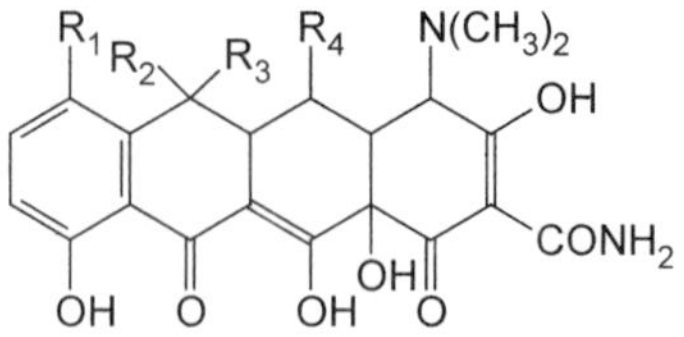

| | R₁ | R₂ | R₃ | R₄ |

	R_1	R_2	R_3	R_4
Tetracycline	H	CH_3	OH	H
Oxytetracycline	H	CH_3	OH	OH
Chlortetracycline	Cl	CH_3	OH	H
Doxycycline	H	CH_3	H	OH

Figure 6.32 *Some tetracycline antibiotics.*

and doxycycline (Figure 6.32) has been described and HPLC-ED used to measure them in pharmaceutical preparations. However, this approach has not been adopted for HPLC tetracyline assays in plasma.[192] Three of these latter compounds have also been measured by CE with reductive fast cyclic voltammetry (FCV),[193] athough the method was again not applied to biological samples. A fused-silica column (80 cm × 50 μm i.d) containing borate buffer (100 mmol L^{-1}) + disodium EDTA (1 mol L^{-1}), pH 8.7, was used (20 kV applied potential). Analyte solutions were introduced by hydrodynamic injection (Δh: 10 cm for 15 s). Detection was at a mercury film electrode deposited on a 25 μm gold microdisk, all three compounds having a reduction peak in the region −1.4 to −1.5 V *vs* SCE. Detection limits (S/N = 2) were in the range 0.7–1.5 μmol L^{-1} (0.3–0.7 mg L^{-1}), although the reproducibility was measured at 50 μmol L^{-1} (22 mg L^{-1}). The RSDs for the migration times were <1% and for the FCV charge were <5%.

9 Antimigraine Drugs

Of the drugs used to treat migraine, only the 5-hydroxytryptamine agonist sumatriptan and related developmental compounds (Figure 6.33) have been measured by HPLC-ED. Andrew et al.[194] measured sumatriptan in plasma and urine using an ODS-modified silica column with methanol-aq. phosphate buffer (8.04 g L^{-1}, pH 7.0) (6 + 4) as eluent and ED (PGEs, E_1: +0.55 V, E_2: +0.80 V *vs* Pd). Sample preparation was by LLE with ethyl acetate-dichloromethane (8 + 2) and back-extraction into aqueous phosphate buffer. No internal standard was used. The LLoQ for a 1 mL plasma sample was 1 μg L^{-1} (RSD = 6.9%). Urine samples were diluted 20-fold with phosphate buffer (pH 7) and 20 μL of the diluate injected directly. The LLoQ for urine was 200 μg L^{-1} (RSD = 8.9%). The same group developed a fully automated procedure based on SPE (ODS-modified silica) and a Zymark robotic system.[195] An ODS-modified silica HPLC column was used with methanol-aq. phosphate buffer (75 mmol L^{-1}, pH 7.0) (65 + 35) as eluent. Detection was at PGEs as described above and the LLoQ for 1 mL serum was 1 μg L^{-1} (RSD = 3.3%).

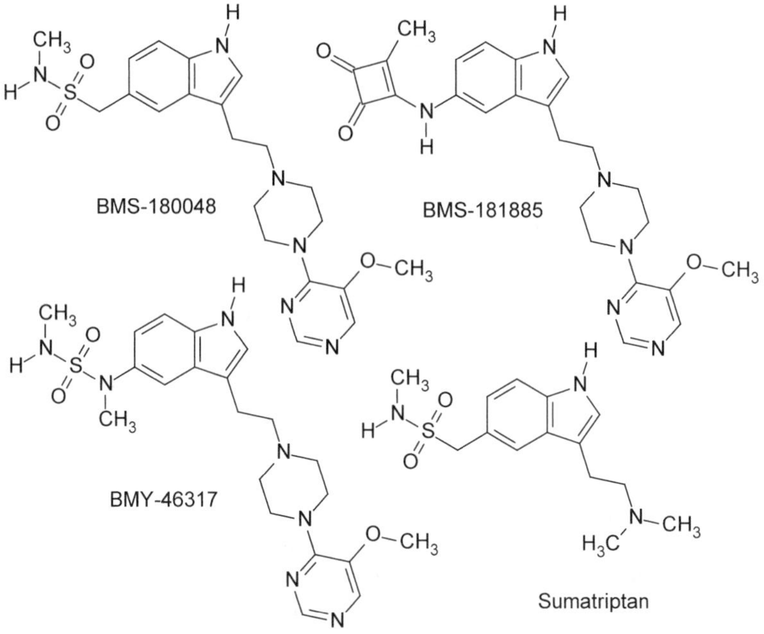

Figure 6.33 *Some 5-hydroxytryptamine agonists.*

Concentrations of BMS-180048 (Figure 6.33) in animal plasma have been reported.[196] BMY-46317 (Figure 6.33) was used as the internal standard and sample preparation was by LLE into 1,2-dichloroethane and back extraction into phosphate buffer (pH 3) containing tetramethylammonium hydroxide (TMAH). HPLC was performed using a nitrile-modified silica column with acetonitrile-methanol-TMAH (1 mol L^{-1}, pH 3)-ammonium phosphate buffer (1 mol L^{-1}, pH 3)-water (15 + 5 + 1 + 1 + 78) as eluent. Detection was at PGEs (E$_1$: +0.5 V, E$_2$: +0.80 V *vs* Pd). A LLoQ of 5 µg L^{-1} was claimed. Srinivas *et al.*[197] used SPE (CN cartridges) to extract BMS-181885 and internal standard (BMY-46317) from monkey plasma. HPLC was with a YMC basic column and acetonitrile-methanol-aq. ammonium acetate (100 mmol L^{-1}, pH 6.0) (2 + 1 + 7) was used as eluent. The PGEs were operated at lower potentials than those used previously (E$_1$: +0.15 V, E$_2$: +0.60 V *vs* Pd). The LLoQ was 2 µg L^{-1} (RSD = 3.9%).

10 Antiparkinsonian Drugs

Of the drugs used to treat Parkinson's disease, only apomorphine and its derivatives, and levodopa (L-DOPA, 3-hydroxy-L-tyrosine), carbidopa (Figure 6.34), and their derivatives and metabolites have been measured by HPLC-ED.

The dopamine agonist apomorphine (Figure 6.34) has been measured in plasma using a cyanopropyl-modified silica column with ED (GCE, +0.77 V *vs* Ag/AgCl).[198] The eluent was acetonitrile-aq. potassium dihydrogen orthophosphate

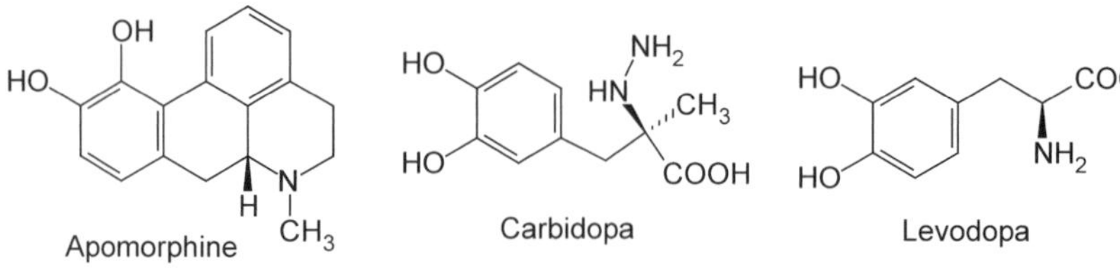

Figure 6.34 *Apomorphine, carbidopa and levodopa.*

$(20\,\text{mmol}\,\text{L}^{-1})$ $(11 + 89)$ containing $0.5\,\text{mmol}\,\text{L}^{-1}$ EDTA adjusted to pH 3.0. Apomorphine was extracted using ethyl acetate and back-extracted into aqueous acid prior to analysis. No internal standard was employed. The LoD was $0.3\,\mu\text{g}\,\text{L}^{-1}$ (1 mL sample). Essink *et al.*[199] used a similar method except that an ODS-modified HPLC column was used and sample preparation was *via* ion-pair LLE of a borate complex using tetraoctylammonium as the counter ion. *N*-Propylnorapomorphine (NPA) was employed as the internal standard. The LoD was $0.5\,\mu\text{g}\,\text{L}^{-1}$ (S/N = 7). An alternative extraction using adsorption of apomorphine onto alumina has been reported.[200] Plasma (0.2 mL), Tris buffer $(2\,\text{mol}\,\text{L}^{-1}$, pH 8.6) and alumina (40 mg) were shaken for 30 min and, after washing, the apomorphine was eluted with PCA in acetonitrile $(0.2\,\text{mL}; 0.2\,\text{mol}\,\text{L}^{-1})$. HPLC was performed using an ODS-modified silica column with acetonitrile-aq. phosphate buffer $(10\,\text{mmol}\,\text{L}^{-1}$, pH 3.6) $(20 + 80)$ as eluent. Detection was at PGEs $(E_1: +0.0\,\text{V}, E_2 +0.35\,\text{V}\ vs$ Pd). The LoD was 20 pg on column corresponding to $80\,\text{ng}\,\text{L}^{-1}$ plasma.

Assays for *R*- and *S*-apomorphine, apocodeine, isoapocodeine, and their sulfate and glucuronide conjugates in the plasma and urine of Parkinsonian patients have been described.[201] The enantiomers were analysed on a Chiralcel OD-R column with acetonitrile-aq. buffer (sodium dihydrogen orthophosphate, $100\,\text{mmol}\,\text{L}^{-1}$; sodium perchlorate, $100\,\text{mmol}\,\text{L}^{-1}$; EDTA, $10\,\text{mg}\,\text{L}^{-1}$; pH 4) $(35 + 65)$ as eluent. Detection was at a GCE $(+0.7\,\text{V}\ vs$ Ag/AgCl). Non-chiral chromatography was performed on an ODS-modified silica column with acetonitrile-aq. phosphate $(100\,\text{mmol}\,\text{L}^{-1}$, pH 3) plus OSA $(20\,\text{mg}\,\text{L}^{-1})$ and EDTA $(10\,\text{mg}\,\text{L}^{-1})$ as eluent. The extraction was similar to that described by Essink *et al.*[199] – complexation with diphenylboronic acid ethanolamine ester and ion pair extraction with tetraoctylammonium bromide. NPA was used as the internal standard and LODs were in the range 1–$2.5\,\mu\text{g}\,\text{L}^{-1}$ (S/N = 3) for the compounds in urine or plasma.

NPA and its potential pro-drug, *R*-10,11-methylenedioxy-*N*-propylnorapomorphine, have been measured in monkey plasma.[202] Samples were prepared by SPE on ODS-modified columns and the residue divided so that the electroactive catechol could be separated (nitrile-modified silica HPLC column) and detected at a GCE $(+0.7\,\text{V}\ vs$ Ag/AgCl). The pro-drug was analysed on an octyl-modified HPLC column with detection at 280 nm. The LoDs were 0.5 and $10\,\mu\text{g}\,\text{L}^{-1}$ for NPA and the methylenedioxy derivative, respectively. The corresponding *N*-methyl homologues were used as internal standards.

As might be expected, methods for measuring exogenous levodopa, L-DOPA methyl ester (LDME) and the DOPA-decarboxylase inhibitor, carbidopa

(S-α-hydrazino-3,4-dihydroxy-α-methylbenzenepropanoic acid), are similar to those developed for endogenous L-DOPA, dopamine and their metabolites. Table 6.2 summarises the eluent composition reported in six papers. All used 5 μm ODS-modified silica columns of various types. The eluents were based on phosphate, phosphate/citrate or acetate buffers containing EDTA and an ion-paring reagent. Manipulation of eluent pH may be useful in obtaining satisfactory resolution of the compounds of interest. This is particularly true of those compounds that contain a carboxylic acid moiety (*e.g.* Figure 1 in Lucerelli *et al.*[203]). A variety of extraction techniques (alumina adsorption, Sephadex ion-exchange, and protein precipitation with PCA) and internal standards (dihydroxybenzylamine (DHBA), *N*-methyldopamine, α-ethyldopa) were employed (Table 6.3). Sagar and Smyth[204] used a column switching technique in which supernatant layers from deproteinated plasma samples were injected onto an octyl-modified silica column. A carbon fibre electrode maintained at $+0.8$ V *vs* Ag/Ag$_3$PO$_4$ was used. Betto *et al.*[205] measured levodopa, carbidopa, and 3,4-dihydroxyphenylacetic acid (DOPAC) at PGEs, but chose to measure 3-*O*-methyl-DOPA by fluorescence (excitation 282 nm, emission 322 nm). In 1990, the same group measured the compound electrochemically.[203] Although some authors quote LoD or LLoQ values, they are of little value for endogenous compounds that will be present in 'blank' plasma. Consequently some authors only give estimates of precision. The lowest concentration and RSD (%) at which precision was measured are given in Table 6.3.

11 Antipsychotics

Compounds considered here include substituted phenothiazines, thioxanthenes and butyrophenones. Not all phenothiazines are antipsychotics, notably promethazine, an antihistamine, and methotrimeprazine (levomepromazine), which is used as a sedative in horses. However, their structural similarities mean that they behave in a similar way analytically. With molar extinction coefficients of *ca* 30,000 L mol^{-1} cm^{-1} in the range 250–260 nm, phenothiazines can be measured using UV detection down to about 1 ng on-column. However, ED offers an increase in sensitivity of an order of magnitude or more. The ring sulfur appears to be the site of oxidation (0.8–1.0 V *vs* Ag/AgCl, Table 4.1) as evidenced by the fact that secondary and primary aliphatic amine, and even *N*-oxide metabolites, show similar responses to the parent compound. However, sulfoxide metabolites do not respond unless they are reduced at the first electrode of a two-electrode system (Figure 3.7).

Tjaden *et al.*[206] used a 'home made' coulometric detector (GCE, $+0.9$ V *vs* Ag/AgCl) to measure perphenazine and fluphenazine (Figure 6.35). A methyl-modified silica column was used with methanol-aq. phosphate (50 mmol L^{-1}, pH 6.9) (53 + 47) containing potassium chloride (7 g L^{-1}) as eluent. Subsequently, cyanopropyl-modified silica columns were used with GCEs and acetonitrile-aq. phosphate buffer[120] or ammonium acetate solution[207–209] as eluents, although Murakami *et al.*[210] employed an ODS-modified column with acetonitrile-THF-pyridine-aq. acetate buffer (0.1 mol L^{-1}, pH 3.5) (69 + 1 + 0.1 + 30) as eluent in the assay of chlorpromazine (Figure 6.35) and levomepromazine. Similarly both

Table 6.2 Comparison of eluents used for analysis of levodopa, carbidopa, LDME and their metabolites

Eluent component	Nissinen and Taskinen[211]	Betto et al.[205]	Titus et al.[212] Plasma	Urine	Lucarelli et al.[203]	Rondelli et al.[213]	Sagar and Smyth[204]
Methanol (%)	8	4	25	22.5	15	8	2
Acetonitrile (%)		4				4	
pH	3.2	3.0	2.8	2.74	2.82	3.2	3.4
Sodium dihydrogen orthophosphate (mmol L^{-1})	100		20	20		50	138*
Citric acid (mmol L^{-1})	20			20			
Sodium acetate (mmol L^{-1})		25			13		
EDTA (mmol L^{-1})	0.15	0.3	0.05		0.54*	0.05	0.1*
Octanesulfonate (mmol L^{-1})	1.25	0.2	4		0.46*		
Heptanesulfonate (mmol L^{-1})						3.5	0.0006*

* calculated from data in paper

Table 6.3 *Extraction techniques and sensitivities for levodopa, carbidopa, LDME and their metabolites*

Compounds	Sensitivity ($\mu g\,L^{-1}$)[a]	Sample	Extraction	Internal Standard	Detection[b]
Nissinen and Taskinen[211]					
Carbidopa	50 (3.6%)	Plasma (1 mL)	Alumina	DHBA	GCE
Levodopa	50 (1.7%)				+0.7 V
DOPAC	50 (3.8%)				
Betto et al.[205]					
Carbidopa	83.1 (3.6%)	Plasma (1 mL)	1. CM-Sephadex	N-methyldopamine	PGEs
Levodopa	18.7 (3.5%)		2. PCA		E_1: +0.25 V
DOPAC	76.4 (5.6%)				E_2: −0.30 V
3-O-methyl-DOPA	78.4 (3.8%)[c]				
Titus et al.[212]					
Carbidopa	25 (5.1%)	Plasma (1 mL)	PCA	α-ethyldopa	GCE
Levodopa	25 (2.5%)				+0.75 V
3-O-methyl-DOPA	25 (1.1%)	Urine (0.1 mL)	Alumina		+0.54 V
Lucarelli et al.[203]					
Carbidopa	370 (6%)	Plasma (1 mL)	MeOH/PCA	N-methyldopamine	PGEs
Levodopa	18.2 (6.7%)				E_1: +0.25 V
DOPAC	38.3 (5.8%)				E_2: −0.35 V
3-O-methyl-DOPA	855 (5.9%)				
Rondelli et al.[213]					
LDME	100 (7.35%)	Plasma (0.5 mL)	PCA	DHBA	PGEs
Levodopa	100 (8.55%)				E_1: +0.35 V
3-O-methyl-DOPA	200 (8.32%)				E_2: −0.35 V
Sagar and Smyth[204]					
Carbidopa	250 (5.34%)	Plasma (not stated)	Octyl-column switching	None	Carbon fibre
Levodopa	250 (5.1%)	Urine (2 mL)			+0.8 V

[a]used for precision (RSD%)
[b]GCEs *vs* Ag/AgCl, PGEs *vs* Pd, carbon fibre *vs* Ag/Ag₃PO₄
[c]fluorescence: excitation 283 nm/emission 322 nm

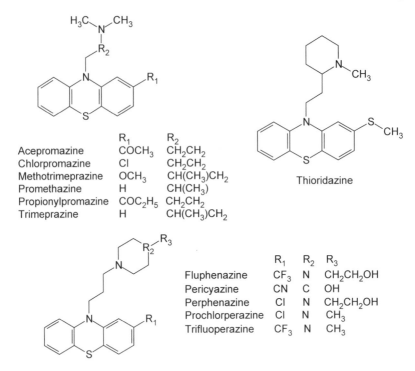

Figure 6.35 *Some phenothiazines.*

cyanopropyl- and ODS-modified silica columns have been used in the measurement of more potent compounds, such as prochlorperazine (Figure 6.35),[214,215] fluphenazine (Figure 6.35)[216,217] and thioridazine (Figure 6.35) and its metabolites.[218,219]

Chlorprothixene (Figure 6.36) has been measured in plasma using LLE (heptane-(3-methylbutanol), 99 + 1) followed by HPLC on a cyanopropyl-modified silica column with acetonitrile-aq. potassium phosphate (0.02 mol L^{-1}, pH 4.5) (60 + 40) as eluent.[220] An internal standard was not used. ED (GCE, +1.0 V *vs* SCE) gave comparable sensitivity to UV absorption (229 nm) (LoD 5 μg L^{-1}, 1 mL sample). Chlorprothixene sulfoxide, however, was considerably less electroactive than the parent compound. Plasma *cis*-thiothixene (Figure 6.36) has been assayed using similar methodology to that described above[221] except that 1. trifluoperazine (Figure 6.35) was the internal standard and 2. use of PGEs (E$_1$: +0.3 V, E$_2$: +0.8 V *vs* Pd) gave a LoD of 0.2 μg L^{-1} (1 mL sample).

Any assay for haloperidol (Figure 6.36) in plasma must be capable of resolving the drug and a metabolite in which the keto group has been reduced to a hydroxyl moiety ('reduced haloperidol'). Korpi *et al.*[222] used ED (GCE, +0.9 V *vs* Ag/AgCl) and reported a LoD of 1–2 μg L^{-1} (2 mL sample). Methods such as those of Midha *et al.*[223] and of Hariharan *et al.*[224] using PGEs have higher sensitivity (0.05 μg L^{-1} using a 1–2 mL sample), but require more complicated sample preparation,

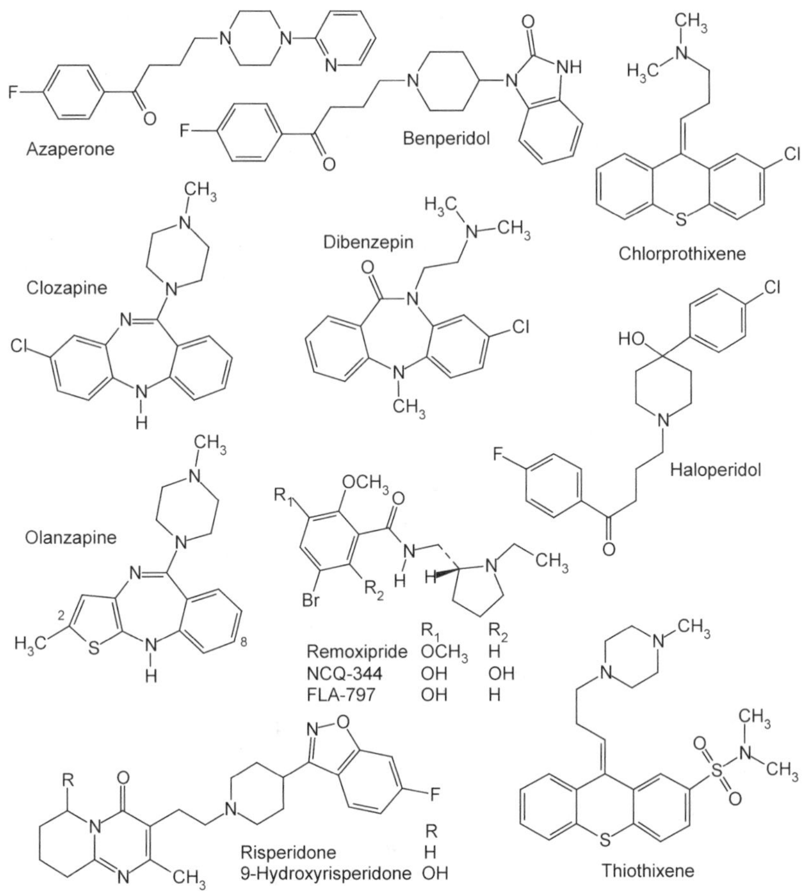

Figure 6.36 *Some non-phenothiazine antipsychotic drugs and metabolites.*

including back-extraction and clean-up procedures. Eddington and Young[225,226] achieved similar sensitivities ($0.1\,\mu g\,L^{-1}$ with a 0.5 mL sample) using SPE (cyanopropyl-modified silica) together with coulometric detection. Pan *et al.*[227] found poor correlations between plasma haloperidol and reduced haloperidol concentrations in patient samples measured using HPLC-UV and HPLC-ED methods. That the ED results (particularly for reduced haloperidol) were generally higher than those obtained using HPLC-UV was attributed to interference from co-administered drugs in the ED method. A related compound benperidol (Figure 6.36) has been assayed following SPE (octyl-modified or sequential ODS- and SCX-modified silica).[228,229]

Clozapine (Figure 6.36) has been measured using an octyl-modified silica column with ED (GCE, +0.7 V *vs* Ag/AgCl).[230] The eluent was acetonitrile-aq. ammonium acetate ($0.25\,mmol\,L^{-1}$) (9 + 1). Sample preparation was by LLE into hexane under alkaline conditions. Dibenzepin (Figure 6.36) was the internal standard. The LoD was $20\,\mu g\,L^{-1}$ ($100\,\mu L$ sample). Clozapine, *N*-desmethylclozapine (norclozapine) and clozapine *N*-oxide have been measured using a GCE ($+0.9$ V *vs* Ag/AgCl) after protein precipitation of patient serum ($200\,\mu L$).[231] An octyl-modified silica column eluted with methanol-acetonitrile-water phosphate buffer ($100\,mmol\,L^{-1}$, pH 6.6) was used. The internal standard was imipramine and the LLoQs for clozapine, norclozapine and clozapine *N*-oxide were 0.5, 0.5, and $5\,\mu g\,L^{-1}$, respectively.

Risperidone (Figure 6.36) and its 9-hydroxy metabolite have been measured in plasma using a cyanopropyl-modified silica analytical column with acetonitrile-aq. phosphate buffer ($50\,mmol\,L^{-1}$, pH 6.5) ($60+40$) as eluent and ED (E_1: $+0.55$ V, E_2: $+0.80$ V *vs* Pd).[232] Sample preparation was *via* a multistep procedure involving both SPE and LLE. Methylrisperidone was the internal standard. The LoD was $2\,\mu g\,L^{-1}$ (1 mL sample). Aravagiri *et al.*[233] reported similar methodology except that sample preparation involved LLE and remoxipride (Figure 6.36) was the internal standard. A LoD of $0.1\,\mu g\,L^{-1}$ (1 mL sample) was claimed. Potential interference from 9-hydroxyrisperidone was not considered, however. Although HPLC-ED methods for remoxipride itself have not been reported, analysis of its unstable quinone metabolite NCQ-344 (Figure 6.36) in human plasma has been described.[234] LLE into hexane-diethyl ether ($1+3$) at pH 7.05 was combined with on-line clean up using a cyanopropyl-modified silica column in a column switching system. A heart-cut from the cyanopropyl column was injected onto an octadecyl-modified silica analytical column using acetonitrile-aq. sodium dihydrogen orthophosphate buffer ($4.42\,g\,L^{-1}$, pH 2.0) containing $0.25\,mmol\,L^{-1}$ *N,N*-dimethylnonylamine, $0.1\,mmol\,L^{-1}$ sodium nonyl sulfate and $0.1\,g\,L^{-1}$ disodium EDTA ($13+37$), as eluent (flow-rate $1.3\,mL\,min^{-1}$). Detection was at a GCE ($+0.70$ V *vs* Ag/AgCl). An internal standard was not used. The absolute recovery was $>95\%$ and the LLoQ $0.04\,\mu g\,L^{-1}$ (RSD 10.0%). The NCQ-344 concentrations in patient sample were 0.04–$0.4\,\mu g\,L^{-1}$ (0.10–$1\,nmol\,L^{-1}$). The remoxipride concentrations in the same samples were 1.8–$7.4\,mg\,L^{-1}$ (5–$20\,\mu mol\,L^{-1}$).

Methods for measuring olanzapine (Figure 6.36) in plasma have been described by Catlow at al.[235] and by Aravagiri *et al.*[236] They both used LY17022 (the 2-ethyl homologue of olanzapine) as the internal standard and obtained similar sensitivity ($0.25\,\mu g\,L^{-1}$). Catlow at al.[235] used SPE with mixed mode (cation exchange and hydrophobic) columns. The analytical column was a YMC basic column and the eluent was acetonitrile-methanol-phosphate buffer ($75\,mmol\,L^{-1}$, pH 7.0) ($13+13+24$). The oxidation current-voltage curve was biphasic, but better results were obtained using the lower potential and working in the redox mode (PGEs, E_1: $+0.2$ V, E_2: -0.2 V *vs* Pd). Aravagiri *et al.*[236] used a nitrile-modified silica column with acetonitrile-methanol-aq. sodium acetate ($130\,mmol\,L^{-1}$, pH 6.8) ($43+3+4$) as eluent. Plasma was extracted with pentane-dichloromethane ($85+15$) and detection was at dual PGEs using the screen mode (E_1: $+0.3$ V, E_2: $+0.93$ V *vs* Pd). Bao and Potts[237] used the HPLC method of Catlow at al.[235] to quantify olanzapine in rat brain homogenates after LLE with cyclohexane-dichloromethane ($85+15$). The 8-methyl substituted analogue (LY170158) was used as the internal standard and the LoD was $0.5\,\mu g\,L^{-1}$. Olanzapine has been extracted from human milk using SPE (mixed mode, ion exchange/reverse phase) and analysed using the

HPLC-ED system described by Catlow *et al.*[235] The LLoQ was 0.25 µg L^{-1} (inter-assay RSD = 8.4%; accuracy 101%) for 1 mL milk.[238]

12 Anxiolytics

This section includes benzodiazepines, which have a variety of clinical uses (*e.g.* as anticonvulsants, muscle relaxants and hypnotics) in addition to reducing anxiety, and non-benzodiazepine anxiolytics. Many benzodiazepines are weak bases (*pKa* values 1.3 to 4.6) and are not readily oxidised unless a second electroactive group is present. However, the 4,5-azomethine moiety can be reduced. Using a dropping mercury electrode (differential pulse mode, −0.82 V *vs* Ag/AgCl), Hackman and Brooks[239] quantified chlordiazepoxide (Figure 6.37) and its desmethyl metabolite. Lloyd and Parry[240,241] refined the method and applied it to several commonly-encountered benzodiazepines. The drugs were analysed using an ODS-modified silica column with methanol-(1-propanol)-aq. phosphate buffer, pH 6 (100 + 7.5 + 80), as eluent. Detection was at a HMDE (−1.2 V *vs* Ag/AgCl). The LoDs were in the range 40–200 pg injected. More recently, a combination of reductive ED at a HMDE (−1.4 V *vs* Ag/AgCl) and UV detection (250 nm) has been described for the detection of benzodiazepines in forensic samples.[242] An octyl-modified silica column was used with acetonitrile-acetate buffer (30 mmol L^{-1}, pH 4.6) (45 + 55) as eluent. LoDs were in the range 2–15 µg L^{-1} for both detection systems. ED was not always the more sensitive, the UV/EC ratios being in the range 0.3–2.7 depending on the compound.

It should be possible to detect 7-nitro-substituted benzodiazepines at carbon electrodes and some of the more complex substituents of some commercially available benzodiazepines should undergo oxidative EC reaction. However, these

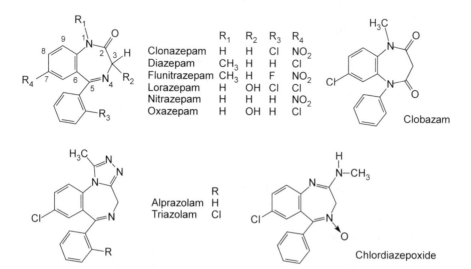

Figure 6.37 *Some representative benzodiazepines.*

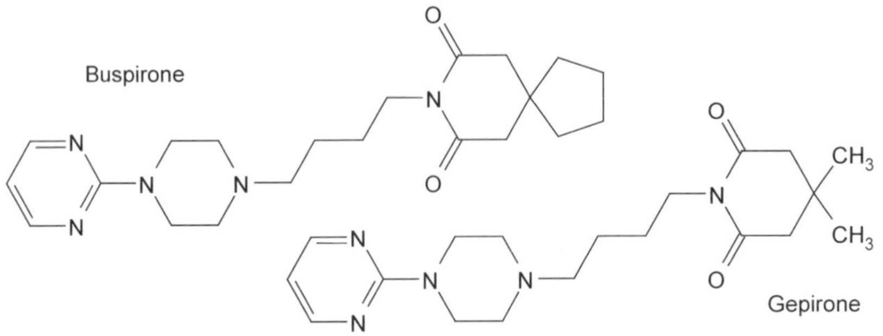

Figure 6.38 *Buspirone and gepirone.*

compounds have well defined UV spectra so HPLC-UV, as for example in the analysis of lorazepam (Figure 6.37),[243] tends to be used. Furthermore, the 7-chloro and 7-nitro substituents impart good electron capturing properties and the high proportion of nitrogen, particularly in the imidazobenzodiazepines, mean that gas chromatography (GC) with electron capture detection (ECD) and/or nitrogen-phosphorus detection (NPD) are suitable alternatives to HPLC-ED.[244]

Buspirone (Figure 6.38) has been measured in plasma by HPLC using a cyanopropyl-modified silica column with acetonitrile-aq. potassium phosphate (40 mmol L^{-1}, pH 6.6) (17 + 33) as eluent and ED (PGEs, E$_1$: +0.55 V, E$_2$: +0.70 V vs Pd).[245] Gepirone (Figure 6.38) was the internal standard. The LoD was 2 µg L^{-1} (2 mL sample). Betto et al.[246] have described the measurement of buspirone and an N-dealkylated metabolite, 1-(2-pyrimidinyl)piperazine (1-PP), in plasma using a cyanopropyl-modified silica analytical column with acetonitrile-aq. potassium phosphate (20 mmol L^{-1}, pH 7.3) (43 + 57) as eluent and ED (PGEs, E$_1$: +0.55 V, E$_2$: +0.70 V vs Pd). Sample preparation was by SPE (carboxymethyl-modified silica column). No internal standard was used. The LoDs were 100 and 70 ng L^{-1} for buspirone and 1-PP, respectively (0.5 mL sample). Odontiadis and Franklin[247] used a similar SPE method for buspirone and 1-PP, but included desipramine (Figure 6.18) as the internal standard. The analytical column was a mixed ODS/CN modified silica with methanol-phosphate buffer (40 mmol L^{-1}, pH 4.5) (74 + 26) as eluent. Detection was at PGEs (E$_1$: +0.75 V, E$_2$: +0.93 V vs Pd) and the LoDs were 0.5 and 2 µg L^{-1} for buspirone and 1-PP, respectively.

13 Cannabinoids

Cannabinoids in cannabis preparations have been measured using an ODS-modified silica column with ED (presumably GCE, +1.2 V vs Ag/AgCl).[248] The eluent was acetonitrile-methanol-aq. sulfuric acid (10 mmol L^{-1}) (16 + 7 + 6). Bourquin and Brenneisen[249] used a similar system to measure 11-carboxy-Δ^9-tetrahydrocannabinol (THC-COOH) in hydrolysed urine. Cannabinol was the internal standard. The reason such a high potential was used in these reports is not clear (no voltammogram was given) since THC-COOH contains a phenolic hydroxyl

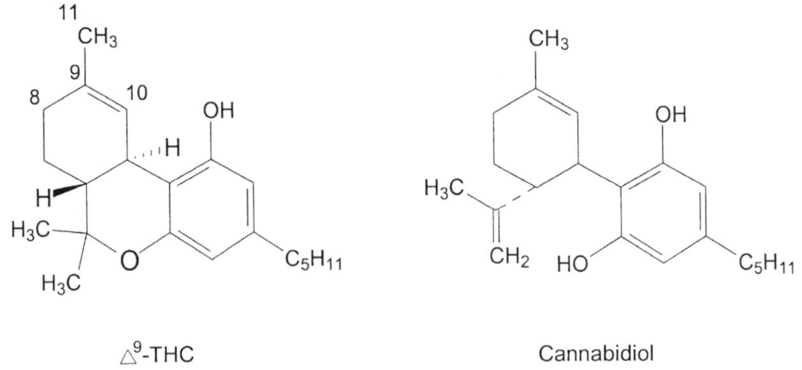

Figure 6.39 *Δ⁹-Tetrahydrocannabinol and cannabidiol.*

(see Figure 6.39). Indeed, Karlsson[250] reported the HPLC of THC-COOH in hydrolysed urine samples *via* direct injection and column switching. Detection was by serial UV (220 nm) and ED (E_1: +0.30 V, E_2: +0.60 V *vs* Pd) detection. No internal standard was used.

Nakahara *et al.*[251] and Nakahara and Cook[252] have also reported the measurement of Δ^9-tetrahydrocannabinol (THC, Figure 6.39) and its major metabolites (11-hydroxy-Δ^9-tetrahydrocannabinol (11-OH-THC) and THC-COOH) in plasma and in urine after alkaline hydrolysis and SPE using ODS-modified silica or sonication, and LLE into diethyl ether. An octyl-modified silica analytical column was used and ED was presumably at a GCE (+1.1 V *vs* Ag/AgCl). Octyl 4-hydroxybenzoate was employed as the internal standard. Gerostamolos and Drummer[253] used similar methodology for these compounds in blood samples obtained post-mortem except that acetonitrile protein precipitation was followed by LLE under both acidic and basic conditions, and no internal standard was employed. ED was at PGEs (E_1: +0.40 V, E_2: +0.68 V *vs* Pd). LoDs of $1\,\mu g\,L^{-1}$ for both THC and 11-OH-THC (1 mL sample) were claimed. THC-COOH has been measured in human urine using a GCE (+0.85 V *vs* Ag/AgCl).[254] The internal standard was prepared by brominating THC-COOH in carbon tetrachloride. Urine was extracted using mixed mode SPE (BondElut Certify II). HPLC was by use of a linear gradient on an ODS-modified silica column. The eluents were 1. methanol-aq. chloroacetic acid ($0.5\,mol\,L^{-1}$, pH 3)-triethylamine-water (628 + 100 + 1.4 + 260.6), and 2. methanol-aq. chloroacetic acid ($0.5\,mol\,L^{-1}$, pH 3)-triethylamine-THF-water (718.9 + 100 + 1.4 + 134.8 + 44.9).

THC concentrations in rat brain have been measured by HPLC-ED.[255] The method was based on that of Nakahara *et al.*[251] An ODS-modified silica column was used with acetonitrile-methanol-aq. sulfuric acid ($10\,mmol\,L^{-1}$) (24 + 21 + 55) as eluent, and detection was at a GCE (+1.2 V *vs* Ag/AgCl). Regions of brain were homogenised in methanol containing 4-dodecylresorcinol as the internal standard. The dried residues were cleaned by LLE (hexane-ethyl acetate, 7 + 3). A sensitivity of 1.5 ng on column (S/N = 2) was claimed.

Backofen *et al.*[256] have suggested the use of capillary HPLC-ED using an ODS-modified silica (3 µm) column (150×0.3 mm i.d.) and acetonitrile containing tetrapropylammonium perchlorate (10 mmol L^{-1}) as eluent for the measurement of cannabinoids in hair. Detection was at a 25 µm Pt microdisk electrode (+1.5 V *vs* Ag wire 'quasi-reference electrode'). A limit of detection of 20 pg THC was reported, although only cannabidiol (Figure 6.39) was discernable on the chromatogram of a methanol extract of a hair sample. Several unassigned peaks were also present. However, a notable claim was the absence of electrode fouling and this paper is of interest as it suggested the use not only of a novel electrode, but also a HPLC system that permits the separation and elution of compounds of widely different polarity within a relatively short time period. Cannabinoids in hair have also been measured using non-aqueous CE-ED.[257] The analysis was performed using a fused silica capillary (95 cm×0.05 mm i.d.) with acetonitrile-methanol (1 + 1) containing sodium hydroxide (5 mmol L^{-1}) as background electrolyte and with a separation voltage of 219 V cm^{-1}. Cannabinol and cannabidiol were well resolved from the THC isomers (Δ^8- and Δ^9-), which were partially resolved from each other. Detection was at a Pt microdisk electrode (25 µm diameter) positioned 25 µm from the end of the capillary and operated at +1.0 V *vs* Ag wire (quasi reference electrode). Samples were injected hydro-dynamically (Δh: 5 cm for 20 s). Hair samples (100 mg) were digested in aq. sodium hydroxide (1 mol L^{-1}) at 95 °C and then extracted using LLE with hexane-ethyl acetate (9 + 1) or SPE (Clean Screen THC columns). The reproducibility (RSD) of signal (peak height) and migration time (average of five injections) of Δ^9-THC (1 mg L^{-1}) were 1.1% and 0.4%, respectively. A LoD for Δ^9-THC of 37 µg L^{-1} (S/N = 2) was claimed.

14 Cardiovascular Drugs

A number of HPLC-ED methods have been published for β-blockers, calcium channel antagonists, direct vasodilators *etc.*, and, for convenience, these compounds are considered together in this section.

14.1 β-Adrenoceptor Antagonists

The HPLC of β-adrenoceptor-blocking drugs has been reviewed.[258] One of the first β-blockers to be assayed by HPLC-ED was sulfinalol,[259] which contains a phenolic hydroxyl moiety (Figure 6.40). A GCE was used (+0.73 V *vs* Ag/AgCl).

The 2-methylindole-containing β-blockers mepindolol and bopindolol (Figure 6.41) have also been measured using ED. Plasma mepindolol was measured down to 1 µg L^{-1} (0.5 mL sample) using ED (GCE, +1.4 V *vs* Ag/AgCl).[260] The drug was extracted into benzene-(3-methyl-1-butanol) (20 + 1) and back-extracted into aqueous acetic acid before analysis using an ODS-modified silica column. Pindolol (Figure 6.41) was the internal standard. Dual PGEs operated in the 'screen' mode (E$_1$: +0.25 V, E$_2$: +0.55 V *vs* Pd) were used for a study of transdermally applied mepindolol.[261] Toluene was the extraction solvent,

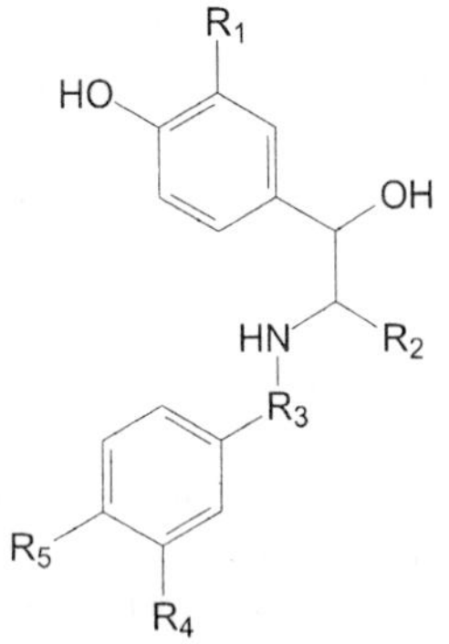

	R₁	R₂	R₃	R₄	R₅
Denopamine	H	H	CH₂CH₂	OCH₃	OCH₃
Isoxsuprine	H	CH₃	CH(CH₃)CH₂O	H	H
Labetalol	CONH₂	H	CH(CH₃)CH₂CH₂	H	H
Nylidrin	H	CH₃	CH(CH₃)CH₂CH₂	H	H
Sulfinalol	SOCH₃	H	CH(CH₃)CH₂CH₂	OCH₃	H

Figure 6.40 *Some cardioactive drugs.*

followed by back-extraction into pH 4 phosphate buffer. An ODS-modified silica column eluted with acetonitrile-aq. citrate buffer-acetate buffer (pH 4) containing pentanesulfonate ($17 \, mmol \, L^{-1}$) ($225 + 775$) was used. Pindolol was the internal standard. The lowest calibration point, $50 \, ng \, L^{-1}$, was taken as the LoD (RSD = 19%, n = 13). The active metabolite of bopindolol, which differs from mepindolol in having a terminal *N*-(*t*-butyl) rather than a terminal *N*-(2-propyl) moiety, has been measured similarly.[262] The LoD was $2.6 \, \mu g \, L^{-1}$. Humbert *et al.*[263] claimed an LoD of $25 \, ng \, L^{-1}$ (1 mL plasma, S/N = 10) 'under favourable chromatographic conditions' for their assay of hydrolysed bopindolol, although they admitted that in routine use a more realistic estimate would be $50 \, ng \, L^{-1}$. The intra-assay RSD at $65 \, ng \, L^{-1}$ was 2.2% (n = 4).

Wang *et al.*[264] showed that labetalol (Figure 6.40) could be measured by HPLC-ED at GCEs (+0.95 V *vs* Ag/AgCl). An ODS-modified silica column was used with acetonitrile-aq. phosphate buffer ($50 \, mmol \, L^{-1}$, pH 6.0) ($35 + 65$) as eluent.

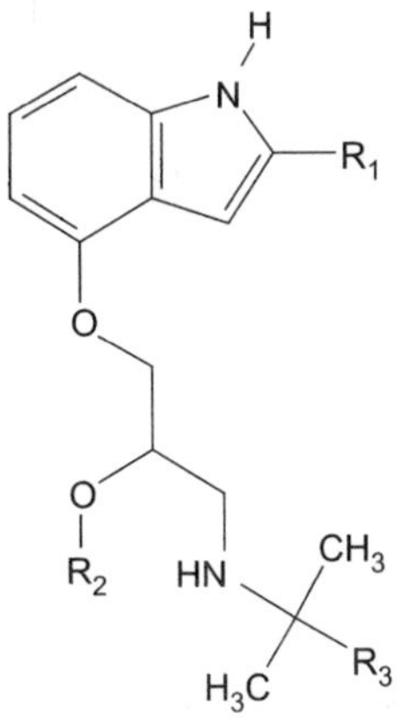

	R₁	R₂	R₃
Bopindolol	CH₃	CO.C₆H₅	CH₃
Mepindolol	CH₃	H	H
Pindolol	H	H	H

Figure 6.41 *Bopindolol, mepindolol and pindolol.*

The method was only applied to 'spiked' urines (6.25–18.75 mg L^{-1}). However, labetalol has since been measured using an ODS-modified silica column with acetonitrile-water (32 + 68) containing acetic acid (1% v/v) as eluent and ED (PGEs E$_1$: +0.40, E$_2$: +1.00 V *vs* Pd).[265] Sample preparation was by LLE into diethyl ether. Propranolol (Figure 6.42) was the internal standard. The LoD was 2.5 μg L^{-1} (2 mL sample).

The non-selective β-blocker timolol (Figure 6.42) has been measured using an ODS-modified silica column with ED (GCE, +1.2 V *vs* Ag/AgCl).[266] The eluent was methanol-aq. sodium dihydrogen orthophosphate (0.2 mol L^{-1})-orthophosphoric acid (88% w/w)-water (500 + 200 + 3 + 297). Sample preparation was by LLE and propranolol was again the internal standard. Plasma timolol has also been measured after application of timolol-containing eye drops. Dual PGEs (E$_1$: +0.5 V, E$_2$: +0.7 V *vs* Pd) were used with an ODS-modified silica column.[267] The eluent was acetonitrile-methanol-aq. disodium hydrogen orthophosphate buffer (40 mmol L^{-1}, pH 3.5) (13 + 5 + 82). Sample preparation was by LLE into 4% (v/v) butanol in hexane. Desmethyltimolol was the internal standard. The LoD was 0.1 μg L^{-1} (1 mL sample, S/N = 3).

Desacetylmetipranolol, the active metabolite of metipranolol (Figure 6.43), has been measured by HPLC-ED after LLE of plasma with dichloromethane.[268] An ODS-modified silica column was used with methanol-phosphoric acid (50 mmol L^{-1}) (4 + 6) containing disodium EDTA (10 mg L^{-1}) as eluent. The final pH was adjusted to 3.0 with aq. sodium hydroxide. Detection was at a GCE (+0.9 V *vs* Ag/AgCl). Pindolol (Figure 6.41) was the internal standard.

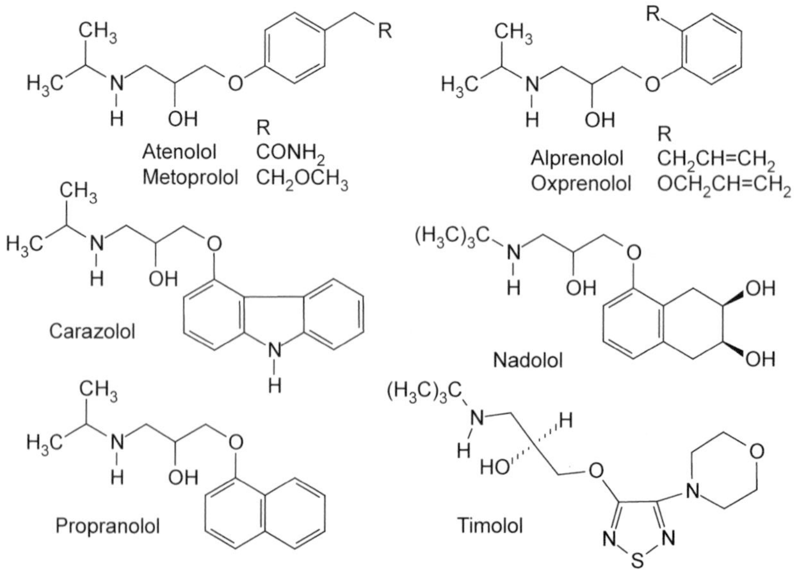

Figure 6.42 *Some β-adrenoceptor blockers.*

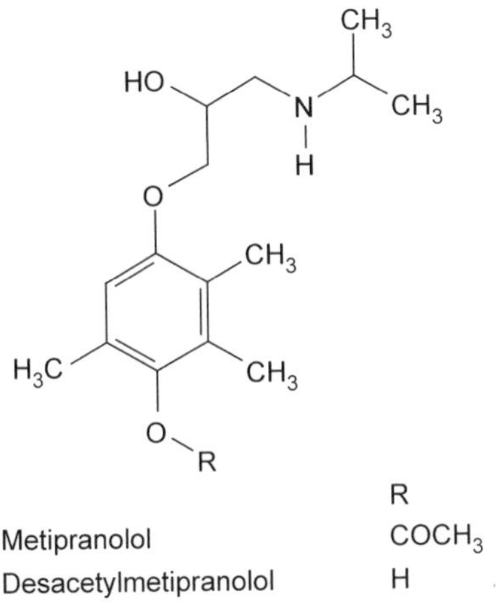

	R
Metipranolol	COCH₃
Desacetylmetipranolol	H

Figure 6.43 *Metipranolol and desacetylmetipranolol.*

A sensitivity of $2 \mu g \, L^{-1}$ (1 mL sample) was claimed; the RSD at $6.0 \mu g \, L^{-1}$ was 6.9%.

An amperometric HPLC method (GCE, $+1.3$ V *vs* Ag/AgCl) for screening six β-blockers (alprenolol, atenolol, metoprolol, nadolol, oxprenolol and timolol – Figure 6.42) in human urine has also been reported,[269] although these compounds seem ill suited to ED. An ODS-modified silica column was used with acetonitrile-aq. phosphate buffer (5 mmol L^{-1}, pH 6.5) (4 + 6) as eluent. Spiked urine (3 mL) was extracted with diethyl ether giving recoveries up to 93% (timolol – recovery for atenolol not stated). Sensitivities of between 15 and $500 \mu g \, L^{-1}$, including $50 \mu g \, L^{-1}$ for atenolol, were claimed.

14.2 Calcium Channel Blockers

Dihydropyridine calcium channel blockers, such as nifedipine and nisoldipine (Figure 6.44), are amenable to EC oxidation.[270] Typical working potentials (GCEs) are $+0.9$ to $+1.0$ V *vs* Ag/AgCl. Suzuki *et al.*[271] obtained a sensitivity of $2 \mu g \, L^{-1}$ for nifedipine (0.5 mL sample). Plasma was extracted with toluene prior to analysis on an ODS-modified silica column using methanol-THF-aq. phosphate buffer (0.05 mol L^{-1}, pH 3.0) (66 + 1 + 33) as eluent. Similar sensitivities were obtained by Huebert *et al.*[272] who used ethyl acetate-hexane (75 + 25) as extraction solvent. Telting-Diaz *et al.*[273] used a carbon fibre electrode ($+1.4$ V *vs* Ag/Ag₃PO₄) to detect nifedipine, nicardipine and pindolol (Figure 6.41) after direct on-column injection (cyanopropyl-modified silica column) with column switching. The eluent was methanol-aq. phosphate buffer (50 mmol L^{-1}, pH 6.7) (4 + 5). LoDs of $15 \mu g \, L^{-1}$ (1 mL plasma, S/N = 3) were claimed for each compound although the method was not applied to patient samples and the RSD for repeated injections (n = 5) of either nifedipine or nicardipine at $100 \mu g \, L^{-1}$ in water was 6.2%. An automated (robotic) sample preparation for nifedipine assay in dog plasma has also

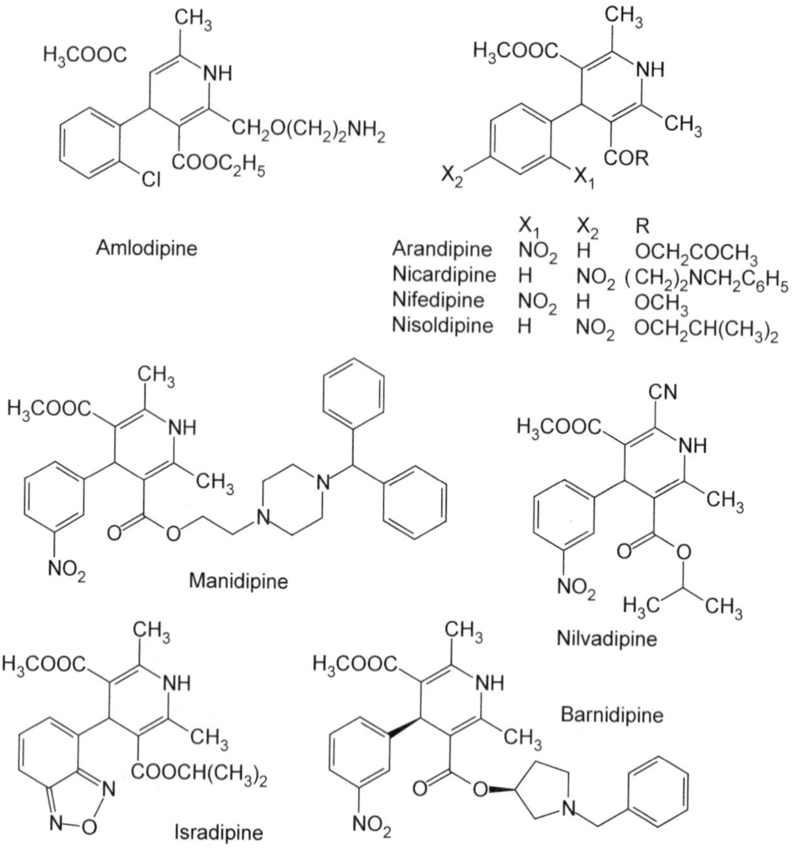

Figure 6.44 *Some dihydropyridine derivatives.*

been described.[274] A nifedipine isomer with the nitro moiety in the 3- rather than the 2- position was used as the internal standard.

Amlodipine (Figure 6.44) has been measured in human serum (1 mL), after diethyl ether extraction, using an ODS-modified silica column with acetonitrile-aq. phosphate buffer (50 mmol L^{-1}, pH 3.1) (35 + 65) containing OSA (sodium salt, 5 mmol L^{-1}) and EDTA (5 mg L^{-1}) as eluent.[275] Detection was at a GCE (+1.0 V vs Ag/AgCl). The LoD was 0.2 μg L^{-1} (S/N = 6). An SPE (C$_2$-column) method for amlodipine has also been described.[276] Detection was at an Antec GCE (+0.95 V vs Ag/AgCl) and an ESA 5020 guard cell (+0.5 V vs Pd) was inserted pre-column. The same group used an α$_1$-acid glycoprotein column with column switching to resolve amlodipine enantiomers.[277]

The dihydropyridine derivative arandipine (MPC-1304, Figure 6.44) and its enol tautomer have been analysed using an ODS-modified silica column with methanol-aq. sodium perchlorate (0.36 mol L^{-1}) (45 + 55) as eluent.[278] Sample preparation was by LLE into toluene and detection was at a GCE (+0.92 V vs Ag/AgCl). Nifedipine was used as the internal standard. A LoD of 0.25 μg L^{-1} was claimed

(1 mL dog plasma), although the precision of the assay was not tested below $5 \mu g L^{-1}$.

Fujitomo *et al.*[279] describe the use of coupled achiral (anion exchange) and chiral (ovomucoid) columns to separate the enantiomers of methyl ($\pm$)-4,7-dihydro-3-isobutyl-6-methyl-4-(3-nitrophenyl)thieno(2,3-*b*)pyridine-5-carboxylate. The eluent was methanol-aq. potassium dihydrogen orthophosphate ($50 \, mmol \, L^{-1}$) ($45 + 55$ for the achiral column and $44 + 56$ for the chiral one). Detection was at PGEs (E_1: $+0.10$ V, E_2: $+0.50$ V *vs* Pd). Sample preparation was by LLE into benzene. The inter-assay RSDs at $1 \mu g L^{-1}$ were 9.1% and 12.7% for the *S*-(+) and *R*-(−) isomers, respectively.

Manidipine and nilvadipine (internal standard) (Figure 6.44) have been analysed on an octyl-modified silica column with detection at $+0.7$ V *vs* Pd (Coulochem 5100A, no further details given).[280] The eluent was acetonitrile-aq. phosphate buffer ($0.1 \, mol \, L^{-1}$, pH 4.5) ($49 + 51$). The complex sample preparation was:

1. SPE on octyl-modified silica columns, the analytes being eluted with 80% (v/v) methanol
2. The dried residue was dissolved in chloroform and extracted on a silica cartridge
3. The residue was dissolved in eluent and subjected to column switching with an ODS pre-column.

The LoD (S/N = 5) was $0.3 \mu g L^{-1}$ (1 mL serum) and the inter-assay RSD at $1 \mu g L^{-1}$ was 7.1%.

Isradipine (Figure 6.44) has been measured in human serum using HPLC-ED by Takamura *et al.*[281] who, despite showing that the 2′,1′,3′-benzoxidazole ring could be reduced at a GCE (−1.4 V *vs* SCE), chose oxidation of the dihydropyridine ring at $+0.6$ V (*vs* Ag/AgCl) for the assay. An Aluspher RP-select B column (polybutadiene-modified alumina) was used with acetonitrile-methanol-Britton Robinson buffer ($0.1 \, mol \, L^{-1}$, pH 11.8) ($6 + 3 + 1$). Barnidipine (Figure 6.44) was used as the internal standard and sample preparation was by LLE into ethyl acetate, giving a LoD of $0.5 \mu g L^{-1}$ (1 mL sample).

Other calcium channel antagonists have been measured by HPLC-ED. Diltiazem (Figure 6.45) and its metabolites, for example, were measured by both UV (237 nm) and ED (GCE, $+1.1$ V *vs* Ag/AgCl) using ODS-modified silica columns and various eluents, some of which were modified by adding amines (DMDA or TEA).[282] *trans*-Diltiazem was used as the internal standard. Adding DMDA gave improved peak shapes, but led to an unacceptable increase in background current and thus UV detection was used for sample analyses. Morishima *et al.*[283] used an ODS-modified silica column with acetonitrile-aq. phosphoric acid ($35 \, mmol \, L^{-1}$) containing dibutylamine ($30 \, mmol \, L^{-1}$), final pH 3 ($45 + 55$) as eluent with ED (PGEs, E_1: $+0.65$ V, E_2: $+0.85$ V *vs* Pd) to measure sesamodil (Figure 6.45) and an *N*-demethylated metabolite in dog plasma. Sample preparation was by SPE (ODS-modified silica column). An analogue of sesamodil was the internal standard. Despite using dibutylamine in the eluent the LoD was 50 pg on column (S/N = 3), corresponding to a LoD of $0.4 \mu g L^{-1}$ (1 mL sample).

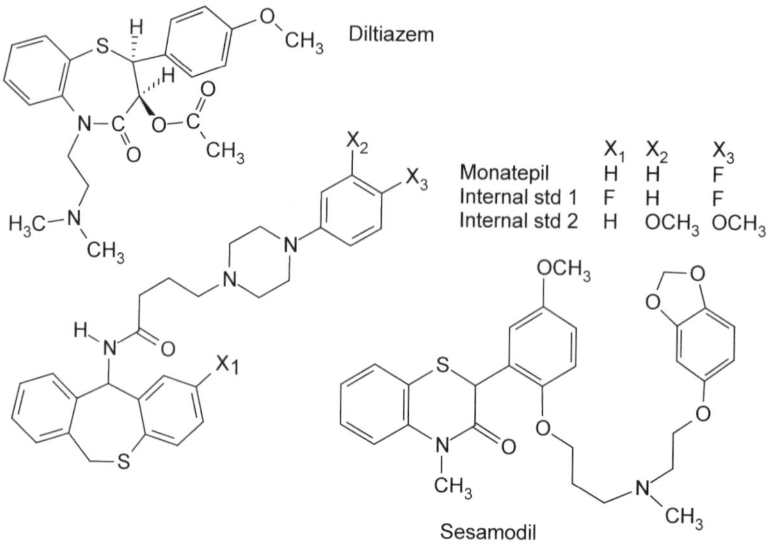

Figure 6.45 *Diltiazem, monatepil and sesamodil.*

	X_1	X_2	X_3
Monatepil	H	H	F
Internal std 1	F	H	F
Internal std 2	H	OCH_3	OCH_3

Monatepil (AJ-2615; Figure 6.45) and its sulfone and diastereoisomeric sulfoxide metabolites have been quantified at a GCE (+0.9 V *vs* Ag/AgCl) after SPE on C_{18} cartridges.[284] An ODS-modified silica column was used with aceto-nitrile-aq. phosphate buffer (50 mmol L^{-1}, pH 6.0) containing disodium EDTA (0.1 mmol L^{-1}) (33 + 67) as eluent. A defluorinated, dimethoxy-monatepil analogue was used as the internal standard (Figure 6.45). The LoDs (0.5 mL sample) were 10 and 5 µg L^{-1} for the parent compound and the three metabolites, respectively. An earlier paper from this same group[285] used a similar HPLC-ED system to measure monatepil alone, except that a fluorinated monatepil analogue was used as the internal standard (Figure 6.45).

14.3 Sympathomimetics

Sympathomimetic agents, particularly catechols and phenols, are suitable candidates for HPLC-ED. β_2-Adrenoceptor agonists have been discussed above (Section 6.2). Etilefrin[286] and epinine (N-methyldopamine, derived from meta-bolism of its di-isobutyroyl ester ibopamine)[287] have been measured using GCEs (+0.95 and +0.65 V *vs* Ag/AgCl, respectively). Etilefrin (Figure 6.46) was extracted from plasma using a cationic SPE column whereas a standard cate-cholamine alumina adsorption method was used for epinine (Figure 6.46). Xamoterol (Figure 6.46), a cardioselective β-adrenoceptor partial agonist, has been determined by HPLC-ED after SPE with C_{18} cartridges.[288] An ODS-modified silica column was used with methanol-THF-aq. perchloric acid (30 mmol L^{-1}) (8 + 0.6 + 91.4) as eluent. Detection was at a GCE (+0.85 V *vs* Ag/AgCl) giving a LoD of 2 µg L^{-1}.

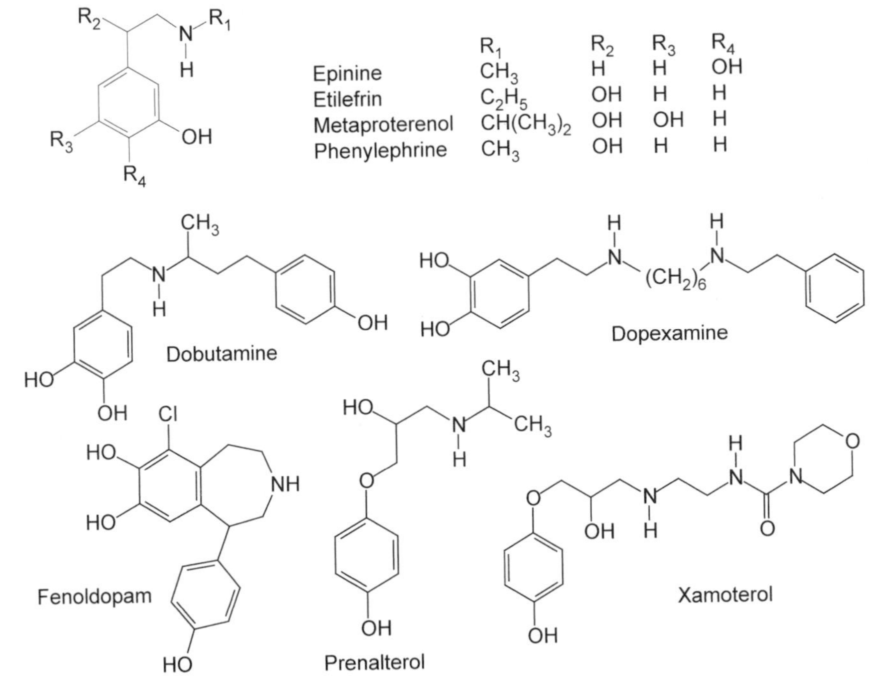
R1 R2 R3 R4
Epinine CH3 H H OH
Etilefrin C2H5 OH H H
Metaproterenol CH(CH3)2 OH OH H
Phenylephrine CH3 OH H H
Dobutamine
Dopexamine
Fenoldopam
Prenalterol
Xamoterol

Figure 6.46 *Some sympathomimetic agents.*

Fenoldopam (Figure 6.46), a potent agonist at peripheral dopamine-1 (D_1) receptors, has been measured using a GCE (+0.65 V *vs* SCE)[289] and using PGEs (E_1: +0.15 V, E_2: −0.15 V *vs* Pd) after LLE into ethyl acetate.[290] The selective β_1-agonist, prenalterol (Figure 6.46) has been measured in plasma using an octyl-modified silica column with methanol-citrate/acetate buffer (pH 3.5) (15 + 85) containing propylamine (0.1 mol L^{-1}) and ED (GCE, +0.70 V *vs* Ag/AgCl).[291] The LoD was 0.5 μg L^{-1} (2 nmol L^{-1}) using a 1 mL sample. The *N*-(1,1-dimethyl)ethyl analogue of prenalterol was selected as the internal standard.

Dobutamine (Figure 6.46) has been measured in plasma (0.1 mL) after protein precipitation with 5% (w/v) trichloracetic acid (0.2 mL).[292] The eluent was aq. sodium dihydrogen orthophosphate (100 mmol L^{-1}), EDTA (0.1 mmol L^{-1}), octanesulfonic acid (3 mmol L^{-1}) and triethylamine (1.5 mL L^{-1})-acetonitrile (80 + 20) adjusted to (presumably final) pH 2.7. The column (150 × 3.9 mm i.d.) contained 4 μm ODS-modified silica. Detection was at PGEs in the redox mode (E_1: +0.35 V; E_2: −0.26 V, *vs* Pd). Intra- and inter-assay coefficients of variation at 0.625, 5 and 10 μg L^{-1} were <6% (n = 6). To measure dobutamine concentrations down to 0.1 μg L^{-1}, a liquid-liquid extraction developed originally for plasma catecholamines was recommended.[293]

Phenylephrine (Figure 6.46) has been measured in plasma using an SCX-modified silica analytical column with methanol-aq. acetate/citrate buffer (0.1 mol L^{-1}, pH 6.2) (7.5 + 92.5) and EC detection (GCE, +0.95 V *vs* Ag/AgCl).[294] Sample preparation was by SPE (ODS-modified silica) after protein precipitation with perchloric acid. Metaproterenol (orciprenaline, Figure 6.46) was the internal standard. The LoD was approximately $1\,\mu g\,L^{-1}$ (2 mL sample). Coulometric assay of phenylephrine in serum (E_1: $+0.6\,V$, E_2: $+0.9\,V$ *vs* Pd) after SPE (phenyl-modified silica columns) with a LLoQ of $0.35\,\mu g\,L^{-1}$ (1 mL serum sample) has been described.[295] An ODS-modified silica column was used with methanol-aq. phosphate buffer ($50\,mmol\,L^{-1}$, pH 3.2) containing HSA ($100\,mg\,L^{-1}$) $(1 + 9)$ as eluent.

Denopamine (Figure 6.40) has been measured in plasma using an ODS-modified silica analytical column with acetonitrile-aq. phosphate buffer ($0.1\,mol\,L^{-1}$) $(27 + 100)$, apparent pH 6.5, as eluent, with ED (GCE, $+0.75\,V$ *vs* Ag/AgCl).[296] Sample preparation was by sequential LLE with chloroform, back-extraction into dilute acid and SPE (ODS-modified silica column). Phenolphthalein was the internal standard. The LoD was $2\,\mu g\,L^{-1}$ (1 mL sample). Baker *et al.*[297] measured dopexamine (Figure 6.46) using an ODS-modified silica column with methanol-aq. sodium heptane sulfonate ($4.04\,g\,L^{-1}$), disodium EDTA ($1\,mmol\,L^{-1}$), orthophosphoric acid ($10\,mL\,L^{-1}$) and diisopropylamine ($22.5\,mL\,L^{-1}$) $(49 + 51)$ as eluent. Whole blood samples were stabilised immediately after collection by the use of dipotassium EDTA and a high concentration of sodium metabisulfite. Sample preparation was by protein precipitation with perchloric acid. Detection was at a GCE ($+0.55\,V$ *vs* Ag/AgCl). The interassay RSD at $5\,\mu g\,L^{-1}$ was 10.6%.

14.4 Other Anti-Hypertensive Agents

The α-adrenoceptor antagonists indoramin and urapidil are both electroactive. Indoramin (Figure 6.47) has been measured using a GCE ($+0.95\,V$ *vs* Ag/AgCl).[298] Although this potential was below the plateau of the hydrodynamic voltammogram, it was chosen to ensure stability at the LoD of $0.5\,\mu g\,L^{-1}$. Sample preparation was relatively simple: alkalinised plasma was extracted with chlorobutane and the solvent evaporated. In contrast, urapidil and some metabolites were measured after direct injection of plasma using column switching.[299] The working potential was $+1.0\,V$ (*vs* Ag/AgCl) and the LoD was $5\,\mu g\,L^{-1}$ (0.1 mL sample).

Hydralazine (Figure 6.47) and its pyruvate- and acetone-hydrazone metabolites have been measured using HPLC-ED (GCE, $+0.9\,V$ *vs* Ag/AgCl). Spiked urine samples were analysed by direct injection to illustrate the potential application of the method.[300] Minoxidil (Figure 6.47), another vasodilator, has been measured using a PGE ($+0.8\,V$ *vs* Pd). The assay was complicated by the reversion of a metabolite, minoxidil *N*-sulfate, to the parent compound.[301] The LoD was $500\,ng\,L^{-1}$ minoxidil. Isoxsuprine (Figure 6.40) has been measured in equine plasma using an ODS-modified silica column with acetonitrile-methanol-aq. phosphate buffer ($30\,mmol\,L^{-1}$, pH 3.0) $(30 + 18 + 52)$ containing octanesulfonic acid ($1.8\,mmol\,L^{-1}$) as eluent, and ED (GCE, $+0.95\,V$ *vs* Ag/AgCl).[302] Sample preparation was by adsorption of isoxsuprine and the internal standard, nylidrin (Figure 6.40), onto Dowex HCR-S ion-exchange resin (20–50 mesh). The LoD was $5\,\mu g\,L^{-1}$ (0.5 mL sample).

The antihypertensive agent pelanserin (Figure 6.47) has been measured in plasma using HPLC-ED (GCE, $+0.85\,V$ *vs* Ag/AgCl).[303] An ODS-modified silica

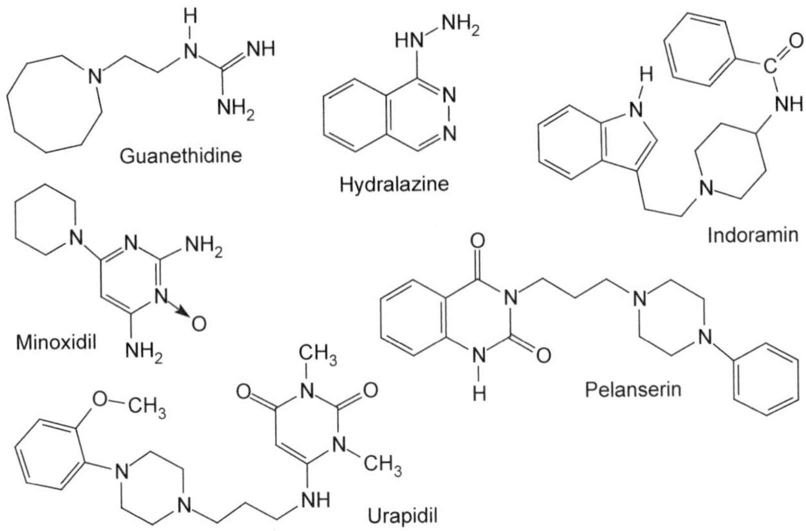

Figure 6.47 *Some antihypertensive agents.*

column was used and the eluent was acetonitrile-aq. acetate buffer (0.1 mol L^{-1}, pH 4.0) (28 + 72) containing EDTA (75 mg L^{-1}). Sample preparation was by LLE into diethyl ether. A pelanserin analogue was the internal standard. A LoD (1 mL sample) of 0.3 µg L^{-1} was claimed. The antihypertensive guanethidine (Figure 6.47) and the diuretic hydrochlorothiazide (Figure 6.50) have been measured simultaneously in pharmaceutical preparations using an ODS-modified silica column and ED (+1.3 V *vs* Ag/AgCl).[304] The eluent was acetonitrile-aq. phosphate buffer (0.05 mol L^{-1}) (30 + 70) containing 20 mmol L^{-1} sodium pentane-sulfonate adjusted to pH 2.5.

The antihypertensive agent L-α-methyldopa (2-methyl-3-(3,4-dihydroxyphenyl)-L-alanine) has been measured in plasma using a Vydac SCX cation-exchange column and ED (+0.54 V *vs* Ag/AgCl).[305] The eluent was aqueous ammonium dihydrogen phosphate (20 mmol L^{-1}, pH 2.55) containing EDTA (0.1 mmol L^{-1}). Sample preparation was by ultrafiltration after treatment with perchloric acid. No internal standard was used. The LoD was 50 µg L^{-1} (1 mL sample). Hoskins and Holliday[306] used an SCX-modified silica column and a GCE (+0.80 V *vs* Ag/AgCl) in the analysis of L-α-methyldopa. The eluent was methanol-water (2 + 8) containing perchloric acid (33.8 mmol L^{-1}), sodium perchlorate (50 mmol L^{-1}) and EDTA (0.2 mmol L^{-1}). 3,4-Dihydroxybenzylamine was the internal standard. Ong *et al.*[307] used a similar approach except that an ODS-modified silica column was used with an eluent that contained citric acid and sodium octanesulfonate.

14.5 Antiarrhythmic Agents

The antiarrhythmic, bisaramil and its metabolite bispidinol (Figure 6.48), have been measured in plasma using an ODS-modified silica column with methanol-aq.

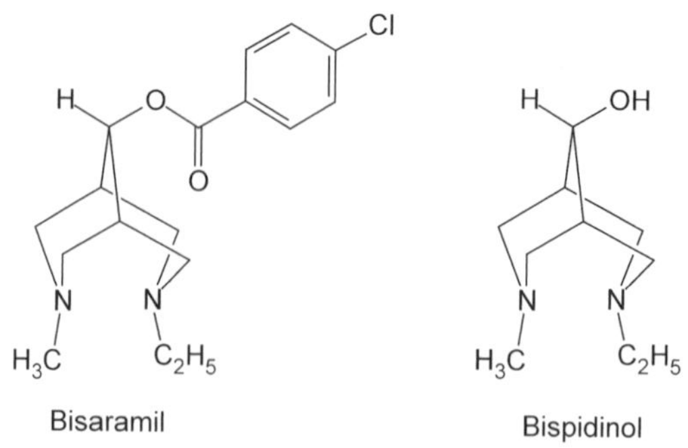

Figure 6.48 *Bisaramil and bispidinol.*

phosphate buffer (0.1 mol L^{-1}, pH 3.2) (35 + 65) (bisaramil) or methanol-aq. phosphate buffer (0.1 mol L^{-1}, pH 3.8) containing sodium heptanesulfonate (2 mmol L^{-1}) (15 + 85) (bispidinol) as eluents, and ED (PGEs, E$_1$: +0.55 V, E$_2$: +0.80 V *vs* Pd).[308] Sample preparation was by LLE into chloroform and involved back-extraction. Analogues of the two compounds were used as internal standards. The LoDs were 0.5 and 1 µg L^{-1} for bisaramil and bispidinol, respectively (0.5 mL sample).

14.6 Angiotensin-Converting Enzyme Inhibitors

The angiotensin-converting enzyme (ACE) inhibitor, captopril (D-2-methyl-3-mercaptopropanoyl-L-proline, Figure 6.49), was derivatised with *N*-(4-dimethyla-minophenyl)maleimide to give a stable product.[309] Detection was at a GCE (+0.9 V *vs* Ag/AgCl). The LoD was 10 µg L^{-1} whole blood (0.5 mL sample). The HPLC-ED of captopril is also discussed in Chapter 5, Section 5.

Lippi *et al.*[310] measured idrapril (Figure 6.49) in urine (human and rat) and tissues (rat) at PGEs (E$_1$: +0.4 V, E$_2$: +0.8 V *vs* Pd) after analysis on an ODS-modified silica column. The eluent for urine and tissue work was acetonitrile-methanol-orthopho-sphoric acid (1% v/v, adjusted to pH 3 with 6 mol L^{-1} aq. sodium hydroxide)

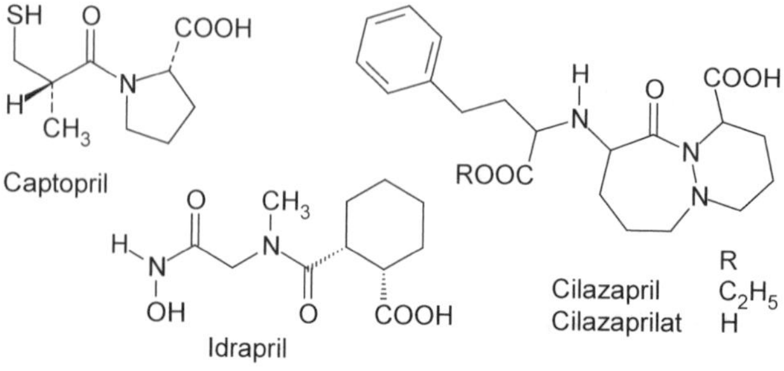

Figure 6.49 *Some angiotensin-converting enzyme inhibitors.*

$(9 + 8 + 83)$ and for plasma was methanol-orthophosphoric acid (1% v/v, pH 3) $(24 + 76)$ for 14 min, after which time the proportion of methanol was increased to elute retained co-extractants. Urine was extracted by SPE (ODS-modified silica cartridges) and plasma and tissue homogenates were deproteinised with acetonitrile. A similar approach was used to measure cilazapril and its active metabolite cilazaprilat (Figure 6.49) in urine.[311] The eluent was methanol-orthophosphoric acid (5 mmol L^{-1}) and detection was at a GCE ($+1.35$ V *vs* Ag/AgCl). Samples (1.5 mL) were extracted by SPE (octyl-modified silica cartridges) giving LoDs of 50 and $40 \, \mu g \, L^{-1}$ for the parent drug and metabolite, respectively.

15 Diuretics

A limited number of diuretics have been assayed by HPLC-ED. Hydrochlorothiazide (Figure 6.50) has been detected at PGEs (E_1: $+0.45$ V; E_2: $+0.63$ V *vs* Pd) after analysis on an ODS-modified silica column with acetonitrile-phosphate buffer $(7.5 \text{ mmol L}^{-1}, \text{pH } 7.3)$ $(1 + 9)$ as eluent.[312] Serum samples (0.2 mL) were prepared by LLE into MTBE. The LLoQ was $5 \, \mu g \, L^{-1}$. HPLC-ED measurements of the loop diuretics frusemide (furosemide), piretanide, bumetanide and torasemide and its carboxylic acid metabolite (Figure 6.50) in urine have been described by Alonso's group (Table 6.4). All the methods used an ODS-modified silica column with acetonitrile-phosphate buffer (5 mmol L^{-1}) as eluent and detection at GCEs.

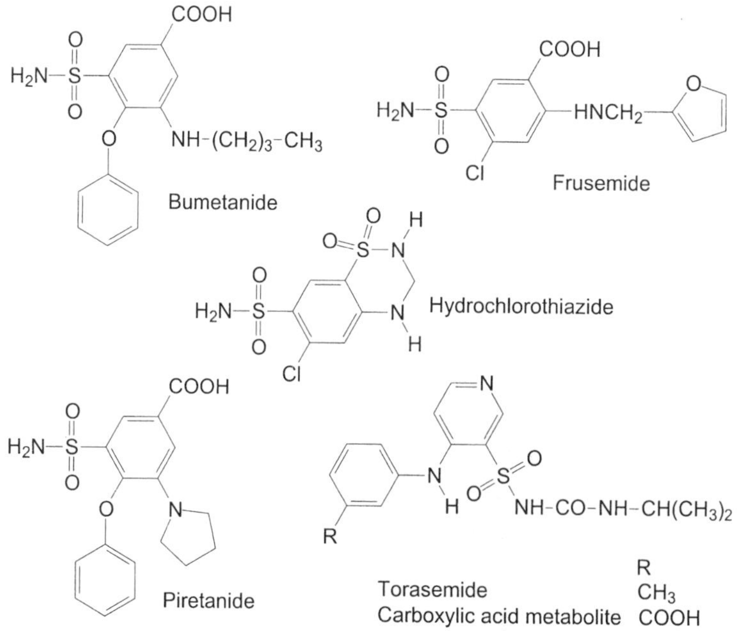

Figure 6.50 *Some diuretic drugs/metabolites.*

Table 6.4 *HPLC-ED measurement of loop diuretics*

Compound	Eluent buffer pH	Acetonitrile:buffer	Potential (V) (vs Ag/AgCl)	Extraction	LLoQ (µg L^{-1})	Reference
Frusemide and piretanide	–	4 + 6	1.2	LLE (ethyl acetate)	both 15	313
Bumetanide	4.0	1 + 1	1.35	LLE (ethyl acetate) SPE – C$_{18}$	0.25	314
Torasemide and carboxylic acid metabolite	3.0	2 + 8	1.3	SPE – C$_2$	8 (metabolite 1)	315

16 Non-Opioid Analgesic/Antipyretic/ Anti-inflammatory Drugs

Paracetamol (acetaminophen, *N*-acetyl-4-aminophenol) was the first drug to be analysed by HPLC-ED (carbon paste electrode, +0.65 V *vs* Ag/AgCl).[316,317] 3-Substituted metabolites, including the 3-cysteinyl and 3-mercapturate conjugates, have also been measured in urine using ED.[318,319] An ODS-modified silica column was used with methanol-glacial acetic acid (0.75% v/v) in aq. potassium dihydrogen phosphate (0.1 mol L^{-1}) (7 + 93) as eluent, and serial UV (248 nm) and EC (CGE, +0.6 V *vs* Ag/AgCl) detection. The sensitivity of the assay was approximately four-fold greater for pure solutions of paracetamol using ED.

A method for plasma paracetamol capable of measuring 10 µg L^{-1} was described by Palmer.[320] The samples were extracted with ethyl acetate before chromatography on an ODS-modified silica column with methanol-aq. sodium acetate (0.1 mol L^{-1}) containing ammonium acetate (1 mmol L^{-1}) (23 + 77) as eluent. The GCE was operated at +0.8 V *vs* Ag/AgCl. A method for measuring paracetamol after therapeutic administration was an incidental finding from a study designed to measure monoamine metabolites in CSF.[321] A method based on the use of PGEs, in the redox mode (E$_1$: +0.25 V, E$_2$: −0.15 V *vs* Pd) has also been described (- Figure 3.12).[322] The high sensitivity given by this approach allowed minimal sample preparation, *viz.* ten-fold dilution of blood or plasma in 5% (w/v) aq. trichloroacetic acid. The LoD (10 µL sample) was 100 µg L^{-1}.

Evans *et al.*[323] reported the HPLC-ED of plasma salicylate (2-hydroxybenzoate). A LoD of 4 ng on column was claimed. The eluent used was methanol-aq. sodium acetate (approximately 60 mmol L^{-1}, pH 6.0) (4 + 46). However, the column used was not stated, a protracted sample purification procedure was needed, no internal standard was used and the applied potential advocated (GCE, +1.35 V *vs* Ag/AgCl) was very high by comparison with other methods involving the ED of phenolic hydroxyl groups.

Nefopam (Figure 6.51) has been measured using an unmodified silica analytical column together with acetonitrile-aq. ammonium nitrate (0.1 mol L^{-1}, pH 9.1) (93 + 7).[324] ED was performed using PGEs (E$_1$: +0.5, E$_2$: +0.8 V *vs* Pd). Sample preparation was by LLE into cyclohexane followed by concentration of the extract

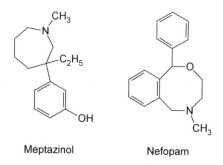

Meptazinol Nefopam

Figure 6.51 *Meptazinol and nefopam.*

on an SPE column (diol-modified silica). *N*-Ethylnornefopam was the internal standard. The LoD was $1 \mu g \, L^{-1}$ (1 mL sample).

Meptazinol (Figure 6.51) has been measured in plasma using a phenylpropyl-modified silica column and ED at a GCE (+1.0 V *vs* Ag/AgCl).[325] The eluent was methanolic ammonium perchlorate (10 mmol L^{-1}, pH 6.7). Sample preparation was by LLE into MTBE. The LoD was $1 \mu g \, L^{-1}$ (0.4 mL specimen, Figure 6.52).

Substituted propionic acid derivatives and related compounds form an important class of non-steroidal anti-inflammatory drugs, but most do not contain readily oxidisable groups. However, pirprofen and diclofenac (Figure 6.53) have been assayed by HPLC-ED. Diclofenac has been measured in plasma using an amino-propyl-modified silica column with acetonitrile-aq. perchloric acid (2.5 mmol L^{-1}) (35 + 65) as eluent after LLE into benzene. ED was at a GCE (+0.9 V *vs* Ag/AgCl).[326] 4-(2,6-Dichloro-4-methoxyphenyl)aminophenylacetic acid was the internal standard. The LoD was $5 \mu g \, L^{-1}$ (2 mL sample). An alternative method for diclofenac in plasma and synovial fluid has been described.[327] LLE into dichloromethane was followed by analysis using an ODS-modified silica column with acetonitrile-aq. sodium acetate (50 mmol L^{-1}, pH 3) (40 + 60) as eluent and ED

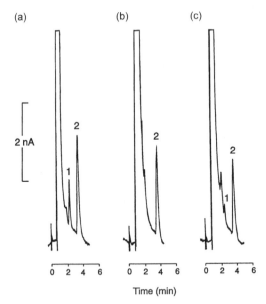

Figure 6.52 *HPLC assay of meptazinol. Column: 125 × 5 (i.d.) mm Spherisorb S5P phenyl-modified silica. Eluent: methanolic ammonium perchlorate (10 mmol L⁻¹, apparent pH 6.7); Flow-rate: 2.0 mL min⁻¹; Detection: V25 grade GCE (+1.0 V vs Ag/AgCl); Injection: 100 mL MTBE extract of (a) Standard solution containing meptazinol (5 mg L⁻¹) in equine plasma, (b) Meptazinol-free human plasma, (c) Plasma sample from volunteer who had ingested meptazinol (meptazinol concentration 1 mg L⁻¹); The initial concentration of the internal standard (imipramine) was 50 mg L⁻¹ in each case; Peaks: 1 = meptazinol, 2 = imipramine.*
(Re-drawn with permission from Flanagan and Ward.[325])

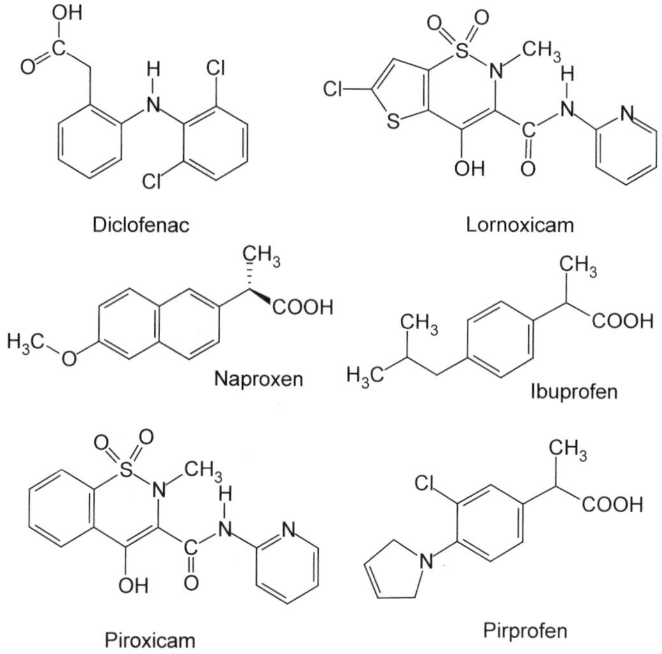

Figure 6.53 *Some nonsteroidal anti-inflammatory drugs.*

(GCE, +0.95 V *vs* Ag/AgCl). 4'-Methoxydiclofenac was the internal standard. As little as $5 \,\mu g\,L^{-1}$ diclofenac could be detected in plasma or synovial fluid (0.5 mL sample) with inter-assay RSDs of <4%. Adoption of SPE (ODS-modified silica columns) extended the method to the measurement of four hydroxylated metabolites in addition to the parent compound.[328] LoDs were $1 \,\mu g\,L^{-1}$ (0.5 mL sample).

Recently, Jin and Zhang[329] described the analysis of diclofenac by CE using a carbon fibre electrode (+0.83 V *vs* SCE) and Pt wire as the auxiliary electrode and also to ground the detector end of the capillary column. A $30.5 \, cm \times 25 \,\mu m$ silica capillary was used with an analysis buffer of disodium hydrogen orthophosphate ($4.9 \, mmol\,L^{-1}$) + sodium dihydrogen orthophosphate ($3.1 \, mmol\,L^{-1}$) (pH 7.0) and a working potential of 10 kV. Samples were introduced by electrokinetic injection (5 kV, 5 s). A LoD of $2.5 \,\mu mol\,L^{-1}$ ($0.74 \, mg\,L^{-1}$) (S/N = 2) was claimed, but precision (RSD 4.7%) was measured at $99 \,\mu mol\,L^{-1}$ ($29 \, mg\,L^{-1}$). Flushing the capillary with double-distilled water, aq. sodium hydroxide ($0.1 \, mol\,L^{-1}$) and double-distilled water before filling with analysis buffer and allowing the electro-osmotic current to stabilise before commencing an analysis gave a reproducibile migration time (RSD 0.8%). The method was applied to 'spiked human urine', but unfortunately diclofenac appeared to co-migrate with an endogenous compound, so that at the lowest dilution tested ($11.7 \, mg\,L^{-1}$) the diclofenac peak was only some 20% larger than the peak due to the endogenous compound.

A similar approach, using LLE into dichloromethane, to that for the HPLC of diclofenac described above was adopted for pirprofen in plasma, synovial fluid and CSF.[330] Not only was pirprofen oxidised at a lower potential (+0.85 V vs Ag/AgCl), but the LoDs were approximately an order of magnitude lower than those for diclofenac. Kuhlmann et al.[331] used a modification of the method described above[328] to measure diclofenac and the local anaesthetic oxybuprocaine (Figure 6.1) in human aqueous humour. Samples (20 μL) were analysed directly using an octyl-modified silica column coated with polyoxyethylene specifically designed for proteinaceous samples. The LoDs were $0.5 \mu g L^{-1}$ and $50 \mu g L^{-1}$ for diclofenac and oxybuprocaine, respectively. Torres-Lopez et al.[332] used a similar approach (GCE, +1.1 V vs Ag/AgCl) to measure diclofenac in a kinetic study in rats. Acidified blood (0.1 mL) was extracted with ethyl acetate. Naproxen (Figure 6.53) was the internal standard. The LoD was $10 \mu g L^{-1}$ (0.1 mL sample).

Lornoxicam (Figure 6.53) and 5'-hydroxylornoxicam have been measured in plasma using an ODS-modified silica column with acetonitrile-methanol-aq. potassium dihydrogen orthophosphate ($50 \, mmol \, L^{-1}$) $(23 + 17 + 60)$ as eluent and ED (PGEs, E_1: +0.40 V, E_2: +0.50 V vs Pd).[333] Piroxicam (Figure 6.53) was the internal standard. The LoDs were 5 and $10 \mu g L^{-1}$ for lornoxicam and 5'-hydroxylornoxicam, respectively (0.1 mL sample). Piroxicam itself has been measured in human plasma and tissues by HPLC-ED.[334] Samples were prepared by LLE (hexane-dichloromethane, $4 + 1$). The analysis employed an ODS-modified silica column with methanol-THF-aq. phosphoric acid ($30 \, mmol \, L^{-1}$) $(32 + 8 + 60)$ containing potassium chloride ($150 \, mg \, L^{-1}$) (final pH adjusted to 2.7) as eluent. Detection was at +0.65 V (details not given, we assume GCE, Ag/AgCl). The LoD was quoted as $0.72 \mu g L^{-1}$ (S/N = 10).

Five ibuprofen (Figure 6.53) metabolites, including three intact glucuronide conjugates, have been detected in human urine ($1 \mu g L^{-1}$, direct injection) using a combination of UV and PAD.[335] The analysis used an ODS-based semi-micro column (250×1.5 mm i.d.) with linear gradient elution from 2–60% acetonitrile (eluent B). Eluent A was aq. potassium dihydrogen orthophosphate ($50 \, mmol \, L^{-1}$, pH 2.45). The UV (210 nm) and PAD (measuring pulse at 0.15 V, cleaning pulses at 0.65 and −0.95 V) were arranged in series, a stream of aq. sodium hydroxide ($0.6 \, mol \, L^{-1}$) being introduced via a T-piece placed between them. From the UV and PAD chromatograms it is difficult to see what advantage is to be gained from using PAD.

Naproxen enantiomers have been resolved using an HPLC stationary phase containing methyl-β-cyclodextrin and amperometric detection (GCE, +1.05 V vs Ag/AgCl).[336] Although the method was developed to measure enantiomeric purity in tablets, it is an interesting example of the combination of chiral separation and ED.

17 Opioid Agonists and Antagonists

The metabolism of diamorphine (heroin, 3,6-diacetylmorphine), morphine and codeine is summarised in Figure 6.54. HPLC is important in the analysis of this

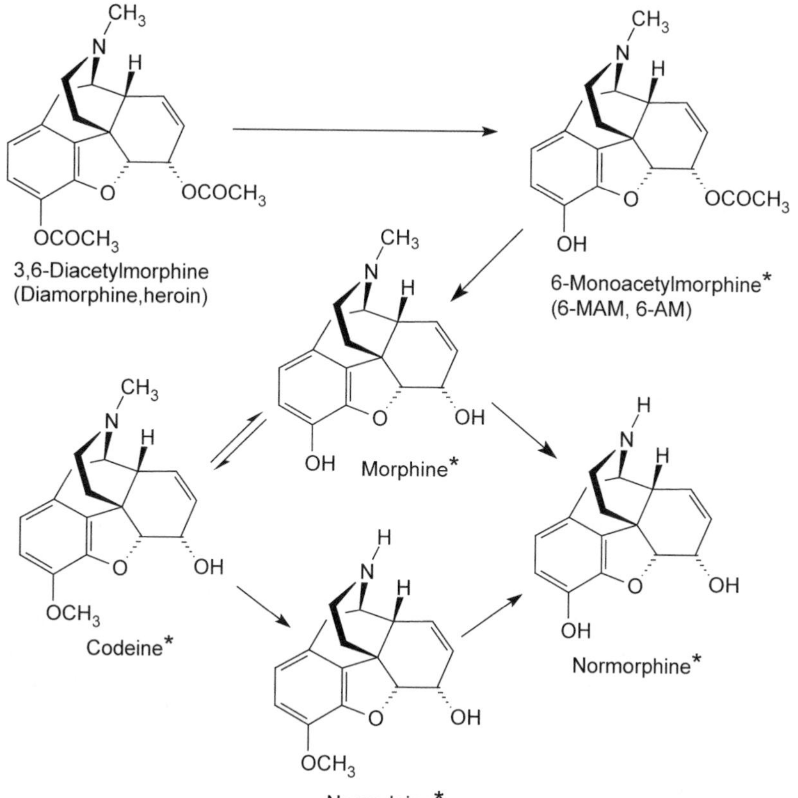

Figure 6.54 *Metabolism of diamorphine (heroin), morphine and codeine (*=metabolised further by conjugation).*

group of compounds since many analytes are hydrophilic, thermally labile or otherwise unsuited to GC.[337] EC oxidation detection has been widely used, especially in the measurement of morphine.[338–344] Generally, detection potentials of +0.8 V *vs* Ag/AgCl have been employed using thin-layer or wall-jet GCEs, whereas potentials between +0.35 and +0.65 V *vs* Pd have been used for the detection of morphine and in some cases morphine-6-glucuronide (M-6-G) and/or hydromorphone (Figure 6.55), using dual PGEs.[345–349] Whelpton[350] has discussed the use of morphine analogues and homologues (nalorphine (Figure 6.58), dextrorphan (the (+)-form of racemorphan, Figure 6.55), *N*-propylnormorphine and *N*-ethylnormorphine) as internal standards in the HPLC-ED of morphine.

Gerostamoulos *et al.*[351] described the use of serial UV (210 nm) and ED (PGEs, E_1: +0.20 V, E_2: +0.55 V *vs* Pd) in the HPLC of 6-monoacetylmorphine (6-MAM), morphine, and codeine in post-mortem urine specimens, whilst Hanisch and Meyer[352] used a GCE (+0.85 V *vs* Ag/AgCl) to measure 6-MAM and morphine in urine after SPE with C_{18}-modified silica cartridges. Rop *et al.*[353] measured 6-MAM and morphine in urine, plasma and whole blood. Samples were

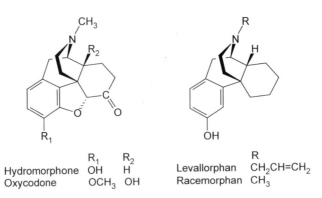

	R₁	R₂		R
Hydromorphone	OH	H	Levallorphan	CH₂CH=CH₂
Oxycodone	OCH₃	OH	Racemorphan	CH₃

Figure 6.55 *Hydromorphone, oxycodone, levallorphan and racemorphan.*

prepared by SPE on C₁₈ cartridges and detection was at a GCE (+0.6 V *vs* Ag/AgCl). Similarly, Gerostamoulos and Drummer[354] used SPE (C₁₈ cartridges) to extract morphine, normorphine, morphine-3-glucuronide (M-3-G) and M-6-G from postmortem blood after disruption of erythrocytes by a freeze-thaw cycle. M-3-G was measured by UV (210) while the other compounds were detected electrochemically (PGEs, E_1: +0.10 V, E_2: +0.6 V *vs* Pd; LoD 10 µg L⁻¹). Rashid *et al.*[355] developed a highly specific solid phase immunoextraction column that showed minimal cross-reactivity to metabolites of morphine and codeine. Morphine detection was at a PGE (+0.45 V *vs* Pd). The RSD at 50 µg L⁻¹ (0.1 mL urine) was 4.7%.

Heybroek *et al.*[356] used PGEs (E_1: +0.4 V, E_2: +0.9 V *vs* Pd) in the analysis of codeine and morphine in plasma and gastric juice. Sample preparation was by on-line extraction (ethyl-modified silica column). Codeine-6-glucuronide was measured as codeine after off-line enzymic hydrolysis. Svensson *et al.*[357] measured morphine, normorphine and M-6-G as metabolites of codeine using PGEs (E_1: +0.25 V, E_2: +0.35 V *vs* Pd) but, because of the low potentials used, the 3-methyl

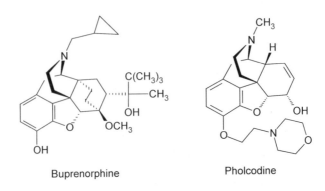

Buprenorphine   Pholcodine

Figure 6.56 *Buprenorphine and pholcodine.*

substituted compounds (codeine, norcodeine, their 6-glucuronides and M-3-G) were quantified by UV (214 nm).

Johansen *et al.*[358] have described the use of serial fluorescence and ED (GCE, +0.35 V *vs* Ag/AgCl) in the analysis of the antitussive pholcodine (3-*O*-morpholinoethylmorphine, Figure 6.56) and morphine, respectively, in urine. Oxycodone (Figure 6.55) has been measured by a similar procedure except that a potential of +1.1 V (*vs* Ag/AgCl) was used and codeine was employed as the internal standard.[359] The LoD was $10 \mu g L^{-1}$ (1 mL sample). In a later paper, the same group used mixed mode SPE (octyl/benzoic acid-modified silica) to extract plasma oxycodone.[360] The LLoQ (0.5 mL sample) was claimed to be $4.5 pg L^{-1}$ $(14.2 nmol L^{-1})$, although it should be noted that the inter-assay RSD at this concentration was 33%.

Buprenorphine (Figure 6.56) and its major metabolite *N*-desalkylbuprenorphine have been measured in urine using a cyanopropyl-modified silica column together with acetonitrile-aq. sodium phosphate $(20 mmol L^{-1}$, pH 3.0) (13 + 87) as eluent.[361] ED was at a GCE (+0.75 V *vs* Ag/AgCl). Sample preparation involved LLE with back-extraction into aqueous acid followed by extraction into toluene under alkaline conditions. The *N*-ethyl analogue of buprenorphine was the internal standard. The LoDs claimed were 250 and $150 ng L^{-1}$ for buprenorphine and *N*-desalkylbuprenorphine, respectively. Schleyer *et al.*[362] have also described a method for these latter compounds using HPLC-ED (PGEs, E_1: +0.16 V, E_2: +0.48 V *vs* Pd). Sample preparation involved SPE (ODS-modified silica column) and column-switching. No internal standard was used. The LoD was $40 ng L^{-1}$ (5 mL sample).

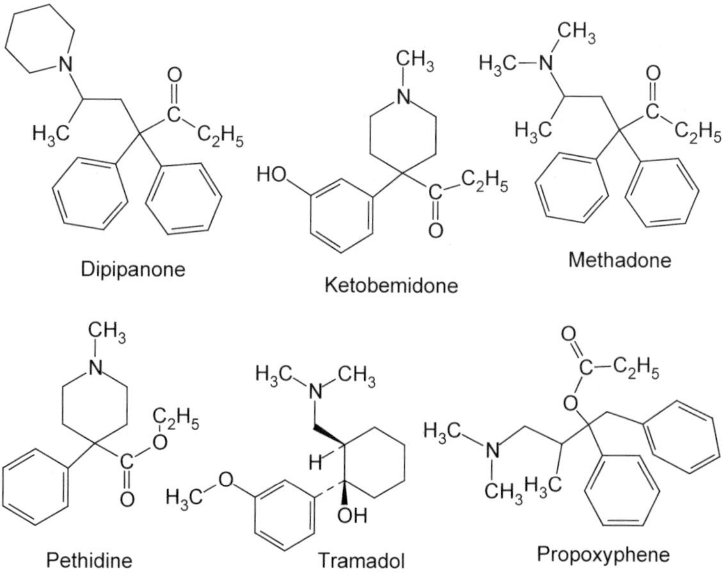

Figure 6.57 *Some synthetic opioid analgesics.*

Many other synthetic compounds (Figure 6.57), such as dextropropoxyphene (D-propoxyphene), dipipanone, ketobemidone methadone and pethidine (meperidine), are potent narcotic analgesics (opioid agonists) and their use is controlled in most countries. Dextropropoxyphene is often formulated together with paracetamol and is extensively metabolised by *N*-demethylation and by other routes. Methadone is metabolised largely by *N*-demethylation and hydroxylation. Dextropropoxyphene and norpropoxyphene have been measured by HPLC-ED (GCE, $+1.2$ V *vs* Ag/AgCl).[363] HPLC was on an unmodified silica column with methanolic ammonium perchlorate (10 mmol L^{-1}, pH 6.7) as eluent. Sample preparation was by LLE into MTBE. The extracts were injected without further treatment. Viloxazine was the internal standard. The LoDs were 20 and 10 μg L^{-1} for norpropoxyphene and dextropropoxyphene, respectively (0.4 mL sample). Methadone and some methadone metabolites may also be measured using this same system. Levorphanol has been measured using an ODS-modified silica column and ED (GCE, $+1.00$ V *vs.* Ag/AgCl).[364] The eluent was acetonitrile-aq. sodium chloride (10 mmol L^{-1}) (30 + 70) containing EDTA (0.1 mmol L^{-1}). Levallorphan (Figure 6.55) was the internal standard. The LoD was approximately 1 μg L^{-1} (1 mL sample). Ketobemidone (Figure 6.57) has been measured using similar methodology.[365] 3-(3-Hydroxyphenyl)-*N*-propylpiperidine was the internal standard.

Tramadol (Figure 6.57) and its 3-*O*-desmethyl metabolite have been measured in rat plasma.[366] Drug, metabolite and internal standard (the *O*-ethyl homologue of tramadol) were extracted into ethyl acetate-hexane (4 + 6). Detection was at dual PGEs, the desmethyltramadol being detected at the first (E$_1$: $+0.61$ V *vs* Pd) and tramadol and the internal standard at the second (E$_2$: $+0.85$ V *vs* Pd). The LoDs (S/N = 3) were 4 μg L^{-1} and 3 μg L^{-1}, for drug and metabolite, respectively (0.1 mL sample).

Nalmefene, nalorphine, naloxone and naltrexone (Figure 6.58) are potent opioid antagonists. Naloxone especially is widely used in therapy. In general, analytical methods similar to those used for narcotic analgesics may be employed.[367] Reid *et al.*[368] measured naloxone in plasma using an ODS-modified silica analytical column with acetonitrile-aq. monochloroacetic acid (0.1 mol L^{-1}) containing octyl sodium sulfate (2.59 mmol L^{-1}) and EDTA (2.39 mmol L^{-1}) (final pH 4.5) (15 + 85) as eluent, and ED (GCE, $+0.89$ V *vs* Ag/AgCl). Sample preparation

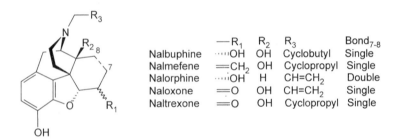

	—R$_1$	R$_2$	R$_3$	Bond$_{7-8}$
Nalbuphine	‥‥OH	OH	Cyclobutyl	Single
Nalmefene	=CH$_2$	OH	Cyclopropyl	Single
Nalorphine	‥‥OH	H	CH=CH$_2$	Double
Naloxone	=O	OH	CH=CH$_2$	Single
Naltrexone	=O	OH	Cyclopropyl	Single

Figure 6.58   *Some opioid partial agonists/antagonists.*

was by SPE (ODS-modified silica column) and naltrexone was the internal standard. The LoD was approximately $20\,\mu g\,L^{-1}$ (0.25 mL sample). A noticeable loss of sensitivity was noted after 25–30 sample injections.

Nalmefene has been measured in plasma using a phenyl-modified silica analytical column with acetonitrile-water $(30 + 70)$ containing sodium pentane-sulfonate $(5\,mmol\,L^{-1})$ and orthophosphoric acid (0.0045% v/v) (final pH 3.1) as eluent, and ED (PGEs, E_1: $+0.23$ V, E_2 $+0.35$ V *vs* Pd).[369] Sample preparation was *via* SPE (cyanopropyl-modified silica column) and nalbuphine was the internal standard. The LoD was $1\,\mu g\,L^{-1}$ (0.7 mL sample).

The opioid agonist/antagonist nalbuphine (Figure 6.58) has been measured using an ODS-modified silica column and ED ($+0.75$ V *vs* Ag/AgCl).[370] The eluent was methanol-aq. potassium phosphate $(10\,mmol\,L^{-1})$ $(45 + 55)$. Sample preparation was by LLE into ethyl acetate-(2-propanol) $(9 + 1)$ after precipitation of plasma protein with perchloric acid. Naloxone (Figure 6.58) was the internal standard. Coulometric detection of nalbuphine was described by Wetzelsberger *et al.*[371] Plasma (3 mL) + naltrexone (internal standard) were extracted with hexane-(3-methyl-2-butanol) $(9 + 1)$. The eluent was prepared from hydrated citric acid (9.5 g), potassium dihydrogen orthophosphate (7.0 g) disodium EDTA (0.08 g), and methanol (680 mL) made up to 2 L with purified water. The HPLC column was ODS-modified silica. Analytes were detected at PGEs in the screen mode (E_1: $+0.03$ V, E_2 $+0.45$ V *vs* Pd). The LoD was claimed to be $<50\,ng\,L^{-1}$, although the RSD at $163\,ng\,L^{-1}$ was 23.2%. Nicolle *et al.*[372] and de Cazanove *et al.*[373] based their HPLC conditions for nalbuphine on the above. Nicolle *et al.*[372] used SPE to extract the drug and internal standard (6-MAM) from plasma (0.5 mL). The LLoQ was $100\,ng\,L^{-1}$ (RSD = 12.4%). de Cazanove *et al.*[373] used LLE into chloroform-(2-propanol) $(98 + 2)$ and achieved a LLoQ of $300\,ng\,L^{-1}$ (RSD = 16.7%) for a 0.5 mL plasma sample.

The quaternary ammonium opioid antagonist, *N*-methylnaltrexone, has been measured in plasma and urine using an ODS-modified silica analytical column $(150 \times 3.2\,mm$ i.d.) and methanol-aq. potassium acetate $(50\,mmol\,L^{-1})$ $(11 + 89)$, (apparent pH 4.5) as eluent.[374] The eluent flow-rate was $0.5\,mL\,min^{-1}$. Detection was at PGEs in the screen mode (E_1: $+0.35$ V, E_2 $+0.70$ V *vs* Pd). Sample preparation was by SPE (weak cation exchanger). Naltrexone (Figure 6.58) was the internal standard. The LoD and LLoQ were 2.0 and $6.0\,\mu g\,L^{-1}$, respectively (RSD at LLoQ 8.0%, accuracy 10.0% (n = 10)). The recovery from plasma was approximately 94%.

18 Miscellaneous Pharmaceuticals

Allopurinol (Figure 6.59), a uricosuric agent, its pharmacologically active metabolite oxypurinol, and uric acid have been measured by HPLC-ED using both dropping mercury ($+0.24$ V *vs* Ag/AgCl) and GCEs ($+1.2$ V *vs* SCE).[375] The analyses were performed using an octyl-modified silica column with methanol-aq. phosphate buffer $(25\,mmol\,L^{-1}$, pH 6) (*ca* $7 + 93$) as eluent. No internal standard was used. The LoD was $0.2\,mg\,L^{-1}$ for both allopurinol and oxypurinol. More

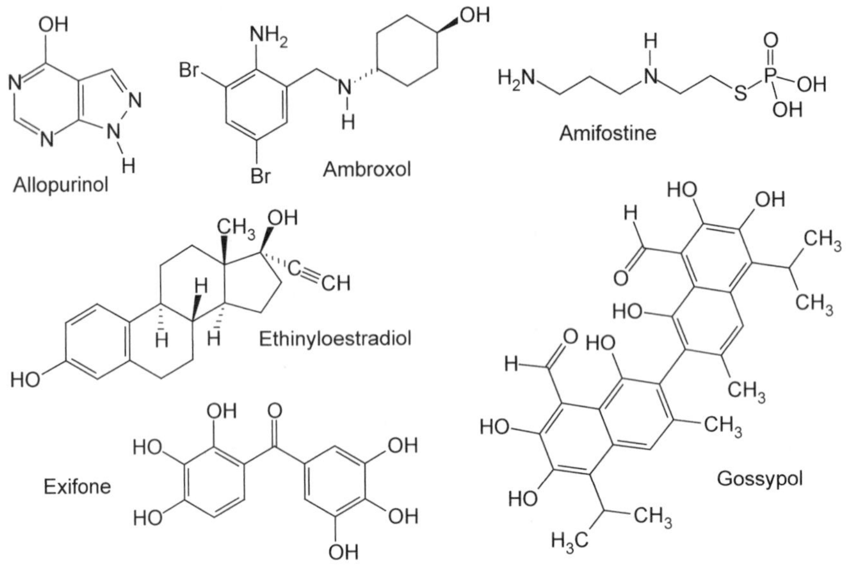

Figure 6.59   *Structural formulae of allopurinol, ambroxol, amifostine, ethinyloestradiol, exifone and gossypol.*

recently, Eisenberg *et al.*[376] used SPE together with HPLC-ED (GCE, +1.07 V *vs* Ag/AgCl) in the analysis of allopurinol and oxypurinol in small (75 μL) volumes of rat plasma. 1,7-Dimethyluric acid was the internal standard. Eisenberg *et al.*[377] have also described the analysis of allopurinol by ED *via* post-column reaction with immobilised xanthine oxidase to give oxypurinol, thereby facilitating ED (+0.85 V *vs* Ag/AgCl). 1-Methylxanthine was the internal standard. The LoD for allopurinol (10 pg on-column) was ten-fold lower using post-column enzyme oxidation than when using direct ED.

The expectorant ambroxol (Figure 6.59) has been measured in plasma using an ODS-modified silica cloumn with methanol-aq. phosphate buffer (0.1 mol L^{-1}, pH 6.5) (65 + 35) as eluent and ED (GCE, +0.85 V *vs* Ag/AgCl).[378] Sample preparation was by LLE into diethyl ether and pelanserin (Figure 6.47) was the internal standard. The LoD was 1 μg L^{-1} (1 mL sample).

The radio- and cyto-protective agent amifostine (*S*-(2-(3-Aminopropylamino)-ethyl)phosphorothioate, Figure 6.59) and its active metabolite (2-(3-dimethylpro-pylamino)ethanediol, WR1065) have been measured in human blood and plasma using HPLC-ED (PGEs: E$_1$: +0.2 V, E$_2$: +0.6 V *vs* Pd).[379] The parent compound was converted to WR1065 and quantified as such with a LLoQ of 50 μg L^{-1} (0.25 μmol L^{-1}) (S/N = 3, 1 mL sample). The analysis was performed using an ODS-modified silica column with methanol-aqueous phase (chloroacetic acid (0.14 mol L^{-1}) + OSA (2.2 mmol L^{-1}) + triethylamine (7.5 mmol L^{-1}) (pH 3.0)) (37 + 63) as eluent. The method was said to minimise WR1065 loss during sample preparation and allowed for the rapid analysis of both compounds on one system.

The 4-hydroxy metabolite of amphetamine (1-(4-hydroxyphenyl)-2-aminopropane) has been measured in brain homogenates by HPLC-ED (GCE, $+0.75$ V *vs* Ag/AgCl).[380] An ODS-modified silica column was used with methanol-THF-aq. citrate buffer (0.1 mol L^{-1}, pH 4.5) (10 + 1 + 89) as eluent. Isoprenaline (Figure 6.2) was the internal standard. The possible phenolic metabolites of MDMA have been studied using a similar system.[381] Amphetamine and one its metabolites, 4-hydroxynorephedrine, have been measured after derivatisation with 2,5-dihydrobenzaldehyde in the presence of BH$_4$ exchange resin.[382] The analytes were measured using an ODS-modified silica column with methanol-aq. phosphate buffer (50 mmol L^{-1}, pH 6.5) (3 + 7) plus triethylamine (0.5% v/v), as eluent. Detection was at a PGE ($+0.6$ V *vs* Pd). Phenylethylamine was used as the internal standard and the LoD was 50 µg L^{-1} (S/N = 3).

Surprisingly Michel *et al.*[383] reported the HPLC of MDMA, MDEA (internal standard) and MDA using an unmodified silica column with methanol-aq. sodium acetate (0.1 mol L^{-1}, pH 4.25) (3 + 10) as eluent and ED (dual GCEs, $+1.24$ V *vs* Ag/AgCl). MDA (primary aliphatic amine) was said to show a greater EC response than MDMA or MDEA (secondary aliphatic amines) and LoDs of 2.7 and 1.6 µg L^{-1} for MDA and MDMA, respectively (0.2 mL samples), were claimed. Given the nature of the electroactive moieties, the high proportion of water in the eluent and the relatively low eluent pH used, all available evidence suggests that primary aliphatic amines should show no EC response whatsoever under these conditions (see Table 4.1)

The γ-aminobutyrate (GABA) mimetic baclofen (4-amino-3-*p*-chlorophenylbutyric acid) has been measured in human plasma after derivatization with OPA-*t*-butanethiol.[384] Baclofen was extracted from plasma (1 mL) using SAX SPE columns. A sample of eluate (borate buffer, pH 10.4, 200 µL) was transferred to an autosampler and derivatised prior to injection. HPLC was performed using an ODS-modified silica column (4 µm; 150×3.9 mm i.d.) with methanol-phosphate buffer (60 mmol L^{-1}, pH 7) (74 + 36) as eluent. Detection was said to be at a 'glassy carbon working cell (Model 5011)' in the screen mode (E$_1$: $+0.2$ V, E$_2$: $+0.7$ V *vs* Pd). Note: the Model 5011 cell has dual PGEs. Intra- and inter-assay RSDs of 6.8 and 7.6%, respectively, were reported at 10 µg L^{-1} (LLoQ). Baclofen has been detected at a GCE ($+0.8$ V *vs* Ag/AgCl) after SPE (propylsulfonic acid, PRS), derivatisation with OPA-sulfite and HPLC on an ODS-modified analytical column using methanol-acetonitrile-phosphate buffer (100 mmol L^{-1}, pH 6.8) (35 + 5 + 65) as eluent.[385] A LLoQ of 5 ng L^{-1} was claimed (0.1 mL plasma sample) and the intra- and inter-batch RSDs were <10% at 10, 50 and 100 ng L^{-1}. Bromobaclofen (BA37061, Novartis) was used as the internal standard.

Ethinyloestradiol (ethynyl estradiol, Figure 6.59) has been measured in rabbit plasma using an ODS-modified silica column with acetonitrile-methanol-aq. phosphate buffer (50 mmol L^{-1}, pH 3.6) (7 + 3 + 10) as eluent and ED (GCE, $+1.0$ V *vs* Ag/AgCl).[386] Sample preparation was by LLE. No internal standard was used. The LoD was 50 ng L^{-1} (1 mL sample). Oestriol, oestradiol and oestrone did not interfere.

The free-radical scavenger exifone (2,3,3′,4,4′,5′-hexahydroxybenzophenone, Figure 6.59) has been measured using an ODS-modified silica column with ED

(PGEs, E_2: $+0.3$ V *vs* Pd).[387] E_1 and the guard cell were not used. The eluent was acetonitrile-aq. orthophosphoric acid $(0.3 \text{ mol L}^{-1}$, pH 2.2) $(15 + 85)$. Sample preparation was by LLE into diethyl ether. An internal standard was not used. The LoD was $0.3 \,\mu g \, L^{-1}$ (1 mL sample).

The polyphenolic binaphthaldehyde gossypol (Figure 6.59), which has been used as a male antifertility agent, has been measured in plasma by HPLC with column switching and ED (GCE, $+0.6$ V *vs* Ag/AgCl).[388] Reduced glutathione was added to stabilise the analyte. The LoD was $5 \,\mu g \, L^{-1}$ (1 mL sample). Gossypol and gossypolone enantiomers have been measured in fish tissues using UV and ED methods[389] (Chapter 6, Section 2).

Mesna and its prodrug, dimesna (BNP7787) (Figure 6.60) have been measured in human plasma and urine.[390] EDTA plasma was mixed with sulfuric acid $(0.33 \text{ mol L}^{-1})$ and sodium hexametaphosphate (50 g L^{-1}) $(1 + 1 + 1)$ to precipitate the protein. Mesna was analysed on an ODS-modified silica column with aq. trisodium citrate (0.1 mol L^{-1}), tetrabutyl ammonium dihydrogen phosphate $(1.0 \text{ mmol L}^{-1})$ and cysteamine $(0.1 \,\mu\text{mol L}^{-1})$, adjusted to pH 3.5 with 85% (w/v) orthophosphoric acid as eluent. Detection was at a gold electrode $(+1.0$ V *vs* Ag/ AgCl). Dimesna was quantified as mesna after pre-column reduction with sodium borohydride. It was claimed that the contribution from mixed disulfides (arising from the reaction of mesna with endogenous thiols) was negligible compared with the concentration of dimesna. The LLoQs (0.1 mL sample) for mesna and dimesna were 0.63 and 1.6, and 1.6 and 0.8 $\mu\text{mol L}^{-1}$ (90 and 450, and 225 and 225 $\mu g \, L^{-1}$) in plasma and in urine, respectively. The same group adopted a similar approach to measure mesna and total mesna (dimesna plus mixed disulfides) in tissues.[391]

The antiemetic metoclopramide (Figure 6.61) has been analysed using an ODS-modified silica column with methanol-phosphate buffer (pH 3) containing sodium chloride (1 mmol L^{-1}) $(3 + 7)$ as eluent.[392] The buffer was prepared by mixing aq. sodium dihydrogen orthophosphate (0.2 mol L^{-1}) and orthophosphoric acid (0.1 mol L^{-1}) $(85 + 15)$ and adjusting to pH 3 with orthophosphoric acid. Detection was a GCE $(+1.1$ V *vs* Ag/AgCl). After addition of the internal standard (methyl 4-hydroxybenzoate), serum (1 mL) was treated with acetonitrile and the supernatant evaporated to dryness. The alkalinised residue was purified by LLE into diethyl ether. The LoD was $0.8 \,\mu g \, L^{-1}$ (S/N = 3) and the LLoQ $2 \,\mu g \, L^{-1}$ (RSD = 14.8%). The inter-assay RSD at $10 \,\mu g \, L^{-1}$ was 6%.

HPLC-ED analysis of nicotine (Figure 6.61) in dog plasma (0.1 mL) after protein precipitation with acetonitrile has been described.[393] An ODS-modified silica column was used with (acetonitrile-methanol, $3 + 1$)-aq. sodium dihydrogen orthophosphate (2 mmol L^{-1}) containing sodium octyl sulfate $(0.25 \text{ mmol L}^{-1})$ $(1 + 19)$ (final pH 3). Detection was at PGEs (E_1: $+0.45$ V, E_2: $+0.75$ V *vs* Pd). A LoD of 20 pg L^{-1} was

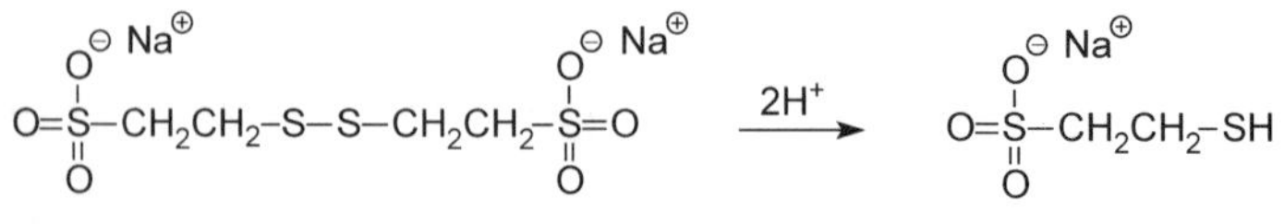

Figure 6.60 *Reduction of dimesna (BNP7787) to mesna with sodium borohydride.*

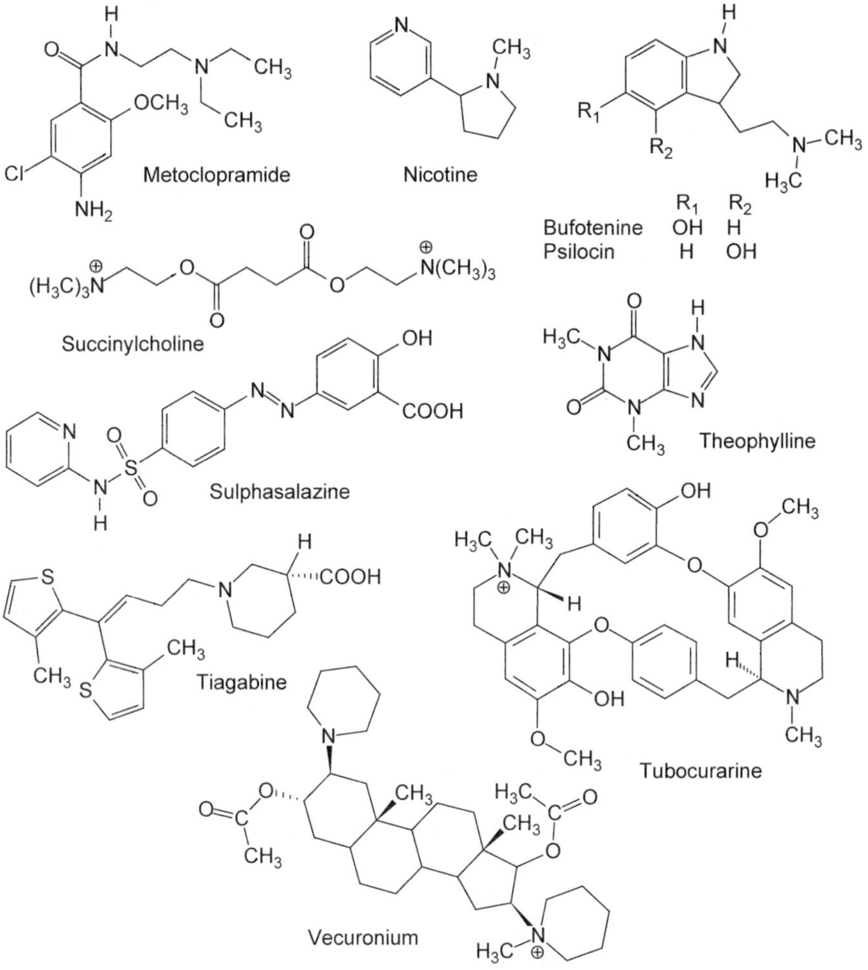

Figure 6.61 *Structural formulae of metoclopramide, nicotine, psilocin and internal standard, succinylcholine, sulfasalazine, theophylline, tiagabine, tubocurarine and vecuronium.*

claimed. HPLC-ED of nicotine in hair has also been described.[394] An ODS-modified silica column was used with acetonitrile-methanol-aq. potassium dihydrogen orthophosphate ($50 \, mmol \, L^{-1}$) ($4 + 2.5 + 93.5$) (final pH 4.8), as eluent. Three-electrode post-column detection (ESA 5021 conditioning cell + ESA 5011 analytical cell) was used (PGEs, conditioning cell: $+0.6 \, V$, E_1: $+0.6 \, V$, E_2: $+0.9 \, V$ *vs* Pd). The potential used for E_2 was chosen to ensure that the internal standard, 2-phenylimidazole, was oxidised. Hair samples were washed with dichloromethane prior to overnight digestion with aq. sodium hydroxide ($1 \, mol \, L^{-1}$) followed by LLE with diethyl ether. An LoD of $50 \, pg \, mg^{-1}$ ($2 \, mg$ sample) was claimed, however the inter-assay RSD at $320 \, pg \, mg^{-1}$ was 19.8% (n = 44).

The hallucinogen psilocybin is hydrolysed *in vivo* to an active metabolite psilocin (4-hydroxy-*N,N*-dimethyltyramine, Figure 6.61) to reveal an electroactive phenol. Several groups have therefore employed HPLC-ED to assay psilocin and other psilocybin metabolites. Kysilka[395] used an ODS-modified silica column with ethanol-citrate/phosphate buffer (pH 2.8) $(1 + 9)$ as eluent to measure this compound in rat urine without sample pretreatment. Detection was at a GCE $(+0.65\ V\ vs\ Ag/AgCl)$ giving a LoD of $0.29\ mg\ L^{-1}$ (10 µL, direct injection). No internal standard was used. LLE (dichloromethane) and on-line SPE (CBA ion-exchange) methods have also been used to measure psilocin in human plasma.[396] An octyl-modified, base deactivated silica column was used in both cases, but each extraction method required different HPLC eluents and detection conditions. The LLE method used acetonitrile-aqueous buffer (sodium acetate $(0.1\ mol\ L^{-1})$, citric acid $(0.1\ mol\ L^{-1})$, disodium EDTA $(0.3\ mmol\ L^{-1})$, pH 4.1) $(17 + 83)$ and detection at $+0.650\ V$ (presumably at a GCE with Ag/AgCl reference). For SPE, the eluent was acetonitrile-aq. phosphate buffer $(150\ mmol\ L^{-1}$, pH 2.3) $(55 + 945)$ containing disodium EDTA (160 µmol) with detection at $+ 0.675\ V$. 5-Hydroxyindole (LLE) or bufotenine (Figure 6.61) (SPE) were used as internal standards. The LLoQ for the SPE method was $10\ µg\ L^{-1}$ (0.4 mL sample) whereas a 2 mL sample was required to achieve the same sensitivity using LLE. Hasler *et al.*[397] investigated the renal excretion profiles of psilocin and psilocin-*O*-glucuronide using an HPLC method based on an earlier plasma psilocin assay.[398] The analysis was performed on an octyl-modified silica column with methanol-ammonium acetate buffer $(0.3\ mol\ L^{-1}$, pH 8.3) $(54 + 46)$ as eluent. Detection was at $+0.15\ V$ *vs* Pd. Urine (6 mL) was freeze-dried and the residue was taken up in methanol (1 mL) and filtered methanol (10 µL portions) injected. The LLoQ was $10\ mg\ L^{-1}$ (RSD = 7.9%, n = 3).

The depolarising neuromuscular blocker succinylcholine (suxamethonium, Figure 6.61), has been quantified by HPLC-ED.[399] A cyanopropyl-modified silica analytical column was used with acetonitrile-methanol-aq. orthophosphoric acid $(30\ mmol\ L^{-1})$ $(35 + 25 + 45)$ (apparent pH 5 (adjusted with ammonium hydroxide)), as eluent. Detection was at PGEs (E_1: $+0.45\ V$, E_2: $+0.75\ V$ *vs* Pd). Plasma was treated with echothiophate to inactive plasma cholinesterases and extracted on methyl-modified silica SPE columns. Pipecurium (an analogue of vecuronium, Figure 6.61) was used as the internal standard. The LLoQ was $250\ µg\ L^{-1}$ (S/N = 3). Although the inter-assay RSD at the LLoQ was 1.2% (n = 5) it was 6.8% at $500\ µg\ L^{-1}$, and 8.3% at $1000\ µg\ L^{-1}$.

The eight principal metabolites of sulfasalazine (salicylazosulfapyridine, Figure 6.61) have been measured in plasma by direct injection HPLC after protein precipitation with methanol.[400] The analytical column was unmodified silica and the eluent was methanol-aq. potassium phosphate $(0.2\ mol\ L^{-1}$, pH 6.5)-water $(45 + 10 + 45)$ containing cetyltrimethylammonium bromide $(3.75\ mmol\ L^{-1})$. Detection was at a GCE $(+0.6\ V\ vs\ Ag/AgCl)$.

The GABA uptake inhibitor tiagabine (Figure 6.61) has been measured in plasma using an ODS-modified silica analytical column with acetonitrile-methanol-water $(37 + 10 + 53)$ containing phosphoric acid and sodium phosphate (both $10\ mmol\ L^{-1}$) and sodium octanesulfonate $(5\ mmol\ L^{-1})$ as eluent, and ED (PGEs,

E_1: +0.50 V, E_2: +0.76 V *vs* Pd).[401] Sample preparation was by SPE (octyl-modified silica columns). The monomethyl analogue of tiagabine was the internal standard. The LoD was $2\,\mu g\,L^{-1}$ (1 mL sample).

The non-depolarising neuromuscular blocking agent vecuronium (Figure 6.61) has been measured using dual GC electrodes in series (+0.65 V and +1.05 V *vs* Ag/AgCl), measurement being performed at the second electrode.[402] Blood (10 mL) was treated with 3% (w/v) PCA to precipitate proteins and extracted with dichloromethane. Presumably this was ion-pair extraction using perchlorate (and possibly endogenous anions) as the counter ion(s). A cyanopropyl modified silica column was used with acetonitrile-aq. phosphate buffer ($100\,mmol\,L^{-1}$, pH 5) (1 + 1) as eluent. Tubocurarine (Figure 6.61) was used as the internal standard and a LoD of $3\,\mu g\,L^{-1}$ was claimed (10 mL sample) when the method was applied to a pilot pharmacokinetic study in man. A similar approach was adopted by Ducharme *et al.*[403] for vecuronium and its 3-α-hydroxy, 17-β-hydroxy, and 3-α,17-β-dihydroxy metabolites: a cyanopropyl-modified silica column was used with acetonitrile-aq. phosphate buffer ($33\,mmol\,L^{-1}$, pH 6.55) (40 + 60) as eluent. Detection was at PGEs (E_1: +0.4 V, E_2: +0.8 V *vs* Pd). Sample preparation was by SPE (methyl-modified silica column). No internal standard was used. LoDs were $4\,mg\,L^{-1}$ for vecuronium and between 8 and $32\,\mu g\,L^{-1}$ for the metabolites (1 mL samples).

Although xanthines such as theophylline (Figure 6.61) are electroactive,[404,405] the relatively high potentials (for example +1.2 V *vs* Ag/AgCl) that are required means that ED offers little advantage over UV detection, particularly if the eluent is chosen to ensure resolution of the caffeine metabolite 1,7-dimethylxanthine.

References

1 M.K. Halbert and R.P. Baldwin, Determination of lidocaine and active metabolites in blood serum by liquid chromatography with electrochemical detection, *J. Chromatogr.*, 1984, **306**, 269–277.

2 G.T. Tucker and M.S. Lennard, Analysis of local anaesthetics in A.S. Curry (ed), *Analytical Methods in Human Toxicology*, Part 1, Macmillan, London, 1984, 159–192.

3 I. Jane, A. McKinnon and R.J. Flanagan, High-performance liquid chromatographic analysis of basic drugs on silica columns using non-aqueous ionic eluents – II. Application of UV, fluorescence and electrochemical oxidation detection, *J. Chromatogr.*, 1985, **323**, 191–225.

4 S.C. Monkman, R. Armstrong, R.J. Flanagan, D.W. Holt and S. Rosevear, High performance liquid chromatographic measurement of lignocaine in tissue samples following transabdominal placental biopsy, *Biomed. Chromatogr.*, 1989, **3**, 88–91.

5 R. Whelpton, P. Dudson, H. Cannell and K. Webster, Determination of prilocaine in human plasma samples using high-performance liquid chromatography with dual-electrode electrochemical detection, *J. Chromatogr.*, 1990, **526**, 215–222.

6 G. Mazzi and M. Schinella, Simple and practical high-performance liquid chromatographic assay of propofol in human blood by phenyl column chromatography with electrochemical detection, *J. Chromatogr.*, 1990, **528**, 537–541.

7 R.E. Uebel, C.A. Wium, A.O. Hawtrey and J. Coetzee, Electrochemical determination of 2,6-diisopropylphenol after high-performance liquid chromatography of extracts from serum, *J. Chromatogr.*, 1990, **526**, 293–295.

8 R.H. Dowrie, W.F. Ebling, J.W. Mandema and D.R. Stanski, High-performance liquid chromatographic assay of propofol in human and rat plasma and fourteen rat tissues using electrochemical detection, *J. Chromatogr. B*, 1996, **678**, 279–288.

9 J. Trocewicz, Z. Suprynowicz and J. Markowicz, Determination of diprivan in urine by a supported liquid membrane technique and liquid chromatography-electrochemical detection, *J. Chromatogr. B*, 1996, **685**, 129–134.

10 P. Favetta, J. Guitton, C.S. Degoute, L. Van Daele and R. Boulieu, High-performance liquid chromatographic assay to detect hydroxylate and conjugate metabolites of propofol in human urine, *J. Chromatogr. B*, 2000, **742**, 25–35.

11 G.B. Park, R.F. Koss, J. Utter, B.A. Mayes and J. Edelson, Determination of isoetharine in plasma by reversed-phase chromatography with amperometric detection, *J. Pharm. Sci.*, 1982, **71**, 932–934.

12 R.C. Causon, R. Desjardins, M.J. Brown and D.S. Davies, Determination of *d*-isoproterenol sulphate by high-performance liquid chromatography with amperometric detection, *J. Chromatogr.*, 1984, **306**, 257–268.

13 M.E. Hadwiger, S.R. Torchia, S. Park, M.E. Biggin and C.E. Lunte, Optimization of the separation and detection of the enantiomers of isoproterenol in microdialysis samples by cyclodextrin-modified capillary electrophoresis using electrochemical detection, *J. Chromatogr. B*, 1996, **681**, 241–249.

14 B. Oosterhuis and C.J. van Boxtel, Determination of salbutamol in human plasma with bimodal high-performance liquid chromatography and a rotated disc amperometric detector, *J. Chromatogr.*, 1982, **232,** 327–334.

15 R.J. Hageman, J.E. Greving, J.H.G. Jonkman and R.A. de Zeeuw, High-performance liquid chromatographic determination of reproterol in plasma using on-line trace enrichment and amperometric detection with a rotating working electrode, *J. Chromatogr.*, 1983, **274**, 239–253.

16 Y.K. Tan and S.J. Soldin, Determination of salbutamol in human serum by reversed-phase high-performance liquid chromatography with amperometric detection, *J. Chromatogr.*, 1984, **311**, 311–317.

17 Y.K. Tan and S.J. Soldin, Analysis of salbutamol enantiomers in human urine by chiral high-performance liquid chromatography and preliminary studies related to the stereoselective disposition kinetics in man, *J. Chromatogr.*, 1987, **422**, 187–195.

18 D.R. Jarvie, A.M. Thompson and E.H. Dyson, Laboratory and clinical features of self-poisoning with salbutamol and terbutaline, *Clin. Chim. Acta.*, 1987, **168**, 313–322.

19 K.M. Sagar, M.T. Kelly and M.R. Smyth, Analysis of terbutaline in human plasma by high-performance liquid chromatography with electrochemical detection using a micro-electrochemical flow cell, *J. Chromatogr.*, 1992, **577**, 109–116.

20 K.M. Sagar, M.T. Kelly and M.R. Smyth, Simultaneous determination of salbutamol and terbutaline at overdose levels in human plasma by high performance liquid chromatography with electrochemical detection, *Biomed. Chromatogr.*, 1993, **7**, 29–33.

21 P.T. McCarthy, S. Atwal, A.P. Sykes and J.G. Ayres, Measurement of terbutaline and salbutamol in plasma by high performance liquid chromatography with fluorescence detection, *Biomed. Chromatogr.*, 1993, **7**, 25–28.

22 S. Bergquist and L.-E. Edholm, Quantitative analysis of terbutaline (Bricanyl®) in human plasma with liquid chromatography and electrochemical detection using on-line enrichment, *J. Liq. Chromatogr.*, 1983, **6**, 559–574.

23 B.-M. Kennedy, A. Blomgren, L.-E. Edholm and C. Roos, Quantitative determination of terbutaline in human plasma after administration of bambuterol using coupled columns and electrochemical detection, *Chromatographia*, 1987, **24**, 895–899.

24 V.L. Herring and J.A. Johnson, Simple method for determination of terbutaline plasma concentration by high-performance liquid chromatography, *J. Chromatogr. B,* 2000, **741**, 307–312.

25 B.T.J. van den Berg, E.J.G. Portier, M. van den Berg, M.C.P. Braat and C.J. van Boxtel, First high-performance liquid chromatography assay of formoterol concentrations in the low-picogram-per-milliliter range, *Ther. Drug Monit.*, 1994, **16**, 196–199.

26 J. Campestrini, J.B. Lecaillon and J. Godbillon, Automated and sensitive method for the determination of formoterol in human plasma by high-performance liquid chromatography and electrochemical detection, *J. Chromatogr. B*, 1997, **704**, 221–229.

27 L.S. Lin, S.N. Caritis and L.K. Wong, Analysis of ritodrine in serum by high-performance liquid chromatography with electrochemical detection, *J. Pharm. Sci.*, 1984, **73**, 131–133.

28 B. Diquet, L. Doare and P. Simon, Determination of clenbuterol in the high nanogram range in plasma of mice by high-performance liquid chromatography with amperometric detection, *J. Chromatogr.*, 1984, **336**, 415–421.

29 H. Hooijerink, R. Schilt, W. Haasnoot and D. Courtheijn, Determination of clenbuterol in urine of calves by high-performance liquid chromatography with in series ultraviolet and electrochemical detection, *J. Pharm. Biomed. Anal.*, 1991, **9**, 485–492.

30 G.A. Qureshi and A. Eriksson, Determination of clenbuterol and mabuterol in equine plasma by ion-pair liquid chromatography with electrochemical detection. Chromatographic and electrochemical characteristics, *J. Chromatogr.*, 1988, **441,** 197–205.

31 B.A. Rashid, P. Kwasowski and D. Stevenson, Solid phase extraction of clenbuterol from plasma using immunoaffinity followed by HPLC, *J. Pharm. Biomed. Anal.*, 1999, **21**, 635–639.

32 A. Koole, J. Bosman, J.P. Franke and R.A. de Zeeuw, Multiresidue analysis of beta2-agonist in human and calf urine using multimodal solid-phase extraction and high-performance liquid chromatography with electrochemical detection, *J. Chromatogr. B*, 1999, **726**, 149–156.

33 M.P. Turberg, J.M. Rodewald and M.R. Coleman, Determination of ractopamine in monkey plasma and swine serum by high-performance liquid chromatography with electrochemical detection, *J. Chromatogr. B*, 1996, **675**, 279–285.

34. P.J. Parsons and A.F. LeRoy, Determination of *cis*-diamminedichloroplatinum(II) in human plasma using ion-pair chromatography with electrochemical detection, *J. Chromatogr.*, 1986, **378**, 395–408.

35 P.J. Parsons, P.F. Morrison and A.F. LeRoy, Determination of platinum-containing drugs in human plasma by liquid chromatography with reductive electrochemical detection, *J. Chromatogr.*, 1987, **385**, 323-335.

36 S.J. Bannister, L.A. Sternson and A.J. Repta, Evaluation of reductive amperometric detection in the liquid chromatographic determination of anti-neoplastic platinum complexes, *J. Chromatogr.*, 1983, **273**, 301–318.

37 W.N. Richmond and R.P. Baldwin, Chloride-assisted electrochemical detection of cis-dichlorodiamineplatinum(II) [cisplatin] after liquid chromatography, *Anal. Chim. Acta*, 1983, **154**, 133-142.

38 F. Elferink, W.J.F. van der Vijgh and H.M. Pinedo, Analysis of antitumour [1,1-bis(aminomethyl)cyclohexane]platinum(II) complexes derived from spiroplatin by high-performance liquid chromatography with differential pulse ampero-metric detection, *J. Chromatogr.*, 1985, **320**, 379–392.

39 I.S. Krull, X.-D. Ding, S. Braverman, C. Selavka, F. Hochberg and L.A. Sternson, Trace analysis for *cis*-platinum anti-cancer drugs *via* LCEC, *J. Chromatogr. Sci.*, 1983, **21**, 166–173.

40 W.J.F. van der Vijgh, H.B.J. van der Lee, G.J. Postma and H.M. Pinedo, Highly sensitive differential pulse amperometric detection of second-generation antitumour platinum compounds in h.p.l.c. eluents, *Chromatographia*, 1983, **17**, 333–336.

41 P. O'Dea, P. Shearan, S. Dunne and M.R. Smyth, Comparison of mercury- and glassy carbon-based electrochemical detection systems for the determination of cisplatin following high-performance liquid chromatographic separation, *Analyst*, 1988, **113**, 1791–1794.

42 L. Silvestro, I. Viano, C. Baiocchi, G. Saini, F. Marmont and R. Ferro, Quantitation of melphalan in plasma of patients by reversed-phase high-performance liquid chromatography with electrochemical detection, *J. Chromatogr.*, 1991, **563**, 443–450.

43 F.M. Rubino, Separation methods for methotrexate, its structural analogues and metabolites, *J. Chromatogr. B*, 2001, **764**, 217–254.

44 F. Palmisano, T.R.I. Cataldi and P.G. Zambonin, Determination of the antineoplastic agent methotrexate in body fluids by high-performance liquid chromatography with electrochemical detection, *J. Chromatogr.*, 1985, **344**, 249–258.

45 J. Dutrieu and Y.A. Delmotte, HPLC determination of methotrexate in serum or plasma optimized with a dual coulometric detector, *Fresenius Z Anal. Chem.*, 1983, **315**, 539–542.

46 J.T. Lin, A.R. Cashmore, M. Baker, R.N. Dreyer, M. Ernstoff, J.C. Marsh, J.R. Bertino, L.R. Whitfield, R. Delap and A. Grillo-Lopez, Phase I studies with trimetrexate: clinical pharmacology, analytical methodology, and pharmacokinetics, *Cancer Res.*, 1987, **47**, 609–616.

47 E.C. Weir, A.R. Cashmore, R.N. Dreyer, M.L. Graham, N. Hsiao, B.A. Moroson, W.L. Sawicki and J.R. Bertino, Pharmacology and toxicity of a potent "nonclassical" 2,4-diamino quinazoline folate antagonist, trimetrexate, in normal dogs, *Cancer Res.*, 1982, **42**, 1696–1702.

48 T.W. Synold, B. Xi, E.M. Newman, F.M. Muggia and J.H. Doroshow, Simple and sensitive method for the quantitative analysis of lometrexol in plasma using high-performance liquid chromatography with electrochemical detection, *J. Chromatogr. B*, 1996, **683**, 245–249.

49 J. Zimák, M. Zimáková and J. Volke, Electrochemical determination of small amounts of mitoxantrone in urine, *Pharmazie*, 1991, **46**, 605–306.

50 C. Akpofure, C.A. Riley, J.A. Sinkule and W.E. Evans, Quantitation of daunorubicin and its metabolites by high-performance liquid chromatography with electrochemical detection, *J. Chromatogr.*, 1982, **232**, 377–383.

51 A.N. Kotake, N.J. Vogelzang, R.A. Larson and N. Choporis, New high-performance liquid chromatographic assay for plasma doxorubicin, *J. Chromatogr.*, 1985, **337**, 194–200.

52 C.A. Riley, W.R. Crom and W.E. Evans, Loop-column extraction and liquid chromatographic analysis of doxorubicin and three metabolites in plasma, *Ther. Drug Monit.*, 1985, **7**, 455–460.

53 C.M. Riley, A.K. Runyan and J. Graham-Pole, Determination of doxorubicin in plasma and urine by high-perfomance liquid chromatography with electrochemical detection (HPLC-EC). Application to the clinical pharmacokinetics of doxorubicin in patients with osteogenic sarcoma, *Anal. Lett.*, 1987, **20**, 97–116.

54 C.M. Riley and A.K. Runyan, High-perfomance liquid chromatography of anthracycline antibiotics with electrochemical detection. Application to the clinical pharmacokinetics of 4′-deoxyrubicin, *J. Pharm. Biomed. Anal.*, 1987, **5**, 33–43.

55 C. Mou, N. Ganju, K.S. Sridhar and A. Krishan, Simultaneous quantitation of plasma doxorubicin and prochlorperazine content by high-performance liquid chromatography, *J. Chromatogr. B*, 1997, **703**, 217–224.

56 R. Ricciarello, S. Pichini, R. Pacifici, I. Altieri, M. Pellegrini, A. Fattorossi and P. Zuccaro, Simultaneous determination of epirubicin, doxorubicin and their principal metabolites in human plasma by high-performance liquid chromatography and electrochemical detection, *J. Chromatogr. B*, 1998, **707**, 219–225.

57 K.E. Choi, J.A. Sinkule, D.S. Han, S.C. McGrath, K.M. Daly and R.A. Larson, High-performance liquid chromatographic assay for mitoxantrone in plasma using electrochemical detection, *J. Chromatogr.*, 1987, **420**, 81–88.

58 A.J. de Vries and K. Nooter, Quantification of mitoxantrone in bone marrow by high-performance liquid chromatography with electrochemical detection, *J. Chromatogr.*, 1991, **563**, 435–442.

59 U.R. Tjaden, J.P. Langenberg, K. Ensing, W.P. van Bennekom, E.A. de Bruijn and A.T. van Oosterom, Determination of mitomycin C in plasma, serum and urine by high-performance liquid chromatography with ultra-violet and electrochemical detection, *J. Chromatogr.*, 1982, **232**, 355–367.

60 M. Barberi-Heyob, H. Rezzoug, J.L. Merlin and F. Guillemin, Sensitive isocratic liquid chromatographic assay for the determination of 5,10,15,20-tetra(m-hydroxyphenyl)chlorin in plasma and tissue with electrochemical detection, *J. Chromatogr. B*, 1997, **688**, 331–338.

61 Q. Wang, H.J. Altermatt, H.B. Ris, B.E. Reynolds, J.C. Stewart, R. Bonnett and C.K. Lim, Determination of 5,10,15,20-tetra-(m-hydroxyphenyl)chlorin in tissues by high performance liquid chromatography, *Biomed. Chromatogr.*, 1993, **7**, 155–157.

62 R. Whelpton, A.T. Michael-Titus, S.S. Basra and M. Grahn, Distribution of temoporfin, a new photosensitizer for the photodynamic therapy of cancer, in a murine tumor model, *Photochem. Photobiol.*, 1995, **61**, 397–401.

63 J.A. Sinkule and W.E. Evans, High-performance liquid chromatographic analysis of the semisynthetic epipodophyllotoxins teniposide and etoposide using electrochemical detection, *J. Pharm. Sci.*, 1984, **73**, 164–168.

64 P. Canal, C. Michel, R. Bugat, G. Soula and M. Carton, Quantification of teniposide in human serum by high-performance liquid chromatography with electrochemical detection, *J. Chromatogr.*, 1986, **375**, 451–456.

65 A. El-Yazigi and C.R. Martin, Improved assay for etoposide in plasma by radial-compression liquid chromatography with electrochemical detection, *Clin. Chem.*, 1987, **33**, 803–805.

66 M.A.J. van Opstal, F.A.L. van der Horst, J.J.M. Holthuis, W.P. van Bennekom and A. Bult, Automated reversed-phase chromatographic analysis of etoposide and teniposide in plasma by using on-line surfactant-mediated sample clean-up and column-switching, *J. Chromatogr.*, 1989, **495,** 139–151.

67 F.A.L. van der Horst, M.A.J. van Opstal, J. Teeuwsen, M.H. Post, J.J.M. Holthuis and U.A.T.h. Brinkman, Comparative study on the determination of the anti-neoplastic drug teniposide in plasma using micellar liquid chromatography and surfactant-mediated plasma clean-up, *J. Chromatogr.*, 1991, **567**, 161–174.

68 R. Zhou, M. Frostvik-Stolt and E. Liliemark, Determination of etoposide in human plasma and leukemic cells by high-performance liquid chromatography with electrochemical detection, *J. Chromatogr. B*, 2001, **757**, 135–141.

69. S. Stremetzne, U. Jaehde and W. Schunack, Determination of the cytotoxic catechol metabolite of etoposide (3′O-demethyletoposide) in human plasma by high-performance liquid chromatography, *J. Chromatogr. B*, 1997, **703**, 209–215.

70 X. Cai, M.H. Woo, M.J. Edick and M.V. Relling, Simultaneous quantitation of etoposide and its catechol metabolite in human plasma using high-performance liquid chromatography with electrochemical detection, *J. Chromatogr. B*, 1999, **728**, 241–250.

71 R. Kitamura, T. Bandoh, M. Tsuda and T. Satoh, Determination of a new podophyllotoxin derivative, TOP-53, and its metabolite in rat plasma and urine by high-performance liquid chromatography with electrochemical detection, *J. Chromatogr. B,* 1997, **690**, 283–288.

72 D.E.M.M. Vendrig, J.J.M. Holthuis, V. Erdélyi-Tóth and A. Hulshoff, Solid-phase extraction of vinblastine and vincristine from plasma and urine: variable drug recoveries due to non-reproducible column packings, *J. Chromatogr.,* 1987, **414**, 91–100.

73 D.E.M.M. Vendrig, J. Teeuwsen and J.J.M. Holthuis, Analysis of vinca alkaloids in plasma and urine using high-performance liquid chromatography with electrochemical detection, *J. Chromatogr.,* 1988, **424**, 83–94.

74 H. Bloemhof, K.N. Van Dijk, S.S.N. De Graaf, D.E.M.M. Vendrig and D.R.A. Uges, Sensitive method for the determination of vincristine in human serum by high-performance liquid chromatography after on-line column-extraction, *J. Chromatogr.,* 1991, **572**, 171–179.

75. P. Koopmans, C.E. Gidding, S.S. de Graaf and D.R. Uges, An automated method for the bioanalysis of vincristine suitable for therapeutic drug monitoring and pharmacokinetic studies in young children, *Ther. Drug Monit.,* 2001, **23**, 406–409.

76 E. Groninger, P. Koopmans, W. Kamps, S. de Graaf and D. Uges, An automated HPLC method to determine intracellular vincristine concentrations in mononuclear cells of children with acute lymphoblastic leukemia, *Ther. Drug Monit.,* 2003, **25**, 441–446.

77 O. van Tellingen, H.R. van der Woude, J.H. Beijnen, K.S.P. Bhushana Rao, W.W. Ten Bokkel Huinink, W.J. Nooyen, High-performance liquid chromatographic bio-analysis and preliminary pharmacokinetics of the experimental antitumour drug vintriptol, *J. Chromatogr.,* 1990, **529**, 329–338.

78 C. Mouchard-Delmas, B. Gourdier and R. Vistelle, Determination of vinorelbine in rabbit plasma by high-performance liquid chromatography with coulometric detection, *J. Chromatogr. B,* 1995, **663**, 390–394.

79 P. Bellon, P. Canal, J. Bernadou and G. Soula, Use of electrochemical detection in the high-performance liquid chromatographic determination of hydroxylated ellipticine derivatives, *J. Chromatogr.,* 1984, **309**, 170–176.

80 I. Švagrová, K. Štulík, V. Pacáková, P. Caliceti and F.M. Veronese, Determination of narciclasine in serum by reversed-phase high-performance liquid chromatography: comparison of amperometric, ultraviolet photometric and fluorescence detection, *J. Chromatogr.,* 1991, **563**, 95–102.

81 R.J. Rucki, A. Ross and S.A. Moros, Application of an electrochemical detector to the determination of procarbazine hydrochloride by high-performance liquid chromatography, *J. Chromatogr.,* 1980, **190**, 359–365.

82 S. Chamart, M. Hanocq, M. Helson, N. Devleeschouwer and G. Leclercq, Determination of 2-methyl derivatives of tamoxifen in cell culture medium using high-performance liquid chromatography and electrochemical detection, *J. Chromatogr.,* 1989, **496**, 365–375.

83 X.-Z. Liang, R.K. Palsmeier and C.E. Lunte, Dual-electrode amperometric detection for the determination of SR4233 and its metabolites with microbore liquid chromatography, *J. Pharm. Biomed. Anal.,* 1995, **14**, 113–119.

84 T. Kuramitsu, K. Takai, R. Ohashi and T. Kuwabara, Determination of the anticancer drug KW-2170, a pyrazoloacridone derivative, and its metabolites in human and dog plasma by high-performance liquid chromatography using an electrochemical detector, *J. Chromatogr. B,* 2002, **768**, 231–237.

85 P. Leroy and A. Nicolas, Determination of atropine in pharmaceutical dosage forms containing vegetal preparations, by high-performance liquid chromatography with U.V. and electrochemical detection, *J. Pharm. Biomed. Anal.,* 1987, **5**, 477–484.

86 R. Whelpton and P.R. Hurst, HPLC-EC analysis of hyoscine (scopolamine) in urine, in *Analysis of Drugs and Metabolites, Including Anti-infective Agents*, R. Reid and I.D. Wilson, (eds) Royal Society of Chemistry, Cambridge, 1990, 279–284.

87 R.R. Brodie, L.F. Chasseaud and A.D. Robbins, Determination of physostigmine in plasma by high-performance liquid chromatography, *J. Chromatogr.,* 1987, **415**, 423–431.

88 N.M. Elsayed, J.R. Ryabik, S. Ferraris, C.R. Wheeler and D.W. Korte, Determination of physostigmine in plasma by high-performance liquid chromatography and fluorescence detection, *Anal. Biochem.,* 1989, **177**, 207–211.

89 M.L. Herold, M.L. Constanzer and B.K. Matuszewski, Determination of picogram levels of heptylphysostigmine in human plasma using high-performance liquid chromatography with fluorescence detection, *J. Chromatogr.,* 1992, **581**, 227–36.

90 R. Whelpton, Analysis of plasma physostigmine concentrations by liquid chromatography, *J. Chromatogr.,* 1983, **272**, 216–220.

91 R. Whelpton and T. Moore, Sensitive liquid chromatographic method for physostigmine in biological fluids using dual-electrode electrochemical detection, *J. Chromatogr.,* 1985, **341**, 361–371.

92 P.R. Hurst and R. Whelpton, Solid phase extraction for an improved assay of physostigmine in biological fluids, *Biomed. Chromatogr.,* 1989, **3**, 226–232.

93 S. Knapp, M.L. Wardlow and L.J. Thal, Sensitive analysis of plasma physostigmine levels using dual-cell electrochemistry in the redox mode, *J. Chromatogr.,* 1990, **526**, 97–107.

94 G.D. Lawrence and N. Yatim, Extraction of physostigmine from biologic fluids and analysis by liquid chromatography with electrochemical detection, *J. Pharmacol. Meth.,* 1990, **24**, 137–143.

95 K. Isaksson and P.T. Kissinger, Determination of physostigmine in plasma by liquid chromatography with dual electrode amperometric detection, *J. Liq. Chromatogr.,* 1987, **10**, 2213–2229.

96 K. Isaksson and P.T. Kissinger, Metabolism of physostigmine in mouse liver microsomal incubations studied by liquid chromatography with dual-electrode amperometric detection, *J. Chromatogr.,* 1987, **419**, 165–175.

97 L.K. Unni, M.E. Hannant, R.E. Becker and E. Giacobini, Determination of physostigmine in plasma and cerebrospinal fluid by liquid chromatography with electrochemical detection, *Clin. Chem.,* 1989, **35**, 292–295.

98 L.K. Unni and R.E. Becker, Determination of heptylphysostigmine in plasma by high-performance liquid chromatography with electrochemical detection, *J. Chromatogr.,* 1992, **573**, 275–281.

99 L. Zecca, D. Radice, A. Mosca and P.G. Pagella, Determination of heptyl-physostigmine in plasma by high-performance liquid chromatography with electrochemical detection, *J. Chromatogr.*, 1993, **615**, 169–173.

100 H.P. Hendrickson, D.O. Scott and C.E. Lunte, Identification of 9-hydroxy-lamine-1,2,3,4-tetrahydroacridine as a hepatic microsomal metabolite of tacrine by high-performance liquid chromatography and electrochemistry, *J. Chromatogr.*, 1989, **487**, 401–408.

101 L.L. Hansen, J.T. Larsen and K. Brosen, Determination of tacrine and its metabolites in human plasma and urine by high-performance liquid chromatography and fluorescence detection, *J. Chromatogr. B,* 1998, **712**, 183–191.

102 C.B. Eap and P. Baumann, Analytical methods for the quantitative determination of selective serotonin reuptake inhibitors for therapeutic drug monitoring purposes in patients, *J. Chromatogr. B,* 1996, **686**, 51–63.

103 E.M. Clement, J. Odontiadis and M. Franklin, Simultaneous measurement of venlafaxine and its major metabolite, oxydesmethylvenlafaxine, in human plasma by high-performance liquid chromatography with coulometric detection and utilisation of solid-phase extraction, *J. Chromatogr. B,* 1998, **705**, 303–308.

104 G. Tournel, N. Houdret, V. Hedouin, M. Deveau, D. Gosset and M. Lhermitte, High-performance liquid chromatographic method to screen and quantitate seven selective serotonin reuptake inhibitors in human serum, *J. Chromatogr. B,* 2001, **761**, 147–158.

105 P. Molander, A. Thomassen, L. Kristoffersen, T. Greibrokk and E. Lundanes, Simultaneous determination of citalopram, fluoxetine, paroxetine and their metabolites in plasma by temperature-programmed packed capillary liquid chromatography with on-column focusing of large injection volumes, *J. Chromatogr. B,* 2002, **766**, 77–87.

106 P.G.L.C. Krugers Dagneaux, C.P.G.G. Loohuis, J.T. Klein Elhorst and T.S. van der Veer, Liquid chromatographic estimation of tranylcypromine in human plasma, *Pharm. Weekbl. Sci.*, 1992, **14**, 46–49.

107 R.F. Suckow and T.B. Cooper, Simultaneous determination of imipramine, desipramine, and their 2-hydroxy metabolites in plasma by ion-pair reversed-phase high-performance liquid chromatography with amperometric detection, *J. Pharm. Sci.*, 1981, **70**, 257–261.

108 E. Koyama, Y. Kikuchi, H. Echizen, K. Chiba and T. Ishizaki, Simultaneous high-performance liquid chromatography-electrochemical detection determination of imipramine, desipramine, their 2-hydroxylated metabolites, and imipramine *N*-oxide in human plasma and urine: preliminary application to oxidation pharmacogenetics, *Ther. Drug Monit.*, 1993, **15**, 224–235.

109 A.G. Chen, Y.K. Wing, H. Chiu, S. Lee, C.N. Chen and K. Chan, Simultaneous determination of imipramine, desipramine and their 2- and 10-hydroxylated metabolites in human plasma and urine by high-performance liquid chromatography, *J. Chromatogr. B,* 1997, **693**, 153–158.

110 J.P. Foglia, D. Sorisio and J.M. Perel, Determination of imipramine, desipramine and their hydroxy metabolites by reversed-phase chromatography with ultra-violet and coulometric detection, *J. Chromatogr.*, 1991, **572**, 247–258.

111 P. Virgili and J.A. Henry, Determination of lofepramine and desipramine using high-performance liquid chromatography and electrochemical detection, *J. Chromatogr.*, 1989, **496**, 228–233.

112 R.F. Suckow and T.B. Cooper, Determination of trimipramine and metabolites in plasma by liquid chromatography with electrochemical detection, *J. Pharm. Sci.*, 1984, **73**, 1745–1748.

113 A.A. Gulaid, G.A. Jahn, C. Maslen and M.J. Dennis, Simultaneous determination of trimipramine and its major metabolites by high-performance liquid chromatography, *J. Chromatogr.*, 1991, **566**, 228–233.

114 O. Spreux-Varoquaux, D. Morin, C. Advenier and M. Pays, Determination of clomipramine and its hydroxylated and demethylated metabolites in plasma and urine by liquid chromatography with electrochemical detection, *J. Chromatogr.*, 1987, **416**, 311–319.

115 R.F. Suckow, T.B. Cooper, F.M. Quitkin and J.W. Stewart, Determination of mianserin and metabolites in plasma by liquid chromatography with electrochemical detection, *J. Pharm. Sci.*, 1982, **71**, 889–892.

116 L.W. Brown, H.K. Hundt and K.J. Swart, Automated high-performance liquid chromatographic method for the determination of mianserin in plasma using electrochemical detection, *J. Chromatogr.*, 1992, **582**, 268–272.

116 R.F. Suckow and T.B. Cooper, Determination of amoxapine and metabo-lites in plasma by liquid chromatography with electrochemical detection, *J. Chromatogr.*, 1985, **338**, 225–229.

117 F.S. Messiha, Determination of carbamazepine by HPLC electrochemical detection and application for estimation of imipramine desipramine, doxepin and nordoxepin, *Alcohol,* 1986, **3**, 135–138.

118 R.F. Suckow, A simultaneous determination of trazodone and its metabolite 1-*m*-chlorophenylpiperazine in plasma by liquid-chromatography with electro-chemical detection, *J. Liq. Chromatogr.*, 1983, **6**, 2195–2208.

119 T. Ohkubo, T. Osanai, K. Sugawara, M. Ishida, K. Otani, K. Mihara and N. Yasui, High-performance liquid chromatographic determination of trazodone and 1-*m*-chlorophenylpiperazine with ultraviolet and electrochemi-cal detector, *J. Pharm. Pharmacol.*, 1995, **47**, 340–344.

120 J.E. Wallace, E.L. Shimek, S. Stavchansky and S.C. Harris, Determination of promethazine and other phenothiazine compounds by liquid chromatography with electrochemical detection, *Anal. Chem.*, 1981, **53**, 960–962.

121 D.E. Leelavathi, D.E. Dressler, E.F. Soffer, S.D. Yachetti and J.A. Knowles, Determination of promethazine in human plasma by automated high-performance liquid chromatography with electrochemical detection and by gas chromatography-mass spectrometry, *J. Chromatogr.*, 1985, **339**, 105–115.

122 A.R. Fox and D.A. McLoughlin, Rapid, sensitive high-performance liquid chromatographic method for the quantification of promethazine in human serum with electrochemical detection, *J. Chromatogr.*, 1993, **631**, 255–259.

123 G. McKay, J.K. Cooper, K.K. Midha, K. Hall and E.M. Hawes, Simple and sensitive high-performance liquid chromatographic procedure with electrochemical detection for the determination of plasma concentrations of

trimeprazine following single oral doses, *J. Chromatogr.*, 1982, **233**, 417–422.

124 O.Y.-P. Hu, E. Gfeller, J.H. Perrin and S.H. Curry, Relative bioavailability of trimeprazine tablets investigated in man using HPLC with electrochemical detection, *J. Pharm. Pharmacol.*, 1986, **38**, 172–176.

125 D.M. Radzik and S.M. Lunte, Application of liquid chromatography/electrochemistry in pharmaceutical and biochemical analysis: A critical review, *CRC Crit. Rev. Anal. Chem.*, 1989, **20**, 317–358.

126 D.W. Hoffman, R.D. Edkins and S.D. Shillcutt, Human metabolism of phenothiazines to sulfoxides determined by a new high performance liquid chromatography-electrochemical detection method, *Biochem. Pharmacol.*, 1988, **37**, 1773–1777.

127 S.R. Vanapalli, S.P. Kambhampati, L. Putcha and D.W. Bourne, A liquid chromatographic method for the simultaneous determination of promethazine and three of its metabolites in plasma using electrochemical and UV detectors, *J. Chromatogr., Sci.*, 2001, **39**, 70–72.

128 R.B. Walker and I. Kanfer, Sensitive high-performance liquid chromatographic determination of cyclizine and its demethylated metabolite, norcyclizine, in biological fluids using coulometric detection, *J. Chromatogr. B*, 1995, **672**, 172–177.

129 A. Marzo and L. Dal Bo, Chromatography as an analytical tool for selected antibiotic classes: a reappraisal addressed to pharmacokinetic applications, *J. Chromatogr. A*, 1998, **812**, 17–34.

130 J. Hoogmartens, Foreword, *J. Chromatogr. A*, 1998, **812**, 1.

131 J.A. Statler, Determination of tobramycin using high-performance liquid chromatography with pulsed amperometric detection, *J. Chromatogr.*, 1990, **527**, 244–246.

132 J. Szúnyog, E. Adams, E. Roets and J. Hoogmartens, Analysis of tobramycin by liquid chromatography with pulsed electrochemical detection, *J. Pharm. Biomed. Anal.*, 2000, **23**, 891–896.

133 E. Adams, R. Schepers, E. Roets and J. Hoogmartens, Determination of neomycin sulfate by liquid chromatography with pulsed electrochemical detection, *J. Chromatogr. A*, 1996, **741**, 233–240.

134 E. Adams, J. Dalle, E. de Bie, I. de Smedt, E. Roets and J. Hoogmartens, Analysis of kanamycin sulfate by liquid chromatography with pulsed electrochemical detection, *J. Chromatogr. A*, 1997, **766**, 133–139.

135 E. Adams, W. Roelants, R. De daepe, E. Roets and J. Hoogmartens, Analysis of gentamicin by liquid chromatography with pulsed electrochemical detection, *J. Pharm. Biomed. Anal.*, 1998, **18**, 689–698.

136 E. Adams, D. Puelings, M. Rafiee, E. Roets and J. Hoogmartens, Determination of netilmicin sulfate by liquid chromatography with pulsed electrochemical detection, *J. Chromatogr. A*, 1998, **812**, 151–157.

137 E. Adams, G. van Vaerenbergh, E. Roets and J. Hoogmartens, Analysis of amikacin by liquid chromatography with pulsed electrochemical detection, *J. Chromatogr. A*, 1998, **819**, 93–97.

138 L.G. McLaughlin and J.D. Henion, Determination of aminoglycoside antibiotics by reversed-phase ion-pair high-performance liquid chromatography coupled with pulsed amperometry and ion spray mass spectrometry, *J. Chromatogr.*, 1992, **591**, 195–206.

139 L. Nordholm and L. Dalgaard, Determination of trimethoprim metabolites including conjugates in urine using high-performance liquid chromatography with combined ultraviolet and electrochemical detection, *J. Chromatogr.*, 1984, **305**, 391–399.

140 L. Nordholm and L. Dalgaard, Assay of trimethoprim in plasma and urine by high-performance liquid chromatography using electrochemical detection, *J. Chromatogr.*, 1982, **233**, 427–431.

141 R. Whelpton, G. Watkins and S.H. Curry, Bratton-Marshall and liquid-chromatographic methods compared for determination of sulfamethazine acetylator status, *Clin. Chem.*, 1981, **27**, 1911–1914.

142 M. Alawi and H. Rüssell, Determination of sulphonamides in milk by HPLC with electrochemical detection, *Fresenius Z Anal. Chem.*, 1981, **307**, 382–384.

143 D.N. Mallett, A.A. Gulaid and M.J. Dennis, High-performance liquid chromatographic method with electrochemical detection for the concomitant assay of sulphadiazine, sulphamerazine and sulphapyridine in plasma, *J. Chromatogr.*, 1988, **428**, 190–195.

144 D. von Baer, A. Momberg, M.E. Carrera, R. Arriagada and M.R. Smyth, Liquid chromatography with amperometric detection of some sulphonamides and their N_4-acetyl-metabolites in serum and urine, *J. Pharm. Biomed. Anal.*, 1991, **9**, 925–928.

145 D.W. Hoffman, K.L. Jones-King, C.L. Ravaris and R.D. Edkins, Electrochemical detection for high-performance liquid chromatography of ketoconazole in plasma and saliva, *Anal. Biochem.*, 1988, **172**, 495–498.

146 M.A. Czech, M. Meltzer and J. Mouskountakis, Determination of Ro 14-4767 (Loceryl) by LC using automated column switching with ultraviolet and electrochemical detection, *J. Pharm. Biomed. Anal.*, 1991, **9**, 1019–1029.

147 F. Schatz and H. Haberl, Analytical methods for the determination of terbinafine and its metabolites in human plasma, milk and urine, *Arzneimittelforschung,* 1989, **39**, 527–532.

148 T.T. Hien and N.J. White, Qinghaosu, *Lancet,* 1993, **341**, 603–608.

149 O.R. Idowu, G. Edwards, S.A. Ward, M.L. Orme and A.M. Breckenridge, Determination of arteether in blood plasma by high-performance liquid chromatography with ultraviolet detection after hydrolysis acid, *J. Chromatogr.*, 1989, **493**, 125–136.

150 D.K. Muhia, E.K. Mberu and W.M. Watkins, Differential extraction of artemether and its metabolite dihydroartemisinin from plasma and determination by high-performance liquid chromatography, *J. Chromatogr., B,* 1994, **660**, 196–199.

151 K.T. Batty, T.M. Davis, L.T. Thu, T.Q. Binh, T.K. Anh and K.F. Ilett, Selective high-performance liquid chromatographic determination of artesunate and

alpha- and beta-dihydroartemisinin in patients with falciparum malaria, *J. Chromatogr. B*, 1996, **677**, 345–350.

152 Z.M. Zhou, J.C. Anders, H. Chung and A.D. Theoharides, Analysis of artesunic acid and dihydroqinghaosu in blood by high-performance liquid chromatography with reductive electrochemical detection, *J. Chromatogr.*, 1987, **414**, 77–90.

153 V. Melendez, J.O. Peggins, T.G. Brewer and A.D. Theoharides, Determination of the antimalarial arteether and its deethylated metabolite dihydroartemisinin in plasma by high-performance liquid chromatography with reductive electrochemical detection, *J. Pharm. Sci.*, 1991, **80**, 132–138.

154 M.V. Nora, G.W. Parkhurst, R.W. Thomas and P.E. Carson, High-performance liquid chromatographic-electrochemical assay method for primaquine in plasma and urine, *J. Chromatogr.*, 1984, **307**, 451–456.

155 R.A. Dean, W. Ochieng, J. Black, S.F. Queener, M.S. Bartlett and N.G. Dumaual, Simultaneous determination of primaquine and carboxyprimaquine in plasma using high-performance liquid chromatography with electrochemical detection, *J. Chromatogr. B*, 1994, **655**, 89–96.

156 Y.S. Endoh, H. Yoshimura, N. Sasaki, Y. Ishihara, H. Sasaki, S. Nakamura, Y. Inoue and M. Nishikawa, High-performance liquid chromatographic determination of pamaquine, primaquine and carboxy primaquine in calf plasma using electrochemical detection, *J. Chromatogr.*, 1992, **579**, 123–129.

157 V. Navaratnam, S.M. Mansor, L.K. Chin, M.N. Mordi, M. Asokan and N.K. Nair, Determination of artemether and dihydroartemisinin in blood plasma by high-performance liquid chromatography for application in clinical pharmacological studies, *J. Chromatogr. B*, 1995, **669**, 289–294.

158 K.L. Chan, K.H. Yuen, S. Jinadasa, K.K. Peh and W.T. Toh, A high-performance liquid chromatography analysis of plasma artemisinin using a GCE for reductive electrochemical detection, *Planta Med.*, 1997, **63**, 66–69.

159 J. Karbwang, K. Na-Bangchang, P. Molunto, V. Banmairuroi and K. Congpuong, Determination of artemether and its major metabolite, dihydroartemisinin, in plasma using high-performance liquid chromatography with electrochemical detection, *J. Chromatogr. B*, 1997, **690**, 259–265.

160 V. Navaratnam, M.N. Mordi and S.M. Mansor, Simultaneous determination of artesunic acid and dihydroartemisinin in blood plasma by high-performance liquid chromatography for application in clinical pharmacological studies, *J. Chromatogr. B*, 1997, **692**, 157–62.

161 N. Sandrenan, A. Sioufi, J. Godbillon, C. Netter, M. Donker and C. van Valkenburg, Determination of artemether and its metabolite, dihydroartemisinin, in plasma by high-performance liquid chromatography and electrochemical detection in the reductive mode, *J. Chromatogr. B Biomed. Sci. Appl.*, 1997, **691**, 145–153.

162 M.A. van Agtmael, J.J. Butter, E.J.G. Portier and C.J. van Boxtel, Validation of an improved reversed-phase high-performance liquid chromatography assay with reductive electrochemical detection for the determination of artemisinin derivatives in man, *Ther. Drug Monit.*, 1998, **20**, 109–116.

163 K. Na-Bangchang, K. Congpuong, L.N. Hung, P. Molunto and J. Karbwang, Simple high-performance liquid chromatographic method with electrochemical detection for the simultaneous determination of artesunate and dihydroartemisinin in biological fluids, *J. Chromatogr. B*, 1998, **708**, 201–207.

164 S.A. Wages, L.C. Patchen and F.C. Churchill, Analysis of blood and urine samples from Macaca mulata for pyronaridine by high-performance liquid chromatography with electrochemical detection, *J. Chromatogr.*, 1990, **527**, 115–126.

165 D.L. Mount, L.C. Patchen, P. Nguyen-Dinh, A.M. Barber, I.K. Schwartz and F.C. Churchill, *J. Chromatogr.*, 1986, **383**, 375–386.

166 D. Blessborn, N. Lindegardh, O. Ericsson, U. Hellgren and Y. Bergqvist, Determination of pyronaridine in whole blood by automated solid phase extraction and high-performance liquid chromatography, *Ther. Drug Monit.*, 2003, **25**, 264–270.

167 J. Moncrieff, Determination of dapsone in serum and saliva using reversed-phase high-performance liquid chromatography with ultraviolet or electrochemical detection, *J. Chromatogr. B*, 1994, **654**, 103–110.

168 A. Loregian, R. Gatti, G. Palu and E.F. De Palo, Separation methods for acyclovir and related antiviral compounds, *J. Chromatogr. B*, 2001, **764**, 289–311.

169 G.C. Visor, S.E. Jackson, R.A. Kenley and G.C. Lee, Electrochemistry of purine derivatives. 1: Direct determination for the antiviral drug 9-[(1,3-dihydroxy-2-propoxy)methyl]guanine by anodic differential pulse voltammetry, *J. Pharm. Sci.*, 1985, **74**, 1078–1081.

170 S. Zhang, Z. Yuan, H. Liu, H. Zou, H. Xiong, Y. Wu, Analysis of acyclovir by high performance capillary electrophoresis with on-column amperometric detection, *Electrophoresis*, 2000, **21**, 2995–2998.

171 R.J. Bopp, J.F. Quay, R.M. Morris, J.F. Stucky and D.J. Miner, Liquid chromatographic analysis of enviradene, a new antiviral agent, in plasma and its application in bioavailability studies in the dog, *J. Pharm. Sci.*, 1985, **74**, 846–850.

172 M.R. Fizzano, L. Valvo, M.L. Dupuis, V. Mennella and M. Cianfriglia, LC determination of indinavir in biological matrices with electrochemical detection, *J. Pharm. Biomed. Anal.*, 2000, **22**, 307–314.

173 K.-J. Pettersson, T. Nordgren and D. Westerlund, Determination of phosphonoformate (foscarnet) in biological fluids by ion-pair reversed-phase liquid chromatography, *J. Chromatogr.*, 1989, **488**, 447-455.

174 M.K. Hassanzadeh, F.T. Aweeka, S. Wu, M.A. Jacobson and J.G. Gambertoglio, Determination of phosphonoformic acid in human plasma and urine by high-performance liquid chromatography with electrochemical detection, *J. Chromatogr.*, 1990, **525**, 133–140.

175 B.B. Ba, A.G. Corniot, D. Ducint, D. Breilh, J. Grellet and M.C. Saux, Determination of phosphonoformate (foscarnet) in calf and human serum by automated solid-phase extraction and high-performance liquid chromatography with amperometric detection, *J. Chromatogr. B.*, 1999, **724**, 127–136.

176 B.B. Ba and M.C. Saux, Separation methods for antiviral phosphorus-containing drugs, *J. Chromatogr. B*, 2001, **764**, 349–362.

177 C.M. Selavka, I.S. Krull and K. Bratin, Analysis for penicillins and cefoperazone by HPLC-photolysis-electrochemical detection (HPLC-hv-EC), *J. Pharm. Biomed. Anal.*, 1986, **4**, 83–93.

178 T. Yamazaki, T. Ishikawa, H. Nakai, M. Miyai, T. Tsubota and K. Asano, Determination of aspoxicillin in broncho-alveolar lavage fluid by high-performance liquid chromatography with photolysis and electrochemical detection, *J. Chromatogr.*, 1993, **615**, 180–185.

179 S. Lihl, A. Rehorek and M. Petz, High-performance liquid chromatographic determination of penicillins by means of automated solid-phase extraction and photochemical degradation with electrochemical detection, *J. Chromatogr. A*, 1996, **729**, 229–235.

180 H. Mascher and C. Kikuta, Determination of amoxicillin in plasma by high-performance liquid chromatography with fluorescence detection after online oxidation, *J. Chromatogr.*, 1990, **506**, 417–421.

181 P. Favetta, J. Guitto, N. Bleyzac, C. Dufresne and J. Bureau, New sensitive assay of vancomycin in human plasma using high-performance liquid chromatography and electrochemical detection, *J. Chromatogr. B*, 2001, **751**, 377–382.

182 I. Kanfer, M.F. Skinner and R.B. Walker, Analysis of macrolide antibiotics, *J. Chromatogr. A*, 1998, **812**, 255–286.

183 R.M. Shepard, G.S. Duthu, R.A. Ferraina and M.A. Mullins, High-performance liquid chromatographic assay with electrochemical detection for azithromycin in serum and tissues, *J. Chromatogr.*, 1991, **565**, 321–337.

184 F. Kees, S. Spangler and M. Wellenhofer, Determination of macrolides in biological matrices by high-performance liquid chromatography with electrochemical detection, *J. Chromatogr. A*, 1998, **812**, 287–293.

185 D.A. Raines, A. Yusuf, M.H. Jabak, W.S. Ahmed, Z.A. Karcioglu and A. El-Yazigi, Simultaneous high-performance liquid chromatography analysis of azithromycin and two of its metabolites in human tears and plasma, *Ther. Drug Monit.*, 1998, **20**, 680–684.

186 C. Taninaka, H. Ohtani, E. Hanada, H. Kotaki, H. Sato and T. Iga, Determination of erythromycin, clarithromycin, roxithromycin, and azithromycin in plasma by high-performance liquid chromatography with amperometric detection, *J. Chromatogr. B*, 2000, **738**, 405–411.

187 I. Niopas and A.C. Daftsios, Determination of clarithromycin in human plasma by HPLC with electrochemical detection: validation and application in pharmacokinetic study, *Biomed. Chromatogr.*, 2001, **15**, 507–508.

188 A. Pappa-Louisi, A. Papageorgiou, A. Zitrou, S. Sotiropoulos, E. Georgarakis and F. Zougrou, Study on the electrochemical detection of the macrolide antibiotics clarithromycin and roxithromycin in reversed-phase high-performance liquid chromatography, *J. Chromatogr. B*, 2001, **755**, 57–64.

189 S. Laakso, M. Scheinin and M. Anttila, Determination of erythromycin base and 2'-acetylerythromycin in human plasma using high-performance liquid chromatography with electrochemical detection, *J. Chromatogr.*, 1990, **526**, 475–486.

190 J. Zhou, Y. Chen and R. Cassidy, Separation and determination of the macrolide antibiotics (erythromycin, spiramycin and oleandomycin) by capillary electrophoresis coupled with fast reductive voltammetric detection, *Electrophoresis,* 2000, **21**, 1349–1353.

191 S. Abou-Khalil, W.H. Abou-Khalil, A.N. Masoud and A.A. Yunis, High-performance liquid chromatographic determination of chloramphenicol and four analogues using reductive and oxidative electrochemical and ultraviolet detection, *J. Chromatogr.,* 1987, **417**, 111–119.

192 W. Hou and E. Wang, Liquid chromatographic determination of tetra-cycline antibiotics at an electrochemically pre-treated GCE, *Analyst,* 1989, **114**, 699–702.

193 J. Zhou, G.C. Gerhardt, A. Baranski and R. Cassidy, Capillary electrophoresis of some tetracycline antibiotics coupled with reductive fast cyclic voltam-metric detection, *J. Chromatogr. A,* 1999, **839**, 193–201.

194 P.D. Andrew, H.L. Birch and D.A. Phillpot, Determination of sumatriptan succinate in plasma and urine by high-performance liquid chromatography with electrochemical detection, *J. Pharm. Sci.,* 1993, **82**, 73–76.

195 M. Dunne and P. Andrew, Fully automated assay for the determination of sumatriptan in human serum using solid-phase extraction and high-performance liquid chromatography with electrochemical detection, *J. Pharm. Biomed. Anal.,* 1996, **14**, 721–726.

196 P.H. Marathe, K.D. Larkin, G.F. Duncan, D.S. Greene and R.H. Barbhaiya, Determination of a novel indolylpiperazine anti-migraine agent in rat, monkey, mouse and rabbit plasma by high-performance liquid chromatography with electrochemical detection, *J. Chromatogr. B,* 1996, **678**, 269–277.

197 N.R. Srinivas, V.R. Shah, A. Schuster, X. Jiang, D.W. Whigan and W.C. Shyu, High-performance liquid chromatographic-electrochemical assay for the qua-ntitation of BMS-181885 in monkey plasma, *J. Chromatogr. B,* 1998, **705**, 317–322.

198 G. Bianchi and M. Landi, Determination of apomorphine in rat plasma and brain by high-performance liquid chromatography with electrochemical detection, *J. Chromatogr.,* 1985, **338**, 230–235.

199 A.W.G. Essink, C.P.G.G. Lohuis, J.T. Klein Elhorst and W.J. Rutten, Selective and quantitative isolation and determination of apomorphine in human plasma, *J. Chromatogr.,* 1991, **570**, 419–24.

200 A. Bolner, L. Barbato, F. Tagliaro, A. Monge, F. Stocchi and G. Nordera, Determination of apomorphine in human plasma by alumina extraction and high-performance liquid chromatography with electrochemical detection, *Forensic. Sci. Int.,* 1997, **89**, 81–91.

201 R. van der Geest, P. Kruger, J.M. Gubbens-Stibbe, T. van Laar, H.E. Bodde and M. Danhof, Assay of *R*-apomorphine, *S*-apomorphine, apocodeine, isoapocodeine and their glucuronide and sulfate conjugates in plasma and urine of patients with Parkinson's disease, *J. Chromatogr. B,* 1997, **702**, 131–141.

202 P. Lampen, J.L. Neumeyer and R.J. Baldessarini, High-performance liquid chromatographic separation and electrochemical or spectrophotometric

determination of R(-)N-n-propylnorapomorphine and R(-)10,11-methylene-dioxy-N-n-propylnorapomorphine in primate plasma, *J. Chromatogr.*, 1988, **426**, 283–294.

203 C. Lucarelli, P. Betto, G. Ricciarello, M. Giambenedetti, C. Corradini, F. Stocchi and F. Belliardo, Simultaneous measurement of L-dopa, its metabolites and carbidopa in plasma of Parkinsonian patients by improved sample pretreatment and high-performance liquid chromatographic determination, *J. Chromatogr.*, 1990, **511**, 167–176.

204 K.A. Sagar and M.R. Smyth, Simultaneous determination of levodopa, carbidopa and their metabolites in human plasma and urine samples using LC-EC, *J. Pharm. Biomed. Anal.*, 2000, **22**, 613–624.

205 P. Betto, G. Ricciarello, M. Giambenedetti, C. Lucarelli, S. Ruggeri and F. Stocchi, Improved high-performance liquid chromatographic analysis with double detection system for L-dopa, its metabolites and carbidopa in plasma of parkinsonian patients under L-dopa therapy, *J. Chromatogr.*, 1988, **459**, 341–349.

206 U.P. Tjaden, J. Lankelma, H. Poppe and R.G. Muusze, Anodic coulometric detection with a GCE in combination with reversed-phase high-performance liquid chromatography. Determination of blood levels of perphenazine and fluphenazine, *J. Chromatogr.*, 1976, **125**, 275–286.

207 S.H. Curry, E.A. Brown, O.Y.-P. Hu and J.H. Perrin, Liquid chromatographic assay of phenothiazine, thioxanthene and butyrophenone neuroleptics and antihistamines in blood and plasma with conventional and radial compression columns and UV and electrochemical detection, *J. Chromatogr.*, 1982, **231**, 361–376.

208 G. McKay, K. Hall, J.K. Cooper, E.M. Hawes and K.K. Midha, Gas chromatographic–mass spectrometric procedure for the quantitation of chlorpromazine in plasma and its comparison with a new high-performance liquid chromatographic assay with electrochemical detection, *J. Chromatogr.*, 1982, **232**, 275–282.

209 J.K. Cooper, G. McKay and K.K. Midha, Subnanogram quantitation of chlorpromazine in plasma by high-performance liquid chromatography with electrochemical detection, *J. Pharm. Sci.*, 1983, **72**, 1259–1262.

210 K. Murakami, K. Murakami, T. Ueno, J. Hijikata, K. Shirasawa and T. Muto, Simultaneous determination of chlorpromazine and levomepromazine in human plasma and urine by high-performance liquid chromatography using electrochemical detection, *J. Chromatogr.*, 1982, **227**, 103–112.

211 E. Nissinen, and J. Taskinen, Simultaneous determination of carbidopa, levodopa and 3,4-dihydroxyphenyl-acetic acid using high-performance liquid chromatography with electrochemical detection, *J. Chromatogr.*, 1982, **231**, 459–462.

212 D.C. Titus, T.F. August, K.C. Yeh, R. Eisenhandler, W.F. Bayne and D.G. Musson, Simultaneous high-performance liquid chromatographic analysis of carbidopa, levodopa and 3-O-methyldopa in plasma and carbidopa, levodopa and dopamine in urine using electrochemical detection, *J. Chromatogr.*, 1990, **534**, 87–100.

213 I. Rondelli, D. Acerbi, F. Mariotti and P. Ventura, Simultaneous determination of levodopa methyl ester, levodopa, 3-O-methyldopa and dopamine in plasma by high-performance liquid chromatography with electrochemical detection, *J. Chromatogr. B,* 1994, **653**, 17–23.

214 M.G. Sankey, J.E. Holt and C.M. Kaye, A simple and sensitive H.P.L.C. method for the assay of prochlorperazine in plasma, *Br. J. Clin. Pharmacol.,* 1982, **13**, 578–580.

215 A. Fowler, W. Taylor and D.N. Bateman, Plasma prochlorperazine assay by high-performance liquid chromatography-electrochemistry, *J. Chromatogr.,* 1986, **380**, 202–205.

216 D.W. Hoffman, R.D. Edkins, S.D. Shillcutt and A. Salama, New high-performance liquid chromatographic method for fluphenazine and metabolites in human plasma, *J. Chromatogr.,* 1987, **414**, 504–509.

217 J.K. Cooper, E.M. Hawes, J.W. Hubbard, G. and K.K. McKay, An ultrasensitive method for the measurement of fluphenazine in plasma by high-performance liquid chromatography with coulometric detection, *Ther. Drug Monit.,* 1989, **11**, 354–360.

218 A.L. Stoll, R.J. Baldessarini, B.M. Cohen and S.P. Finklestein, Assay of plasma thioridazine and metabolites by high-performance liquid chromatography with amperometric detection, *J. Chromatogr.,* 1984, **307**, 457–463.

219 C.N. Svendsen and E.D. Bird, HPLC with electrochemical detection to measure chlorpromazine, thioridazine and metabolites in human brain, *Psychopharmacology (Berlin),* 1986, **90**, 316–321.

220 M.A. Brooks, G. DiDonato and H.P. Blumenthal, Determination of chlorprothixene and its sulfoxide metabolite in plasma by high-performance liquid chromatography with ultraviolet and amperometric detection, *J. Chromatogr.,* 1985, **337**, 351–362.

221 M. Hariharan, T. VanNoord, E.K. Kindt and R. Tandon, A simple, sensitive liquid chromatographic assay of cis-thiothixene in plasma with coulometric detection, *Ther. Drug Monit.,* 1991, **13**, 79–85.

222 E.R. Korpi, B.H. Phelps, H. Granger, W.-H. Chang, M. Linnoila, J.L. Meek and R.J. Wyatt, Simultaneous determination of haloperidol and its reduced metabolite in serum and plasma by isocratic liquid chromatography with electrochemical detection, *Clin. Chem.,* 1983, **29**, 624–628.

223 K.K. Midha, J.K. Cooper, E.M. Hawes, J.W. Hubbard, E.D. Korchinski and G. McKay, An ultrasensitive method for the measurement of haloperidol and reduced haloperidol in plasma by high-performance liquid chromatography with coulometric detection, *Ther. Drug Monit.,* 1988, **10**, 177–183.

224 M. Hariharan, E.K. Kindt, T. VanNoord and R. Tandon, An improved sensitive assay for simultaneous determination of plasma haloperidol and reduced haloperidol levels by liquid chromatography using a coulometric detector, *Ther. Drug Monit.,* 1989, **11**, 701–707.

225 H.D. Eddington and D. Young, Sensitive electrochemical high-performance liquid chromatography assay for the simultaneous determination of haloperidol and reduced haloperidol, *J. Pharm. Sci.,* 1988, **77**, 541–543.

226 H.D. Eddington and D. Young Biliary excretion of reduced haloperidol glucuronide, *Psychopharmacology (Berl.)* 1990, **100**, 46–48.

227 L. Pan, M.T. Rosseel and F.M. Belpaire, Comparison of two high-performance liquid chromatographic methods for monitoring plasma concentrations of haloperidol and reduced haloperidol, *Ther. Drug Monit.*, 1998, **20**, 224–230.

228 M. Furlanut, A. Perosa, P. Benetello and G. Colombo, Electrochemical detection of benperidol in serum for drug monitoring in humans, *Ther. Drug. Monit.*, 1987, **9**, 343–346.

229 S. Süss, W. Seiler, C. Hiemke, G. Schöllnhammer, H. Wetzel and A. Hillert, Determination of benperidol and its reduced metabolite in human plasma by high-performance liquid chromatography and electrochemical detection, *J. Chromatogr.*, 1991, 565, 363–373.

230 C. Humpel, C. Haring and A. Saria, Rapid and sensitive determination of clozapine in human plasma using high-performance liquid chromatography and amperometric detection, *J. Chromatogr.*, 1989, **491**, 235–239.

231 E. Schulz, C. Fleischhaker and H. Remschmidt, Determination of clozapine and its major metabolites in serum samples of adolescent schizophrenic patients by high-performance liquid chromatography. Data from a prospective clinical trial, *Pharmacopsychiatry,* 1995, **28**, 20–25.

232 J.P. Le Moing, S. Edouard and J.C. Levron, Determination of risperidone and 9-hydroxyrisperidone in human plasma by high-performance liquid chromatography with electrochemical detection, *J. Chromatogr.*, 1993, **614**, 333–339.

233 M. Aravagiri, S.R. Marder, T. Van Putten and K.K. Midha, Determination of risperidone in plasma by high-performance liquid chromatography with electrochemical detection: application to therapeutic drug monitoring in schizophrenic patients, *J. Pharm. Sci.*, 1993, **82**, 447–449.

234 L.B. Nilsson, High sensitivity determination of the remoxipride hydroquinone metabolite NCQ-344 in plasma by coupled column reversed-phase liquid chromatography and electrochemical detection, *Biomed. Chromatogr.*, 1998, **12**, 65–68.

235 J.T. Catlow, R.D. Barton, M. Clemens, T.A. Gillespie, M. Goodwin and S.P. Swanson, Analysis of olanzapine in human plasma utilizing reversed-phase high-performance liquid chromatography with electrochemical detection, *J. Chromatogr. B*, 1995, **668**, 85–90.

236 M. Aravagiri, D. Ames, W.C. Wirshing and S.R. Marder, Plasma level monitoring of olanzapine in patients with schizophrenia: Determination by high-performance liquid chromatography with electrochemical detection, *Ther. Drug. Monit.*, 1997, **19**, 307–313.

237 J. Bao and B.D. Potts, Quantitative determination of olanzapine in rat brain tissue by high-performance liquid chromatography with electrochemical detection, *J. Chromatogr. B*, 2001, **752**, 61–67.

238 S.C. Kasper, E.L. Mattiuz, S.P. Swanson, J.A. Chiu, J.T. Johnson and C.O. Garner, Determination of olanzapine in human breast milk by high-performance liquid chromatography with electrochemical detection, *J. Chromatogr. B,* 1999, **726**, 203–209.

239 M.R. Hackman and M.A. Brooks, Differential pulse amperometric detection of drugs in plasma using a dropping mercury electrode as a high-performance liquid chromatographic detector, *J. Chromatogr.*, 1981, **222**, 179–190.

240 J.B.F. Lloyd and D.A. Parry, Detection and determination of common benzodiazepines and their metabolites in blood samples of forensic science interest. Microcolumn cleanup and high-performance liquid chromatography with reductive electrochemical detection at a pendent mercury drop electrode, *J. Chromatogr.*, 1988, **449**, 281–297.

241 J.B.F. Lloyd and D.A. Parry, Forensic applications of the determination of benzodiazepines in blood samples by microcolumn cleanup and high-performance liquid chromatography with reductive mode electrochemical detection, *J. Anal. Toxicol.*, 1989, **13**, 163–168.

242 M. Wilhelm, H.J. Battista and D. Obendorf, HPLC with simultaneous UV and reductive electrochemical detection at the hanging mercury drop electrode: A highly sensitive and selective tool for the determination of benzodiazepines in forensic samples, *J. Anal. Toxicol.*, 2001, **25**, 250–257.

243 J.M. Egan and D.R. Abernethy, Lorazepam analysis using liquid chromatography: improved sensitivity for single-dose pharmacokinetic studies, *J. Chromatogr.*, 1986, **380**, 196–201.

244 Y. Gaillard, J.P. Gay-Montchamp and M. Ollagnier, Simultaneous screening and quantitation of alpidem, zolpidem, buspirone and benzodiazepines by dual-channel gas chromatography using electron-capture and nitrogen-phosphorus detection after solid-phase extraction, *J. Chromatogr.*, 1993, **622**, 197–208.

245 M. Franklin, Determination of plasma buspirone by high-performance liquid chromatography with coulometric detection, *J. Chromatogr.*, 1990, **526**, 590–596.

246 P. Betto, A. Meneguz, G. Ricciarello and S. Pichini, Simultaneous high-performance liquid chromatographic analysis of buspirone and its metabolite 1-(2-pyrimidinyl)-piperazine in plasma using electrochemical detection, *J. Chromatogr.*, 1992, **575**, 117–121.

247 J. Odontiadis and M. Franklin, Simultaneous quantitation of buspirone and its major metabolite 1-(2-pyrimidinyl)piperazine in human plasma by high-performance liquid chromatography with coulometric detection, *J. Pharm. Biomed. Anal.*, 1996, **14**, 347–351.

248 Y. Nakahara and H. Sekine, Studies on confirmation of cannabis use. I. Determination of the cannabinoid contents in marijuana cigarette, tar, and ash using high performance liquid chromatography with electrochemical detection, *J. Anal. Toxicol.*, 1985, **9**, 121–124.

249 D. Bourquin and R. Brenneisen, Confirmation of cannabis abuse by the determination of 11-nor-delta 9-tetrahydrocannabinol-9-carboxylic acid in urine with high-performance liquid chromatography and electrochemical detection, *J. Chromatogr.*, 1987, **414**, 187–191.

250 L. Karlsson, Direct injection of urine on a high-performance liquid chromatographic column-switching system for determination of delta 9-tetrahydrocannabinol-11-oic acid with both ultraviolet and electrochemical detection, *J. Chromatogr.* 1987, **417**, 309–317.

251 Y. Nakahara, H. Sekine, C.E. Cook, Confirmation of cannabis use. II. Determination of tetrahydrocannabinol metabolites in urine and plasma by HPLC with ECD, *J. Anal. Toxicol.*, 1989, **13**, 22–24.

252 Y. Nakahara and C.E. Cook, Confirmation of cannabis use. III. Simultaneous quantitation of six metabolites of delta 9-tetrahydrocannabinol in plasma by high-performance liquid chromatography with electrochemical detection, *J. Chromatogr.*, 1988, **434**, 247–252.

253 J. Gerostamoulos and O.H. Drummer, Incidence of psychoactive cannabinoids in drivers killed in motor vehicle accidents, *J. Forensic Sci.*, 1993, **38**, 649–656.

254 D.H. Fisher, M.I. Broudy and L.M. Fisher, Quantification of 9-carboxy-11-nor-delta 9-tetrahydrocannabinol in urine using brominated 9-carboxy-11-nor-delta 9-tetrahydrocannabinol as the internal standard and high-performance liquid chromatography with electrochemical detection, *Biomed Chromatogr.*, 1996, **10**, 161–166.

255 E.C. Nyoni, B.R. Sitaram and D.A. Taylor, Determination of delta 9-tetrahydrocannabinol levels in brain tissue using high-performance liquid chromatography with electrochemical detection, *J. Chromatogr. B*, 1996, **679**, 79–84.

256 U. Backofen, W. Hoffmann and F.M. Matysik, Determination of cannabinoids by capillary liquid chromatography with electrochemical detection, *Biomed. Chromatogr.*, 2000, **141**, 49–52.

257 U. Backofen, F.M. Matysik and C.E. Lunte, Determination of cannabinoids in hair using high-pH* non-aqueous electrolytes and electrochemical detection. Some aspects of sensitivity and selectivity, *J. Chromatogr. A*, 2002, **942**, 259–269.

258 L. Šoltés, High-performance liquid chromatographic determination of beta-adrenoceptor blocking agents in body fluids, *Biomed. Chromatogr.*, 1989, **3**, 139–152.

259 G.B. Park, R.K. Koss, S.K. O'Neil, G.P. Palace and J. Edelson, Determination of sulfinalol hydrochloride in human plasma and urine by liquid chromatography with amperometric detection, *Anal. Chem.*, 1981, **53**, 604–606.

260 W. Krause, Determination of plasma mepindolol levels by high-performance liquid chromatography and electrochemical detection, *J. Chromatogr.*, 1980, **181**, 67–75.

261 I. Meineke, S. Hartnack and C. de Mey, Determination of mepindolol in plasma after transdermal drug delivery by high-performance liquid chromatography with electrochemical detection, *J. Chromatogr.*, 1990, **526**, 240–245.

262 S.L. Perkins, B. Tattrie, P.M. Johnson and E.Z. Rabin, Analytical problems encountered during high-performance liquid chromatographic separation and coulometric detection of bopindolol metabolites in human plasma, *Ther. Drug Monit.*, 1988, **10**, 480–485.

263 H. Humbert, J. Denouel and H.P. Keller, Column liquid chromatographic determination of hydrolysed bopindolol, in the picogram per millilitre range in plasma, using cartridge extraction and dual electrochemical detection, *J. Chromatogr.*, 1987, **422**, 205–215.

264 J. Wang, M. Bonakdar and B.K. Deshmukh, Measurement of labetalol by high-performance liquid chromatography with electrochemical detection, *J. Chromatogr.*, 1985, **34**, 412–415.

265 D.R. Abernethy, E.L. Todd, J.L. Egan and G. Carrum, Labetalol analysis in human plasma using liquid chromatography with electrochemical detection. Application to pharmacokinetic studies, *J. Liq. Chromatogr.*, 1986, **9**, 2153–2163.

266 M.R. Gregg and D.B. Jack, Determination of timolol in plasma and breast milk using high-performance liquid chromatography with electrochemical detection, *J. Chromatogr.*, 1984, **305**, 244–249.

267 H. He, T.I. Edeki and A.J.J. Wood, Determination of low plasma timolol concentrations following topical application of timolol eye drops in humans by high-performance liquid chromatography with electrochemical detection, *J. Chromatogr. B,* 1994, **661**, 351–356.

268 M. Tkaczykova and L. Safarik, High-performance liquid chromatography of deacetylmetipranolol in plasma, *J. Pharm. Biomed. Anal.*, 1989, **7**, 1805–1810.

269 M.I. Maguregui, R.M. Alonso and R.M. Jiménez, High-performance liquid chromatography with amperometric detection applied to the screening of beta-blockers in human urine, *J. Chromatogr. B,* 1995, **674**, 85–91.

270 A. Álvarez-Lueje, L. Naranjo, L.J. Núñez-Vergara and J.A. Squella, Electrochemical study of nisoldipine: analytical application in pharmaceutical forms and photodegradation, *J. Pharm. Biomed. Anal.*, 1998, **16**, 853–862.

271 H. Suzuki, S. Fujiwara, S. Kondo and I. Sugimoto, Determination of nifedipine in human plasma by high-performance liquid chromatography with electrochemical detection, *J. Chromatogr.*, 1985, 341, 341–347.

272 N.D. Huebert, M. Spedding and K.D. Haegele, Quantitative analysis of the dihydropyridines, 3-(2-furoyl)-5-methoxycarbonyl-2,6-dimethyl-4-(2-nitro-phenyl)-1,4-dihydropyridine and nifedipine, by high-performance liquid chromatography with electrochemical detection, *J. Chromatogr.*, 1986, **353**, 175–180.

273 M. Telting-Diaz, M.T. Kelly, C. Hua and M.R. Smyth, High-performance liquid chromatographic determination of nifedipine, nicardipine and pindolol using a carbon fibre flow-through amperometric detector, *J. Pharm. Biomed. Anal.*, 1991, **9**, 889–893.

274 V. Horváth, A. Hrabéczy-Páll, Z. Niegreisz, E. Kocsi, G. Horvai, L. Gödörházy, A. Tolokán, I. Klebovich and K. Balogh-Nemes, Sensitive high-performance liquid chromatographic determination of nifedipine in dog plasma using an automated sample preparation system with laboratory robot, *J. Chromatogr.*, 1996, **686**, 211–219.

275 K. Shimooka, Y. Sawada and H. Tatematsu, Analysis of amlodipine in serum by a sensitive high-performance liquid chromatographic method with amperometric detection, *J. Pharm. Biomed. Anal.*, 1989, **7**, 1267–1272.

276 M. Josefsson, A.L. Zackrisson and B. Norlander, Sensitive high-performance liquid chromatographic analysis of amlodipine in human plasma with amperometric detection and a single-step solid-phase sample preparation, *J. Chromatogr. B,* 1995, **672**, 310–313.

277 M. Josefsson and B. Norlander, Coupled-column chromatography on a Chiral-AGP phase for determination of amlodipine enantiomers in human plasma: An HPLC assay with electrochemical detection, *J. Pharm. Biomed. Anal.*, 1996, **15**, 267–277.

278 Y. Iida, Y. Kinouchi, Y. Takeichi, T. Imai and M. Otagiri, Simultaneous determination of a new dihydropyridine calcium antagonist (MPC-1304) and its metabolite in dog plasma by high-performance liquid chromatography with electrochemical detection, *J. Chromatogr.*, 1991, **571**, 277–282.

279 H. Fujitomo, I. Nishino, K. Ueno and T. Umeda, Determination of the enantiomers of a new 1,4-dihydropyridine calcium antagonist in dog plasma by achiral/chiral coupled high-performance liquid chromatography with electrochemical detection, *J. Pharm. Sci.*, 1993, **82**, 319–322.

280 T. Ohkubo, T. Uno and K. Sugawara, Liquid chromatographic determination of manidipine in serum with electrochemical detection, *J. Chromatogr. B Biomed. Appl.*, 1996, **687**, 413–418.

281 K. Takamura, F. Kusu, H. Abdel-Wadood, N. El-Rabbat, G. Saleh and I. Refaat, Redox properties of isradipine and its electrochemical detection in the HPLC determination of the compound in human serum, *Biomed. Chromatogr.*, 2000, **14**, 453–458.

282 P. Höglund and L.G. Nilsson, Liquid chromatographic determination of diltiazem and its metabolites using *trans* isomers as internal standards, with dynamic modification of the solid phase by addition of an amine to the mobile phase, *J. Chromatogr.*, 1987, **414**, 109–120.

283 K. Morishima, R. Tahara and M. Horiuchi, Determination of a new calcium antagonist, sesamodil fumarate (SD-3211), and its metabolite in plasma by liquid chromatography with electrochemical detection, *J. Chromatogr.*, 1990, **527**, 381–388.

284 M. Kurono, K. Suzuki, K., Yoshida and S. Naruto, Simultaneous determination of a novel calcium entry blocker, monatepil maleate, and its metabolites in rat plasma by means of solid-phase extraction and reversed-phase liquid chromatography with electrochemical detection, *J. Chromatogr. B*, 1997, **689**, 427–432.

285 M. Kurono, K. Yoshida, S. Arakawa and S. Naruto, Determination of a novel calcium entry blocker, AJ-2615, in plasma using solid-phase extraction and high-performance liquid chromatography with electrochemical detection, *J. Chromatogr.*, 1990, **532**, 175–180.

286 K. Kojima, M. Yamanaka, Y. Nakanishi and S. Arakawa, High-performance liquid chromatographic determination of etilefrine in human plasma using combined solid-phase and organic solvent extraction and electrochemical detection, *J. Chromatogr.*, 1990, **525**, 210–217.

287 R. Gifford, W.C. Randolph, F.C. Heineman and J.A. Ziemniak, Analysis of epinine and its metabolites in man after oral administration of its pro-drug ibopamine using high-performance liquid chromatography with electrochemical detection, *J. Chromatogr.*, 1986, **381**, 83–93.

288 P.C. Davis, Determination of xamoterol in human plasma by high-performance liquid chromatography with electrochemical detection, *J. Chromatogr.*, 1987, **417**, 233–235.

289 V.L. Osborne and W.C. Randolph, Analysis of SKandF 82526 in plasma by high-performance liquid chromatography with electrochemical detection, *J. Chromatogr.*, 1983, **255**, 491–496.

290 V.K. Boppana, K.M. Dolce, M.J. Cyronak and J.A. Ziemniak, Simplified procedures for the determination of fenoldopam and its metabolites in human plasma by high-performance liquid chromatography with electrochemical detection: comparison of manual and robotic sample preparation methods, *J. Chromatogr.*, 1989, **487**, 385–399.

291 P.-O. Lagerström, P. Carlebom, A.F. Clarke and D.B. Jack, Determination of prenalterol in plasma and urine by liquid chromatography with electrochemical detection, *J. Chromatogr.*, 1984, **307**, 230–234.

292 H. Husseini, V. Mitrovic and M. Schlepper, Rapid and sensitive assay of dobutamine in plasma by high-performance liquid chromatography and electrochemical detection, *J. Chromatogr.*, 1993, **620**, 164–168.

293 F. Smedes, J.C. Kraak and H. Poppe, Simple and fast solvent extraction system for selective and quantitative isolation of adrenaline, noradrenaline and dopamine from plasma and urine, *J. Chromatogr.*, 1982, **231**, 25–39.

294 A. Martinsson, S. Bevegard and P. Hjemdahl, Analysis of phenylephrine in plasma: initial data about the concentration-effect relationship, *Eur. J. Clin. Pharmacol.*, 1986, **30**, 427–431.

295 V. Vuma and I. Kanfer, High-performance liquid chromatographic determination of phenylephrine in human serum with coulometric detection, *J. Chromatogr. B*, 1996, **678**, 245–252.

296 K. Tagawa, T. Ueki, M. Mizobe, K. Noda and M. Samejima, Determination of denopamine in human and dog plasma by high-performance liquid chromatography with electrochemical detection, *J. Chromatogr.*, 1990, **529**, 500–506.

297 P.R. Baker, J.J. Gardner, W.J. Lockley and D. Wilkinson, Determination of dopexamine hydrochloride in human blood by high-performance liquid chromatography with electrochemical detection, *J. Chromatogr. B*, 1995, **667**, 283–290.

298 D.E. Leelavathi, E.F. Soffer, D.E. Dressler and J. Knowles, Liquid chromatographic assay using electrochemical detection for the quantitation of indoramin in human plasma, *J. Pharm. Sci.*, 1986, **75**, 421–423.

299 K. Zech and R. Huber, Determination of urapidil and its metabolites in human serum and urine: comparison of liquid-liquid and fully automated liquid-solid extraction, *J. Chromatogr.*, 1986, **353**, 351–360.

300 K. Ravichandran and R.P. Baldwin, Determination of hydralazine and metabolites in urine by liquid chromatography with electrochemical detection, *J. Chromatogr.*, 1985, **343**, 99–108.

301 G. Carrum, D.R. Abernethy, M. Sadhukhan and C.E. Wright, Minoxidil analysis in human plasma using high-performance liquid chromatography with electrochemical detection, Application to pharmacokinetic studies, *J. Chromatogr.*, 1986, **381**, 127–135.

302 A. Hashem and B. Lubczyk, Determination of isoxsuprine in equine plasma by high-performance liquid chromatography with electrochemical detection, *J. Chromatogr.*, 1991, **563**, 216–223.

303 F.J. Flores-Murrieta, G. Castañeda-Hernández and E., Hong, Determination of pelanserin in human plasma by high-performance liquid chromatography with amperometric detection, *J. Chromatogr.*, 1990, **528**, 501–508.

304 J.T. Stewart and S.S. Clark, Liquid chromatographic determination of guanethidine salts and hydrochlorothiazide using electrochemical detection and ion-pair techniques, *J. Pharm. Sci.*, 1986, **75**, 413–415.

305 M.J. Cooper, R.F. O'Dea and B.L. Mirkin, Determination of methyldopa and metabolites in human serum by high-performance liquid chromatography with electrochemical detection, *J. Chromatogr.*, 1979, **162**, 601–604.

306 J.A. Hoskins and S.B. Holliday, Determination of alpha-methyldopa and methyldopate in human breast milk and plasma by ion-exchange chromatography using electrochemical detection, *J. Chromatogr.*, 1982, **230**, 162–167.

307 H. Ong, S. Sved and N. Beaudoin., Assay and stability of alpha-methyldopa in man using high-performance liquid chromatography with electrochemical detection, *J. Chromatogr.*, 1982, **229,** 433–438.

308 E. Matsushima and T. Sato, Determination of bisaramil and its metabolite in plasma using high-performance liquid chromatography with electrochemical detection, *J. Chromatogr.*, 1992, **573**, 339–344.

309 K. Shimada, M. Tanaka, T. Nambara, Y. Imai, K. Abe and K. Yoshinaga, Determination of captopril in human blood by high-performance liquid chromatography with electrochemical detection, *J. Chromatogr.*, 1982, **227**, 445–451.

310 A. Lippi, M. Criscuoli, G. Sardelli and A. Subissi, High-performance liquid chromatographic method with electrochemical detection for the determination of idrapril, a novel angiotensin-converting enzyme inhibitor, in biological matrices, *J. Chromatogr. B,* 1994, **660**, 127–134.

311 J.A. Prieto, R.M. Jimenez and R.M. Alonso, Quantitative determination of the angiotensin-converting enzyme inhibitor cilazapril and its active metabolite cilazaprilat in pharmaceuticals and urine by high-performance liquid chromatography with amperometric detection, *J. Chromatogr. B,* 1998, **714,** 285–292.

312 K. Richter, R. Oertel and W. Kirch, New sensitive method for the determination of hydrochlorothiazide in human serum by high-performance liquid chromatography with electrochemical detection, *J. Chromatogr. A,* 1996, **729**, 293-296.

313 M.B. Barroso, R.M. Jiménez, R.M. Alonso and E. Ortiz, Determination of piretanide and furosemide in pharmaceuticals and human urine by high-performance liquid chromatography with amperometric detection, *J. Chromatogr. B,* 1996, **675**, 303–312.

314 M.J. Legorburu, R.M. Alonso, R.M. Jiménez and E. Ortiz, Quantitative determination of the loop diuretic bumetanide in urine and pharmaceuticals by high-performance liquid chromatography with amperometric detection, *J. Chromatogr. Sci.*, 2001, **39**, 425–430.

315 M.B. Barroso, R.M. Alonso and R.M. Jiménez, Simultaneous determination of torasemide and its major metabolite M5 in human urine by high-performance liquid chromatography-electrochemical detection, *J. Chromatogr. Sci.*, 2001, **39**, 491–496.

316 R.M. Riggin, A.L. Schmidt and P.T. Kissinger, Determination of acetamino-
     phen in pharmaceutical preparations and body fluids by high-performance
     liquid chromatography with electrochemical detection, *J. Pharm. Sci.*, 1975,
     64, 680–683.
317 D.J. Miner and P.T. Kissinger, Trace determination of acetaminophen in
     serum, *J. Pharm. Sci.*, 1979, **68**, 96–97.
318 M. Hamilton and P.T. Kissinger, Determination of acetaminophen metabolites
     in urine by liquid chromatography/electrochemistry, *Anal. Biochem.*, 1982,
     125, 143–148.
319 J.M. Wilson, J.T. Slattery, A.J. Forte and S.D. Nelsonz, Analysis of acetamino-
     phen metabolites in urine by high-performance liquid chromatography with UV
     and amperometric detection, *J. Chromatogr.*, 1982, **227**, 453–462.
320 J.L. Palmer, Novel method of sample preparation for the determination of
     paracetamol in plasma by high-performance liquid chromatography with
     electrochemical detection, *J. Chromatogr.*, 1986, **382,** 338–342.
321 F.X. Walsh, P.J. Langlais and E.D. Bird, Liquid-chromatographic identifica-
     tion of acetaminophen in cerebrospinal fluid with use of electrochemical
     detection, *Clin. Chem.*, 1982, **28**, 382–383.
322 R. Whelpton, K. Fernandes, K.A. Wilkinson and D.R. Goldhill, Determination
     of paracetamol (acetaminophen) in blood and plasma using high performance
     liquid chromatography with dual electrode coulometric quantification in the
     redox mode, *Biomed. Chromatogr.*, 1993, **7**, 90–93.
323 D. Evans, J.P. Hart and G. Rees, Voltammetric behaviour of salicylic acid at a
     GCE and its determination in serum using liquid chromatography with
     amperometric detection, *Analyst,* 1991, **116**, 803–806.
324 L.C. Burton, N.J. Loftus, D.W. Vere and R. Whelpton, Determination of
     plasma nefopam by liquid chromatography and electrochemical detection,
     J. Chromatogr., 1990, **526**, 159–168.
325 R.J. Flanagan and N.M. Ward, HPLC measurement of meptazinol in plasma
     using electrochemical oxidation detection, *Biomed. Chromatogr.*, 1986, **1**,
     180–182.
326 F. Plavšić and J. Čulig, Determination of serum diclofenac by high-
     performance liquid chromatography by electromechanical detection, *Hum.
     Toxicol.*, 1985, **4**, 317–322.
327 L. Zecca and P. Ferrario, Determination of diclofenac in plasma and synovial
     fluid by high-performance liquid chromatography with electrochemical
     detection, *J. Chromatogr.*, 1989, **495**, 303–308.
328 L. Zecca, P. Ferrario and P. Costi, Determination of diclofenac and its
     metabolites in plasma and cerebrospinal fluid by high-performance liquid
     chromatography with electrochemical detection, *J. Chromatogr.*, 1991, **567**,
     425–432.
329 W. Jin and J. Zhang, Determination of diclofenac sodium by capillary zone
     electrophoresis with electrochemical detection, *J. Chromatogr. A,* 2000, **868**,
     101–107.
330 L. Zecca, P. Ferrario, R. Pirola, F. Zambotti, N. Zonta and F. Fraschini,
     Analysis of pirprofen in cerebrospinal fluid, plasma, and synovial fluid

by high-performance liquid chromatography with electrochemical detection, *J. Pharm. Sci.*, 1989, **78**, 776–779.

331 O. Kuhlmann, G. Stoldt, H.-G. Struck and G.-J. Krauss, Simultaneous determination of diclofenac and oxybuprocaine in human aqueous humor with HPLC and electrochemical detection, *J. Pharm. Biomed. Anal.*, 1998, **17**, 1351–1356.

332 J.E. Torres-Lopez, M.B. Robles, J. Perez-Urizar, F.J. Flores-Murrieta and V. Granados-Soto, Determination of diclofenac in micro-whole blood samples by high-performance liquid chromatography with electrochemical detection, Application in a pharmacokinetic study, *Arzneimittelforschung,* 1997, **47**, 1040–1043.

333 T. Suwa, H. Urano, Y. Shinohara and J. Kokatsu, Simultaneous high-performance liquid chromatographic determination of lornoxicam and its 5'-hydroxy metabolite in human plasma using electrochemical detection, *J. Chromatogr.*, 1993, **617**, 105–110.

334 A.D. de Jager, H. Ellis, H.K.L. Hundt, K.J. Swart and A.F. Hundt, High-performance liquid chromatographic determination with amperometric detection of piroxicam in human plasma and tissues, *J. Chromatogr. B,* 1999, **729**, 183–189.

335 T. Kimura, O. Shirota and Y. Ohtsu, Analysis of ibuprofen metabolites by semi-microcolumn liquid chromatography with ultraviolet absorption and pulsed amperometric detectors, *J. Pharm. Biomed. Anal.*, 1997, **15**, 1521–1526.

336 L.O. Healy, J.P. Murrihy, A. Tan, D. Cocker, M. McEnery and J.D. Glennon, Enantiomeric separation of *R,S*-naproxen by conventional and nano-liquid chromatography with methyl-beta-cyclodextrin as a mobile phase additive, *J. Chromatogr. A,* 2001, **924**, 459–464.

337 F. Tagliaro, D. Franchi, R. Dorizzi and M. Marigo, High-performance liquid chromatographic determination of morphine in biological samples: an overview of separation methods and detection techniques, *J. Chromatogr.*, 1989, **488**, 215–228.

338 M.W. White, Determination of morphine and its major metabolite, morphine-3-glucuronide, in blood by high-performance liquid chromatography with electrochemical detection, *J. Chromatogr.*, 1979, **178**, 229–240.

339 J.E. Wallace, S.C. Harris and M.W. Peek, Determination of morphine by liquid chromatography with electrochemical detection, *Anal. Chem.*, 1980, **52**, 1328–1330.

340 R.D. Todd, S.M. Muldoon and R.L. Watson, Determination of morphine in cerebrospinal fluid and plasma by high-performance liquid chromatography with electrochemical detection, *J. Chromatogr.*, 1982, **232**, 101–110.

341 K.R. Bedford and P.C. White, Improved method for the simultaneous determination of morphine, codeine and dihydrocodeine in blood by high-performance liquid chromatography with electrochemical detection, *J. Chromatogr.*, 1985, **347**, 398–404.

342 J.-O. Svensson, Determination of morphine, morphine-6-glucuronide and normorphine in plasma and urine with high-performance liquid chromatography and electrochemical detection, *J. Chromatogr.*, 1986, **375**, 174–178.

343 J. Zoer, P. Virgili and J.A. Henry, High-performance liquid chromatographic assay for morphine with electrochemical detection using an unmodified silica column with a non-aqueous ionic eluent, *J. Chromatogr.*, 1986, **382**, 189–197.

344 F. Tagliaro, G. Carli, R. Dorizzi and M. Marigo, Direct injection high-performance liquid chromatographic assay of morphine with electrochemical detection, a polymeric column and an alkaline eluent, *J. Chromatogr.*, 1990, **507**, 253–258.

345 J.G. Besner, C. Band, J.J. Rondeau, L. Yamlahi, G. Caille, F. Varin and J. Stewart, Determination of opiates and other basic drugs by high-performance liquid chromatography with electrochemical detection, *J. Pharm. Biomed. Anal.*, 1989, **7**, 1811–1817.

346 A.I. Bouquillon, D. Freeman and D.E. Moulin, Simultaneous solid-phase extrmaction and chromatographic analysis of morphine and hydromorphone in plasma by high-performance liquid chromatography with electrochemical detection, *J. Chromatogr.*, 1992, **577**, 354–357.

347 C.F. McLean, L.E. Mather, J. Odontiadis and P.A. Sloan, Improved method for morphine determination in biological fluids and tissues: rapid, sensitive and selective, *J. Pharm. Pharmacol.*, 1990, **42**, 669–671.

348 J.L. Mason, S.P. Ashmore and A.R. Aitkenhead, Simple method for the determination of morphine and its active glucuronide metabolite in human plasma by high-performance liquid chromatography with electrochemical detection, *J. Chromatogr.*, 1991, **570**, 191–197.

349 C.P.W.G.M. Verwey-Van Wissen, P.M. Koopman-Kimenai and T.B. Vree, Direct determination of codeine, norcodeine, morphine and normorphine with their corresponding *O*-glucuronide conjugates by high-performance liquid chromatography with electrochemical detection, *J. Chromatogr.*, 1991, **570**, 309–320.

350 R. Whelpton, Morphine analogues as internal standards in liquid chromatographic assay, in *Methods of Morphine Estimation in Biological Fluids and the Concept of Free Morphine,* J.F.B. Stuart (ed) Academic Press/Royal Society of Medicine, London, 1983, 15–20.

351 J. Gerostamoulos, K. Cramp, I.M. McIntyre and O.H. Drummer, Simultaneous determination of 6-monoacetylmorphine, morphine and codeine in urine using high-performance liquid chromatography with combined ultraviolet and electrochemical detection, *J. Chromatogr.*, 1993, **617**, 152–156.

352 W. Hanisch and L.V. Meyer, Determination of the heroin metabolite 6-monoacetyl-morphine in urine by high-performance liquid chromatography with electrochemical detection, *J. Anal. Toxicol.*, 1993, **17**, 48–50.

353 P.P. Rop, F. Grimaldi, J. Burle, M.N. De Saint Leger and A. Viala, Determination of 6-monoacetylmorphine and morphine in plasma, whole blood and urine using high-performance liquid chromatography with electrochemical detection, *J. Chromatogr. B,* 1994, **661**, 245–253.

354 J. Gerostamoulos and O.H. Drummer, Solid phase extraction of morphine and its metabolites from postmortem blood. *Forensic Sci. Int.*, 1995, **77**, 53–63.

355 B.A. Rashid, G.W. Aherne, M.F. Katmeh, P. Kwasowski and D. Stevenson, Determination of morphine in urine by solid-phase immunoextraction and

high-performance liquid chromatography with electrochemical detection, *J. Chromatogr. A*, 1998, **797**, 245–250.

356 W.M. Heybroek, M. Caulfield, A. Johnston and P. Turner, Automatic on-line extraction coupled with electrochemical detection as an improved method for the HPLC co-analysis of codeine and morphine in plasma and gastric juice, *J. Pharm. Biomed. Anal.*, 1990, **8**, 1021–1027.

357 J.O. Svensson, Q.Y. Yue and J. Sawe, Determination of codeine and metabolites in plasma and urine using ion-pair high-performance liquid chromatography, *J. Chromatogr. B*, 1995, **674**, 49–55.

358 M. Johansen, F. Tønnesen and K.E. Rasmussen, Column-switching high-performance liquid chromatographic detection of pholcodine and its metabolites in urine with fluorescence and electrochemical detection, *J. Chromatogr.*, 1992, **573**, 283–288.

359 M.T. Smith, J.A. Watt, G.P. Mapp and T. Cramond, Quantitation of oxycodone in human plasma using high-performance liquid chromatography with electrochemical detection, *Ther. Drug Monit.*, 1991, **13**, 126–130.

360 A.W. Wright, J.A. Lawrence, M. Iu, T. Cramond and M.T. Smith, Solid-phase extraction method with high-performance liquid chromatography and electrochemical detection for the quantitative analysis of oxycodone in human plasma, *J. Chromatogr. B*, 1998, **712**, 169–175.

361 L. Debrabandere, M. van Boven and P. Daenens, High-performance liquid chromatography with electrochemical detection of buprenorphine and its major metabolite in urine, *J. Chromatogr.*, 1991, **564**, 557–566.

362 E. Schleyer, R. Lohmann, C. Rolf, A. Gralow, C.C. Kaufmann, M. Unterhalt and W. Hiddemann, Column-switching solid-phase trace-enrichment high-performance liquid chromatographic method for measurement of buprenorphine and norbuprenorphine in human plasma and urine by electrochemical detection, *J. Chromatogr.*, 1993, **614**, 275–283.

363 R.J. Flanagan, J.D. Ramsey and I. Jane, Measurement of dextropropoxyphene and nordextropropoxyphene in biological fluids, *Hum. Toxicol.*, 1984, **3**, 103S–114S.

364 R. Lucek and R. Dixon, Quantitation of levorphanol in plasma using high-performance liquid chromatography with electrochemical detection, *J. Chromatogr.*, 1985, **341**, 239–243.

365 P.-Å. Hynning, P. Anderson, U. Bondesson and L.O. Boréus, Determination of ketobemidone in plasma by high-performance liquid chromatography with electrochemical detection, *J. Chromatogr.*, 1986, **375**, 207–211.

366 M. Valle, J.M. Pavon, R. Calvo, M.A. Campanero and I.F. Troconiz, Simultaneous determination of tramadol and its major active metabolite *O*-demethyltramadol by high-performance liquid chromatography with electrochemical detection, *J. Chromatogr. B*, 1999, **724**, 83–89.

367 E.F. O'Connor, S.W.T. Cheng and W.G. North, Simultaneous extraction and chromatographic analysis of morphine, dilaudid, naltrexone and naloxone in biological fluids by high-performance liquid chromatography with electrochemical detection, *J. Chromatogr.*, 1989, **491**, 240–247.

368 R.W. Reid, A. Deakin and D.J. Leehey, Measurement of naloxone in plasma using high-performance liquid chromatography with electrochemical detection, *J. Chromatogr.*, 1993, **614**, 117–122.

369 J.Z. Chou, H. Albeck and M.J. Kreek, Determination of nalmefene in plasma by high-performance liquid chromatography with electrochemical detection and its application in pharmacokinetic studies, *J. Chromatogr.*, 1993, **613**, 359–364.

370 C.L. Lake, C.A. DiFazio, E.N. Duckworth, J.C. Moscicki, J.S. Engle and C.G. Durbin, High-performance liquid chromatographic analysis of plasma levels of nalbuphine in cardiac surgical patients, *J. Chromatogr.*, 1982, **233**, 410–416.

371 N. Wetzelsberger, P.W. Lucker and W. Erking, Internally standardized method for the determination of nalbuphine in human plasma by means of high performance liquid chromatography with electrochemical coulometric detection, *Arzneimittelforschung*, 1988, **38**, 768–771.

372 E. Nicolle, S. Veitl, C. Guimier and G. Bessard, Modified method of nalbuphine determination in plasma: validation and application to pharmacokinetics of the rectal route, *J. Chromatogr. B*, 1997, **690**, 89–97.

373 F. de Cazanove, J.M. Kinowski, M. Audran, A. Rochette and F. Bressolle, Determination of nalbuphine in human plasma by high-performance liquid chromatography with electrochemical detection, Application to a pharmacokinetic study. *J. Chromatogr. B*, 1997, **690**, 203–210.

374 J. Osinski, A. Wang, J.A. Wu, J.F. Foss and C.S. Yuan, Determination of methylnaltrexone in clinical samples by solid-phase extraction and high-performance liquid chromatography for a pharmacokinetics study, *J. Chromatogr. B*, 2002, **780**, 251–259.

375 F. Palmisano, E. Desimoni and P.G. Zambonin, High-performance liquid chromatography with polarographic and voltammetric anodic detection: simultaneous determination of allopurinol, oxipurinol and uric acid in body fluids, *J. Chromatogr.*, 1984, **306**, 205–214.

376 E.J. Eisenberg, P. Conzentino, G.G. Liversidge and K.C. Cundy, Determination of allopurinol and oxypurinol in rat plasma, intestinal wash, and bile by high-performance liquid chromatography with electrochemical detection (HPLC/EC) following automated solid phase extraction, *Pharm. Res.*, 1991, **8**, 653–655.

377 E.J. Eisenberg, P. Conzentino, G.G. Liversidge and K.C. Cundy, Simultaneous determination of allopurinol and oxypurinol by liquid chromatography using immobilized xanthine oxidase with electrochemical detection, *J. Chromatogr.*, 1990, **530**, 65–73.

378 F.J. Flores-Murrieta, C. Hoyo-Vadillo, E. Hong and G. Castañeda-Hernández, Assay of ambroxol in human plasma by high-performance liquid chromatography with amperometric detection, *J. Chromatogr.*, 1989, **490**, 464–464.

379 F. Bai, M.N. Kirstein, S.K. Hanna and C.F. Stewart, New liquid chromatographic assay with electrochemical detection for the measurement of amifostine and WR1065, *J. Chromatogr. B*, 2002, **772**, 257–265.

380 K. Ishikawa, J.L. Martinez and J.L. McGaugh, Simple determination of *p*-hydroxyamphetamine by high-performance liquid chromatography with electrochemical detection, *J. Chromatogr.*, 1984, **306**, 394–397.

381 M.Y. Yousif, R.L. Fitzgerald, N. Narasimhachari, J.A. Rosecrans, R.V. Blanke and R.A. Glennon, Identification of metabolites of 3,4-methylenedioxy-methamphetamine in rats, *Drug Alcohol Depend.*, 1990, **26**, 127–135.

382 N.A. Santagati, G. Ferrara, A. Marrazzo and G. Ronsisvalle, Simultaneous determination of amphetamine and one of its metabolites by HPLC with electrochemical detection, *J. Pharm. Biomed. Anal.*, 2002, **30**, 247–255.

383 R.E. Michel, A.B. Rege and W.J. George, High-pressure liquid chromato-graphy/electrochemical detection method for monitoring MDA and MDMA in whole blood and other biological tissues, *J. Neurosci. Methods,* 1993, **50**, 61–66.

384 L. Millerioux, M. Brault, V. Gualano and A. Mignot, High-performance liquid chromatographic determination of baclofen in human plasma, *J. Chromatogr. A,* 1996, **729**, 309–314.

385 H.E. Wiersma, C.J. van Boxtel, J.J. Butter, W.M. van Aalderen, T. Omari and M.A. Benninga, Pharmacokinetics of a single oral dose of baclofen in pediatric patients with gastroesophageal reflux disease, *Ther. Drug Monit.*, 2003, **25**, 93–98.

386 N. Fernández, J.J. García, M.J. Diez, M.T. Terán and M. Sierra, Rapid high-performance liquid chromatographic assay of ethynyloestradiol in rabbit plasma, *J. Chromatogr.*, 1993, **619**, 143–147.

387 J.-J. Descombe, G. Doumont and M. Picard, Determination of exifone in human plasma and urine by high-performance liquid chromatography with electrochemical detection, *J. Chromatogr.*, 1989, **496**, 345–353.

388 M.-Z. Wang, D.-F. Wu and Y.-W. Yu, High-performance liquid chromato-graphy with electrochemical detection of gossypol in human plasma, *J. Chromatogr.*, 1985, **343**, 387–396.

389 K.J. Lee and K. Dabrowski, High-performance liquid chromatographic deter-mination of gossypol and gossypolone enantiomers in fish tissues using simultaneous electrochemical and ultraviolet detectors, *J. Chromatogr. B,* 2002, **779**, 313–319.

390 M. Verschraagen, T.H. Zwiers, P.E. de Koning, J. Welink and W.J. van der Vijgh, Quantification of BNP7787 (dimesna) and its metabolite mesna in human plasma and urine by high-performance liquid chromatography with electrochemical detection, *J. Chromatogr. B,* 2001, **753**, 293–302.

391 M. Verschraagen, M. Bosma, T.H. Zwiers, E. Torun and W.J. van der Vijgh, Quantification of mesna and total mesna in kidney tissue by high-performance liquid chromatography with electrochemical detection, *J. Chromatogr. B,* 2003, **783**, 33–42.

392 H. Lamparczyk, A. Chmielewska, L. Konieczna, A. Plenis and P.K. Zarzycki, RP-HPLC method with electrochemical detection for the determination of metoclopramide in serum and its use in pharmacokinetic studies, *Biomed. Chromatogr.*, 2001, **15**, 513–517.

393 C.Y. Chien, J.N. Diana and P.A. Crooks, High-performance liquid chromatography with electrochemical detection for the determination of nicotine in plasma, *J. Pharm. Sci*, 1988, **77**, 277–279.

394 G.N. Mahoney and W. Al-Delaimy, Measurement of nicotine in hair by reversed-phase high-performance liquid chromatography with electrochemical detection, *J. Chromatogr. B*, 2001, **753**, 179–187.

395 R. Kysilka, Determination of psilocin in rat urine by high-performance liquid chromatography with electrochemical detection, *J. Chromatogr.*, 1990, **534**, 287–290.

396 H. Lindenblatt, E. Krämer, P. Holzmann-Erens, E. Gouzoulis-Mayfrank and K.-A. Kovar, Quantitation of psilocin in human plasma by high-performance liquid chromatography and electrochemical detection: comparison of liquid-liquid extraction with automated on-line solid-phase extraction, *J. Chromatogr. B*, 1998, **709**, 255–263.

397 F. Hasler, D. Bourquin, R. Brenneisen and F.X. Vollenweider, Renal excretion profiles of psilocin following oral administration of psilocybin: a controlled study in man, *J. Pharm. Biomed. Anal.*, 2002, **30**, 331–339.

398 F. Hasler, D. Bourquin, R. Brenneisen, T. Bär and F.X. Vollenweider, Determination of psilocin and 4-hydroxyindole-3-acetic acid in plasma by HPLC-ECD and pharmacokinetic profiles of oral and intravenous psilocybin in man, *Pharm. Acta Helv.*, 1997, **72**, 175–184.

399 H. Gao, S. Roy, F. Donati and F. Varin, Determination of succinylcholine in human plasma by high-performance liquid chromatography with electrochemical detection, *J. Chromatogr. B*, 1998, **718**, 129–134.

400 S.H. Hansen, Simple and rapid method for the simultaneous determination of the eight main metabolites and conjugates of sulphasalazine in human plasma, urine and faeces using dynamically modified silica, *J. Chromatogr.*, 1989, **491**, 175–185.

401 L.E. Gustavson and S. Chu, High-performance liquid chromatographic procedure for the determination of tiagabine concentrations in human plasma using electrochemical detection, *J. Chromatogr.*, 1992, **574**, 313–318.

402 O.Y.-P. Hu, C.-H. Chou, W. Ho and S.-T. Ho, Determination of vecuronium in blood by HPLC with UV and electrochemical detection: A pilot study in man, *Proc. Nat. Sci. Republic China B: Life Sci.*, 1991, **15**, 186–190.

403 J. Ducharme, F. Varin, D.R. Bevan, F. Donati and Y. Théorêt, High-performance liquid chromatography-electrochemical detection of vecuronium and its metabolites in human plasma, *J. Chromatogr.*, 1992, **573**, 79–86.

404 E.C. Lewis and D.C. Johnson, Evaluation of a glassy-carbon electrode for amperometric detection of selected methylxanthines in serum after their separation by reversed-phase chromatography, *Clin. Chem.*, 1978, **24**, 1711–1719.

405 M.S. Greenberg and W.J. Mayer, High-performance liquid chromatographic determination of theophylline and its derivatives with electrochemical detection, *J. Chromatogr.*, 1979, **169**, 321–327.

CHAPTER 7

Applications of HPLC-ED in Toxicology and Related Areas

Although the principal area of application of HPLC-ED has been in the analysis of naturally-occurring analytes, such as catecholamines, and pharmaceuticals in biological samples, HPLC-ED has also been applied to the analysis of pesticides and other analytes of interest to the toxicologist. This section details some of these additional applications.

1 Ammatoxins

Urine concentrations of α-amanitin (Figure 7.1, from *Amanita phalloides)* have been measured by HPLC-ED.[1] Sample preparation was by mixed mode SPE (Bond-Elut Certify) and the extracts were analysed using an ODS-modified silica column with acetonitrile-phosphate buffer (5 mmol L^{-1}, pH 7.2) (1 + 9) as eluent. Detection was at PGEs in the screen mode (E$_1$: +0.05 V, E$_2$: +0.5 V *vs* Pd). The intra-assay RSD at 10 μg L^{-1} was 3.6%, but the accuracy at this concentration was only 77%. Furthermore the calibration curve appears to be non-linear, although linear calibration was applied.

2 Aromatic Amines and Nitro-compounds

Felice *et al.*[2] used a GCE (+0.7 to +0.9 V *vs* Ag/AgCl) in the analysis of polycyclic aromatic amines, such as 2-aminonaphthalene, 4-aminobiphenyl, and 2-aminoanthracene, in rodent skin samples after topical application of these compounds. The HPLC system used consisted of an ODS-modified silica column with acetonitrile-aq. citrate/perchlorate buffer (7 + 3 or thereabouts) as eluent – the buffer composition and the proportion of acetonitrile were varied in different experiments. A LoD of 0.1 pmol on column could be expected.

Heterocyclic aromatic amines, such as 2-amino-3-methylimidazo(4,5-f)quinoline, arise from amino acids, proteins, *etc.* during cooking. Billedeau *et al.*[3] used a 'short-chain alkyl-modified' (SynChropak SCD-100) analytical column and ED (GCE, +0.6–0.9 V *vs* Ag/AgCl) in the analysis of these compounds. The eluent was acetonitrile-aq. ammonium acetate (50 mmol L^{-1}, pH 5.5–6.5) (25 + 75). No results of sample analyses were reported, however.

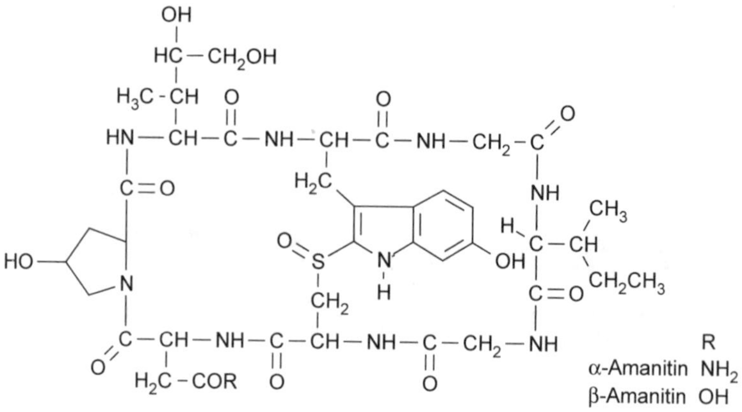

Figure 7.1 *Amanitin.*

Nitrated polycyclic aromatic hydrocarbons, such as 1-nitropyrene, 1-nitrosopyrene and 1,3-dinitropyrene, have been measured in car exhaust deposits by HPLC with chemiluminescence detection after on-line EC reduction (PGEs, −1.6 V *vs* Pd).[4] LoDs of fmol on-column were claimed.

3 Algal and Fungal Toxins

The trichothecene mycotoxins deoxynivalenol (vomitoxin), nivalenol and fusarenon-X (Figure 7.2) have been measured in wheat extracts using HPLC with post-column photolysis (Hg vapour lamp, 2.3–3 min residence time) and ED (parallel GCEs, +1.10 V and +0.85 V *vs* Ag/AgCl).[5] An ODS-modified silica analytical column was used with methanol-aq. sodium chloride (50 mmol L^{-1}) (15 + 85) as eluent. LoDs were 1–2 ng on-column.

The mycotoxin agaritine (β-*N*-(γ-L(+)glutamyl)-4-hydroxymethylphenylhydrazine, Figure 7.3) has been measured in cultivated mushrooms using an ODS-modified silica column with methanol-aq. potassium hydrogen phosphate (50 mmol L^{-1}) (2.5 + 97.5) as eluent and ED (PGEs, E$_1$: +0.6 V *vs* Pd).[6] Sample preparation (fresh produce) was by homogenisation and methanol extraction. No internal standard was used. The LoD was 5 mg kg^{-1} dry weight.

Saxitoxin (Figure 7.3) and several other algal toxins that are known to sometimes contaminate shellfish have been measured by fluorescence (excitation

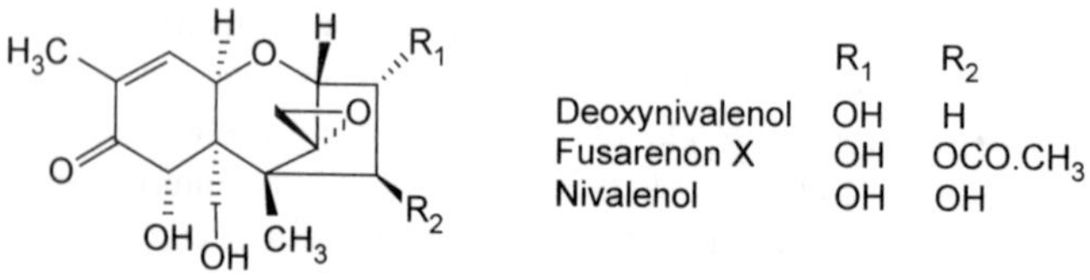

	R$_1$	R$_2$
Deoxynivalenol	OH	H
Fusarenon X	OH	OCO.CH$_3$
Nivalenol	OH	OH

Figure 7.2 *Some trichothecene mycotoxins.*

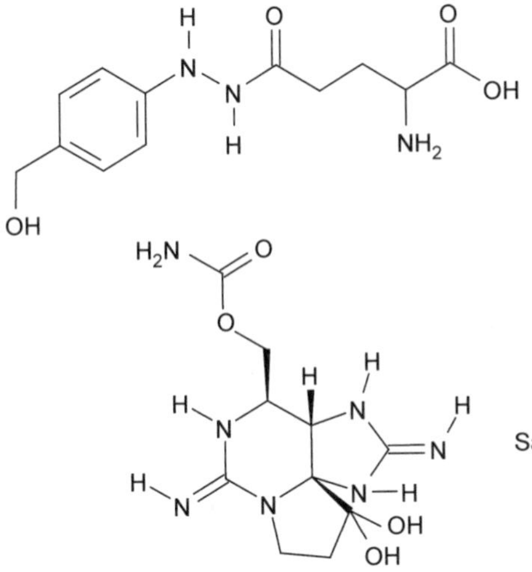

Figure 7.3 *Agaritine and saxitoxin.*

330 nm; emission 395 nm) and mass spectrometric detection after electrochemical oxidation in an ESA 5020 guard cell (PGE +1.05 V *vs* Pd).[7] Chromatographic analysis was achieved using anion-exchange and cation-exchange columns connected in series with gradient elution (20–450 mmol L^{-1} aq. ammonium acetate, pH 6.9). LoDs (S/N = 3) for saxitoxin were claimed to be 0.03 ng (fluorescence) and 0.5 ng (MS). The method was applied to the measurement of the toxins in mussels.

4 Pesticides and Related Compounds

Chlorophenols, such as 2,4-dichlorophenol, occur as impurities in chlorophenoxy herbicide formulations at concentrations of about 1% (w/w) of the herbicides. Åkerblom and Lindgren[8] used serial UV (280 nm) and ED (carbon paste electrode, +1.05 V *vs* Ag/AgCl) for the herbicides and chlorophenols, respectively. The analytical column was ODS-modified silica and the eluent was methanol-aq. acetic acid (0.1 mol L^{-1}) (9 + 11 to 11 + 9 depending on the formulation under analysis). This approach could prove useful for the analysis of these compounds in biological specimens.

The insecticide carbaryl (1-naphthyl-*N*-methylcarbamate, Figure 7.4) has been measured on filter paper exposed during aerial spraying using an ODS-modified silica column and ED (GCE, +0.75 V *vs* Ag/AgCl) after pre-column alkaline hydrolysis to 1-naphthol.[9] The eluent was acetonitrile-acetic acid-aq. potassium chloride (0.5 mol L^{-1}) (50 + 1 + 49). 2-Naphthol was the internal standard. Post-column hydrolysis with sodium hydroxide of carbaryl and some other *N*-methyl-carbamate pesticides has also been employed in measuring crop residues of these compounds.[10] The analytical column used was octyl-modified silica. Elution was

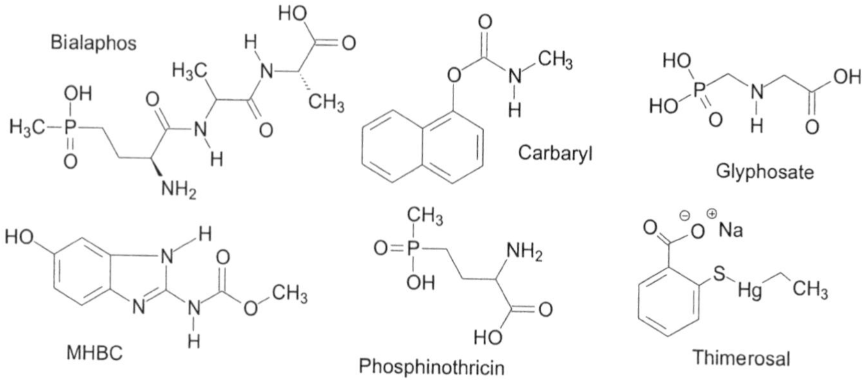

Figure 7.4 *Some pesticides and related compounds measured by HPLC-ED (MHBC =
methyl 5-hydroxy-2-benzimidazole carbamate, glufosinate is the ammonium
salt of phosphinothricin).*

by means of an acetonitrile-water gradient and ED was at PGEs (E_1: +0.20 V, E_2:
+0.60 V *vs* Pd).

Methyl 5-hydroxy-2-benzimidazole carbamate (MHBC, Figure 7.4), a urinary
metabolite of the pesticides carbendazium, benomyl and thiophanate-methyl,
has been measured using an ODS-modified silica analytical column with methanol-
aq. ammonium acetate (approximately 60 mmol L^{-1}, pH 8) (27 + 73) as eluent
and ED (PGE, +0.22 V *vs* Pd).[11] Sample preparation was by a complex procedure
involving SPE (SCX-modified silica) of hydrolysed specimens. No internal standard
was used. The LoD was 5 µg L^{-1} (1 mL sample).

The HPLC-ED of the herbicides glyphosate, glufosinate (the ammonium salt of
phosphinothricin), and bialaphos (Figure 7.4) using integrated pulsed amperometric
detection (IPAD) has been described.[12] The analytes were analysed using an
IonPac AS15 anion-exchange column with gradient elution (purified water/aq.
sodium hydroxide (0.2 mol L^{-1})). Detection was by adsorption onto a gold
working electrode at −0.3 V and oxidation at +0.23 V *vs* Ag/AgCl. The electrode surface was cleaned and reactivated by applying potentials of −0.2 V
and +0.63 V, before being returned to −0.3 V. LoDs of 50, 20, and 65 µg L^{-1}
(S/N = 3) were claimed for glyphosate, glufosinate and bialaphos, respectively.
The method was applied to 'spiked' human urine and plasma and to urine from a
mouse dosed with glyphosate (750 mg kg^{-1}).

The antiseptic thimerosal (sodium ethylmercury thiosalicylate, Figure 7.4) and
its major degradation products thiosalicylic acid and 2,2'-dithiodibenzoic acid
have been studied in pharmaceutical preparations using HPLC-ED (ODS-modified
silica or anion-exchange resin).[13] Glassy carbon or carbon paste electrodes in
wall-jet or thin-layer configuration were used (+0.9 V *vs* Ag/AgCl (+1.2 V for
2,2'-dithiodibenzoic acid)) in addition to PGEs (+0.8 V *vs* Pd, all analytes).
LoDs (PGE) ranged from 2 to 20 µg L^{-1}. A similar approach was adopted by Kang
and Kim[14] to measure methylparaben and propylparaben (methyl and propyl
4-hydroxybenzoate), and thimerosal, again in pharmaceutical preparations. Several

ODS-based columns and eluents of pH 2.5–6.0 were evaluated. Detection was at a GCE (+1.25 V *vs* Ag/AgCl).

5 Veterinary Drug and Other Food Residues

The use of HPLC-ED in the analysis of the anabolic steroids and metabolites, diethylstilbestrol, taleranol, zearalenol, zearalenone and zeranol (Figure 7.5), in mammalian tissue has been discussed.[15] Diethylstilbestrol has been measured in animal tissue using an ODS-modified silica column with methanol-aq. phosphate buffer (50 mmol L^{-1}, pH 3.5) (67 + 33) as eluent and ED (GCE, +0.9 V *vs* Ag/AgCl).[16] Sample homogenisation was followed by LLE into MTBE, back-extraction into aqueous sodium hydroxide and SPE (ODS-modified silica). The LoD was approximately 0.5 μg kg^{-1} wet weight (10 g sample). Clenbuterol assay in calf urine was discussed above (Chapter 6, Section 2).

The tranquillisers acepromazine, azaperone, chlorpromazine, haloperidol, propionylpromazine (Figure 6.35) and xylazine (Figure 7.6), the azaperone metabolite/tautomer azaperol and the β-adrenoreceptor blocker carazolol (Figure 6.42) have been measured in animal tissue using a methyl-modified silica analytical column with acetonitrile-aq. ammonium acetate (1.54 g L^{-1}) (1 + 1) as eluent with ED (PGEs, E$_1$: +0.4 V, E$_2$: +0.7 V *vs* Pd).[17] Homogenised tissue was extracted using SPE (ODS-modified silica). LoDs were 2 μg kg^{-1} wet weight (5 g samples).

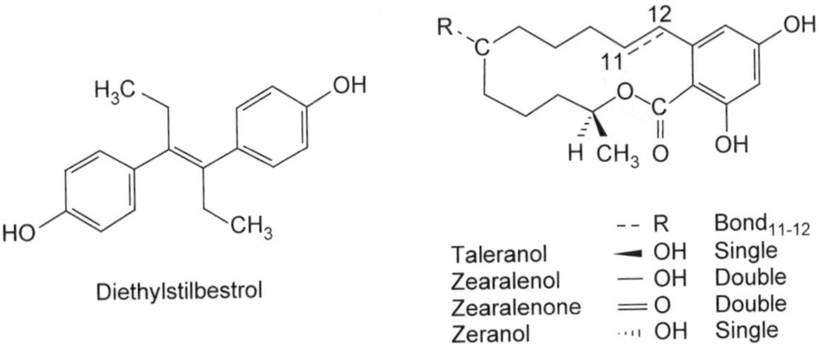

	R	Bond$_{11-12}$
Taleranol	◄ OH	Single
Zearalenol	— OH	Double
Zearalenone	= O	Double
Zeranol	⋯ OH	Single

Figure 7.5 *Some anabolic steroids.*

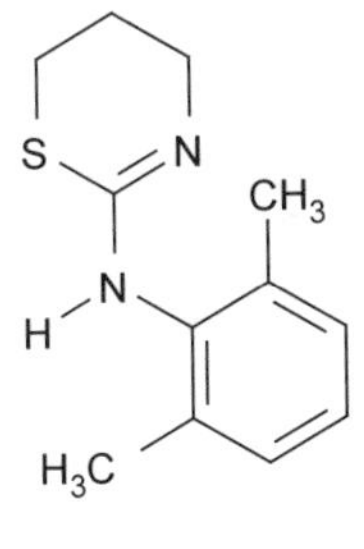

Figure 7.6 *Xylazine.*

Gossypol (Figure 6.59) and gossypolone (gossypol in which the hydoxyl moieties on the two inner aromatic rings are replaced by quinones) enantiomers have been measured in fish using HPLC with EC and UV detection.[18] The analytes were derivatised prior to analysis on an ODS-modified silica column with acetonitrile-water (8 + 1) containing aq. potassium dihydrogen orthophosphate (2 mmol L^{-1}), final pH 3.0 adjusted with orthophosphoric acid, as eluent. Blood, seminal plasma and kidney homogenates were extracted with dimethylformamide-(R(-)-2-amino-1-propanol)-acetic acid (88 + 2 + 10) and heated (90–95 °C, 30 min). Detection was at a GCE (+0.85 V *vs* Ag/AgCl). LoDs (S/N = 3) using ED were 2.5 and 50 μg L^{-1} for gossypol and gossypolone, respectively, some 16 and 6 times lower than those obtained by UV. Sample sizes and extraction volumes were not given.

A number of penicillin antibiotics, including amoxicillin, ampicillin, cloxacillin and penicillin G (benzylpenicillin, Figure 6.29), have been measured in milk using an ODS-modified silica column with gradient elution (acetonitrile-methanol-aq. acetate buffer (20 mmol L^{-1}) and PAD (serial dual Au electrodes, Ag/AgCl reference).[19] Sample preparation was by preconcentration of the analytes on the analytical column after protein precipitation with acetonitrile and fat extraction using dichloromethane-hexane (1 + 1). LoDs were of the order of 0.2 μmol L^{-1}. A similar approach was adopted by Dasenbrock and LaCourse[20] for the measurement of cephapirin and ampicillin in milk. Isocratic elution with acetonitrile-aq. acetate buffer (0.5 mol L^{-1}, pH 3.75)-water (4 + 20 + 76) and an octyl-modified silica column was used. Detection was by IPAD at a gold electrode; four sweeps between +0.05 to +1.15 V *vs* Ag/AgCl were integrated. The LoD (both compounds) was 5 mg L^{-1} (10 mL sample).

Three nitrofuran derivatives, nitrofurantoin, furazolidone and furaltadone (Figure 7.7), have been measured in cow's milk after deproteinisation with TCA and SPE (ODS-modified silica).[21] The analysis was performed using an ODS-modifed silica column with 0.5% (v/v) acetic acid in acetonitrile-aq. sodium

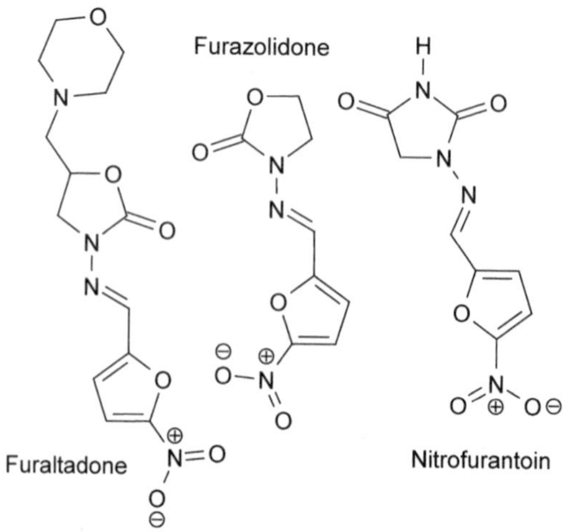

Figure 7.7 *Furaltadone, furzaolidone and nitrofurantoin.*

perchlorate (0.1 mol L^{-1}) $(28 + 72)$. Oxygen was removed by bubbling with nitrogen. Three electrode PGE detection was described (ESA 5021 conditioning cell and ESA 5011 high sensitivity analytical cell), but only one working potential, -0.6 V *vs* Pd, was quoted. LoDs of 4–6 µg L^{-1} were claimed (50 mL sample). Interestingly, this same group used a very similar approach to measure nitro-pesticides in river water.[22] The conditioning cell was operated at -0.6 V *vs* Pd 'to contribute to the elimination of oxygen in the mobile phase'. The analytical cell was operated in the redox mode (E_1: -1.3 V E_2: $+0.8$ V *vs* Pd), the signals from both electrodes being used to quantify the compounds.

References

1 C. Defendenti, E. Bonacina, M. Mauroni and L. Gelosa, Validation of a high performance liquid chromatographic method for alpha-amanitin determination in urine, *Forensic Sci. Int.*, 1998, **92**, 59–68.

2 L.J. Felice, R.E. Schirmer, D.L. Springer and C.V. Veverka, Determination of polycyclic aromatic amines in skin by liquid chromatography with electrochemical detection, *J. Chromatogr.*, 1986, **354**, 442–448.

3 S.M. Billedeau, M.S. Bryant and C.L. Holder, Analysis of heterocyclic amines using reversed-phase high performance liquid chromatography with electrochemical detection, *LC-GC Intl.*, 1991, **4**, 38–41.

4 N. Imaizumi, K. Hayakawa, Y. Suzuki and M. Miyazaki, Determination of nitrated pyrenes and their derivatives by high performance liquid chromatography with chemiluminescence detection after online electrochemical reduction, *Biomed. Chromatogr.*, 1990, **4**, 108–112.

5 W.L. Childress, I.S. Krull and C.M. Selavka, Determination of deoxynivalenol (DON, vomitoxin) in wheat by high-performance liquid chromatography with photolysis and electrochemical detection (HPLC-hv-EC), *J. Chromatogr. Sci.*, 1990, **28**, 76–82.

6 M. Sharman, A.L. Patey and J. Gilbert, A survey of the occurrence of agaritine in U.K. cultivated mushrooms and processed mushroom products, *Food Addit. Contam.*, 1990, **7**, 649–656.

7 E. Jaime, C. Hummert, P. Hess and B. Luckas, Determination of paralytic shellfish poisoning toxins by high-performance ion-exchange chromatography, *J. Chromatogr. A*, 2001, **929**, 43–49.

8 M. Åkerblom and B. Lindgren, Simultaneous determination of active ingredients and chlorophenol impurities in phenoxy acid herbicide formulations by high-performance liquid chromatography with ultra-violet and electrochemical detection, *J. Chromatogr.*, 1983, **258**, 302–306.

9 S. Kawai, K. Goto, K. Kano and T. Kubota, Determination of carbaryl by high-performance liquid chromatography with electrochemical detection, *J. Chromatogr.*, 1988, **442**, 451–454.

10 R.T. Krause, High-performance liquid chromatographic determination of aryl *N*-methylcarbamate residues using post-column hydrolysis electrochemical detection, *J. Chromatogr.*, 1988, **442**, 333–343.

11 L.H. Leenheers, R. Engel, W.E.T. Spruit, W.J.A. Meuling and M.J.M. Jongen, Determination of methyl 5-hydroxy-2-benzimidazole carbamate in urine by high-performance liquid chromatography with electrochemical detection, *J. Chromatogr.*, 1993, **613**, 89–94.

12 K. Sato, J.-Y. Jin, T. Takeuchi, T. Miwa, K. Suenami, Y. Takekoshi and K. Susumu, Integrated pulsed amperometric detection of glufosinate, bialaphos and glyphosate at gold electrodes in anion-exchange chromatography, *J. Chromatogr. A.*, 2001, **919**, 313–320.

13 M. del Pilar da Silva, J.R. Procopio and L. Hernández, Evaluation of the capability of different chromatographic systems for the monitoring of thimerosal and its degradation products by high-performance liquid chromatography with amperometric detection, *J. Chromatogr. A*, 1993, **653**, 267–273.

14 S.H. Kang and H. Kim, Simultaneous determination of methylparaben, propylparaben and thimerosal by high-performance liquid chromatography and electrochemical detection, *J. Pharm. Biomed. Anal.*, 1997, **15**, 1359–1364.

15 A. Laganà and A. Marino, General and selective isolation procedure for high-performance liquid chromatographic determination of anabolic steroids in tissues, *J. Chromatogr.*, 1991, **588**, 89–98.

16 Th. Reuvers, E. Perogordo and R. Jiménez, Rapid screening method for the determination of diethylstilbestrol in edible animal tissue by column liquid chromatography with electrochemical detection, *J. Chromatogr.*, 1991, **564**, 477–484.

17 M.D. Rose and G. Shearer, Determination of tranquilisers and carazolol residues in animal tissues using high-performance liquid chromatography with electrochemical detection, *J. Chromatogr.*, 1992, **624**, 471–477.

18 K. Lee and K. Dabrowski, High-performance liquid chromatographic determination of gossypol and gossypolone enantiomers in fish tissues using simultaneous electrochemical and ultraviolet detectors, *J. Chromatogr. B*, 2002, **779**, 313.

19 E. Kirchmann, R.L. Earley and L.E. Welch, The electrochemical detection of penicillins in milk, *J. Liq. Chromatogr.*, 1994, **17**, 1755–1772.

20 C.O. Dasenbrock and W.R. LaCourse, Assay for cephapirin and ampicillin in raw milk by high-performance liquid chromatography-integrated pulsed amperometric detection, *Anal. Chem.*, 1998, **70**, 2415–2420.

21 T. Galeano Díaz, A. Guiberteau Cabanillas, M.I. Acedo Valenzuela, C.A. Correa and F. Salinas, Determination of nitrofurantoin, furazolidone and furaltadone in milk by high-performance liquid chromatography with electrochemical detection, *J. Chromatogr. A*, 1997, **764**, 243-248.

22 T. Galeano-Díaz, A. Guiberteau-Cabanillas, N. Mora-Díez, P. Parrilla-Vázquez and F. Salinas-López, Rapid and sensitive determination of 4-nitrophenol, 3-methyl-4-nitrophenol, 4,6-dinitro-*o*-cresol, parathion-methyl, fenitrothion, and parathion-ethyl by liquid chromatography with electrochemical detection, *J. Agric. Food Chem.*, 2000, **48**, 4508–4513.

APPENDIX

Non-aqueous Ionic Eluent Systems for Basic Drugs

Efficient performance can be obtained for many basic drugs on unmodified silica columns using methanol containing an ionic modifier as the eluent (Figure 4.2).[1-3] A feature of such silica column/non-aqueous ionic eluent systems is that EC detection can be used since the eluent contains a supporting electrolyte. Indeed, higher oxidation potentials may be used than with conventional aqueous eluents while maintaining good sensitivity, since methanol is more resistant to oxidation than water.

Perchloric acid (0.01 or 0.02% v/v, *ca* 1 or 2 mmol L^{-1}) is a useful ionic modifier if a strongly acidic eluent is required. A methanolic ammonium perchlorate (10 mmol L^{-1})-methanolic sodium hydroxide (0.1 mol L^{-1}) (999 + 1) eluent (final apparent pH 6.7) balances retention, peak shape and EC response for many analytes, especially secondary and tertiary aliphatic amines.[4] Under these conditions a V25 grade (2500 °C pyrolysed) GCE (Le Carbone) in a wall-jet assembly gives a good response for secondary aliphatic amines at an applied potential of +1.2 V (*vs* Ag/AgCl) at pH 6.7. The sensitivity towards other analytes is influenced primarily by the electroactive moieties present and the applied potential (Table 4.1). As with ED in general, loss of response occurrs with the V25 electrode in routine use. This is especially noticeable with secondary aliphatic amines, particularly if extracts of specimens obtained postmortem or other 'dirty' samples have been analysed. Repolishing the electrode is required to fully restore the response.[4] The standing current is a useful indicator of performance during routine operation – the values that should be attainable at different applied potentials are given in Table 4.1. Loss of response is normally accompanied by a decreased current while excessive noise due to, for example, a contaminated eluent is often paralleled by an increased current.

Jane *et al.*[4] have recorded retention (125×5 (i.d.) mm Spherisorb S5W Silica column) and relative detector response data (EC (V25 grade GCE, +1.2 V *vs* Ag/AgCl) and UV absorption (254 nm)) for 462 basic drugs and quaternary ammonium compounds using methanolic ammonium perchlorate (10 mmol L^{-1}, pH 6.7) as eluent. In addition to their use in qualitative analyses, these tables provide information on retention and EC response if quantitative analyses are contemplated. Even for compounds where the EC/UV ratio reported is small, the added selectivity of the EC detector may prove useful for analytes derived from biological samples.

The data reported by Jane *et al.*[4] were compiled by one analyst using a single column and detector arrangement at *ca* 22 °C. System performance with each new batch of eluent was checked by analysis of a standard drug mixture. Retention times were measured relative to the retention time of imipramine with each eluent batch. Retention and relative detector response data for 208 compounds had been compiled previously in a different laboratory employing different Spherisorb S5W silica columns, different analysts and different detector arrangements over a period of several years (Jane and McKinnon, unpublished). Comparisons of retention (k) and relative detector response ($EC_{(+1.2\ V\ vs\ Ag/AgCl)}/UV_{(254\ nm)}$) data in this latter report with the results published by Jane *et al.*[4] are shown in Figure A.1. If some outliers (Table A.1) were excluded from the calculations, good correlations ($r = 0.97$) between the two sets of data ($\log_{10}$) were obtained. A similar correlation ($r = 0.96$) was obtained if retention times relative to imipramine were compared.

In some cases the major discrepancies between the two sets of results (Figure A.1, Table A.1) can be easily explained. With dioxaphetyl butyrate and promethazine, data on *in vitro* decomposition products were recorded by Jane and McKinnon – minor peaks with identical retention and relative detector response characteristics were observed during the definitive study.[4] With brompheniramine, chlorpheniramine and pheniramine the k values reported by Jane and McKinnon were approximately three times those established subsequently, suggesting a systematic error, such as eluent ionic strength – the relative detector responses were similar suggesting the same compounds had been studied (Table A.1). With the remaining outliers either the k values or the relative detector responses were similar, again suggesting that the same compounds had been studied. However, some of the data recorded in the earlier work are clearly inconsistent. Dipipanone and methadone, for example, have similar UV absorption to dextropropoxyphene (254 nm) and contain the same electroactive moiety (tertiary aliphatic amine) yet the EC/UV ratios recorded in the earlier, unpublished work were 23, 92 and 1350 mA au^{-1}, respectively. The corresponding values reported by Jane *et al.*[4] were 280, 670 and 1200 mA au^{-1}.

Overall the comparison thus showed reasonable reproducibility not only of retention data, but also of EC response between laboratories. However, it is misleading to include ED response data from work on the V25 electrode (+1.2 V *vs* Ag/AgCl) used with the methanolic ammonium perchlorate-modified (10 mmol L^{-1}, pH 6.7) eluent[4] in a single list of 'electrochemically detectable substances' at a 'GCE' without further qualification.[5] This is because commercially available glassy carbon working electrodes, including those supplied by Hewlett-Packard, are manufactured from a variety of starting materials. Such electrodes invariably give only a poor response to tertiary aliphatic amines, and either a minimal or no

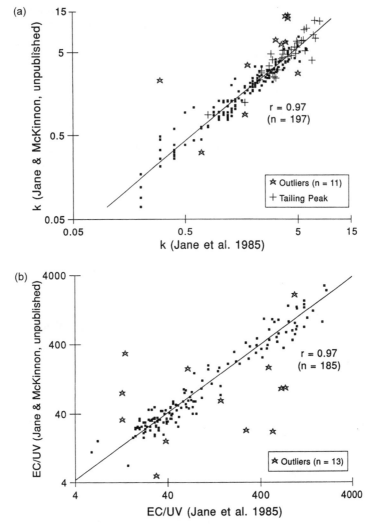

Figure A.1 *Comparison of retention (k) and relative detector response (EC (V25 grade GCE, +1.2 V vs Ag/AgCl) and UV absorption (254 nm)) (mA au^{-1}) for basic drugs recorded by Jane* et al.[4] *and by Jane and McKinnon (unpublished). Details of the data not included in the regression calculation ('outliers') are given in Table 4.2.*

response to secondary amines even at +1.4 V *vs* Ag/AgCl and eluent flow-rates of 1–2 mL min^{-1}, possibly due to the relatively small surface area available on the electrodes. Evidence for this latter view is that a PGE (large surface area) gives an excellent response to secondary and tertiary aliphatic amines using a methanolic ammonium perchlorate (5 mmol L^{-1}, pH 6.7)-modified eluent at an applied potential of +1.0 V (*vs* Pd) (Figure A.2).

Table A.1 *Comparison of retention and relative detector response data for basic drugs (Jane et al.[4] (A) and Jane and McKinnon, unpublished (B)): Data excluded from the regression analysis (Figure A.1) as "outliers". (N.B. Data in italics not shown in Figure A.1)*

Compound	k		EC/UV Ratio (mA au^{-1})	
	A	B	A	B
Retention Data Comparison (Figure A.1 (a))				
Antazoline	1.8	3.5	72	63
Benzoylecgonine	*0.9*	(†)	*32*	–
Brompheniramine	4.1	13.4	75	50
Chlorpheniramine	3.9	13.2	80	60
Dioxaphetyl butyrate	0.3	2.3	540	22
Ecgonine	*1.1*	(†)	(‡)	–
Meclozine	0.7	0.3	220	230
Mepyramine	3.9	6.5	24	26
Methylphenidate	1.7	0.9	29	50
Morpheridine	*1.6*	(†)	*2800*	–
Narceine	*0.7*	(†)	*38*	–
Pheniramine	4.1	12.4	76	60
Promethazine	5.0	2.8	38	16
Protokylol	*3.1*	(†)	*270*	–
Tripelennamine	3.6	6	45	40
Triprolidine	3.2	6.9	29	19
Detector Response Ratio Comparison (Figure A.1 (b))				
Alprenolol	1.2	0.8	13	80
Buclizine	0.7	0.4	490	186
Bufotenine	3.1	2.8	940	2100
Clemastine	3.7	3.1	740	94
Dipipanone	2.2	2.7	280	23
Dioxaphetyl butyrate	0.3	2.3	540	22
Ephedrine	1.0	1.3	150	62
Hyoscine	*1.1*	*1.2*	*940*	(‡)
Labetalol	*1.7*	*1.2*	*250*	(‡)
Maprotiline	2.2	2.1	30	5
Mescaline	1.3	2.0	13	33

(Continued)

Table A.1 *(Continued)*

Compound	k		EC/UV Ratio (mA au^{-1})	
	A	B	A	B
Methadone	2.2	2.4	670	92
Oxprenolol	1.3	1.0	14	300
Promethazine	5.0	2.8	38	16
Propranolol	1.3	0.9	66	180

Key: (†) Not eluted
    (‡) No UV absorption (254 nm)

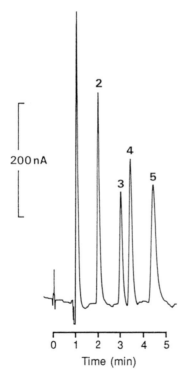

Figure A.2 *Analysis of a mixture of basic drugs using the ESA Coulochem II (PGEs, E_1 and E_2 both +1.0 V vs Pd). Column: 100×4.6 (i.d.) mm Spherisorb S5W silica; Eluent: methanolic ammonium perchlorate (5 mmol L^{-1}, pH 6.7); Flow-rate: 1.3 mL min^{-1}; Injection: 20 µL methanolic solution of nortriptyline (2), amitriptyline (3), imipramine (4) and methdilazine (5) (all 100 mg L^{-1}).*

References

1 R.J. Flanagan and I. Jane, High-performance liquid chromatographic analysis of basic drugs on silica columns using non-aqueous ionic eluents – I. Factors influencing retention, peak shape and detector response, *J. Chromatogr.,* 1985, **323**, 173–189.
2 M.B. Thomas and T.A. Last, Voltammetric electrochemical detection for normal-phase high-performance liquid chromatography, *Anal. Chem.,* 1988, **60**, 2158–2161.
3 R.J. Flanagan, HPLC of psychotropic and related drugs in *High Performance Liquid Chromatography in Neuroscience Research*, R.B. Holman, A.J. Cross and M.H. Joseph (eds), Wiley, Chichester, 1993, 321–356.
4 I. Jane, A. McKinnon and R.J. Flanagan, High-performance liquid chromatographic analysis of basic drugs on silica columns using non-aqueous ionic eluents – II. Application of UV, fluorescence and electrochemical oxidation detection, *J. Chromatogr.,* 1985, **323**, 191–225.
5 Hewlett-Packard, *Electrochemical Detection in High Performance-Liquid Chromatography – A Practical Primer.* Hewlett-Packard, Waldbronn, Germany, 1989.

Subject Index